Drugs and Foods from Little-Known Plants

2693

ETHNOBOTANICAL AND ECOLOGICAL STUDIES
IN THE NORTHWESTERN AMAZON BASIN

Sci. name *Unonopsis veneficiorum*
(Mart.) R. E. Fr. vel aff.
Tribal name
Kofán: ï tesï ja'ndi
Locality Santa Rosa on Río Guamué
Putumayo, Colombia
Habitat swamp – Spanish: vega;
Kofán: andEkï
Uses of root:
Curare ingredient
Remarks
Small tree 20 ft ±

Coll. Homer V. Pinkley   No. 558
Date 11-26-'66 Det. James W. Walker

Herbarium specimen of *Unonopsis veneficiorum* (Annonaceae) on which is based the
modern ethnotoxicological note of the Kofán Indian use of the plant in curare.
H. V. Pinkley 558. Courtesy: Botanical Museum of Harvard University.

Drugs and Foods from Little-Known Plants

Notes in Harvard University Herbaria

By SIRI VON REIS ALTSCHUL

Harvard University Press   Cambridge, Massachusetts   1973

© Copyright 1973 by the President and Fellows of Harvard College

All rights reserved

Preparation of this volume has been aided by grants from the National Institutes of Health and the National Library of Medicine

Library of Congress Catalog Card Number 72-85145

SBN 674-21676-8

Printed in the United States of America

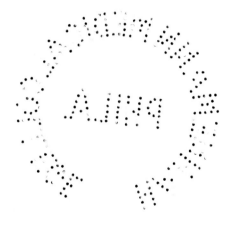

To the observant and devoted botanical collectors
who made possible this book and the researches that may follow it

Contents

Acknowledgments

Thanks are due many people for helping to make this volume possible, but first of all to my esteemed senior colleague, Professor Richard Evans Schultes, Director of the Botanical Museum of Harvard University. His early enthusiasm for the project led to its being sponsored by the Botanical Museum, an institution known, among other activities, for its research in economic botany. It was Professor Schultes' continuing interest and guidance which encouraged those of us engaged in the project when difficulties and frustrations threatened.

Professor Paul C. Mangelsdorf, Director Emeritus of the Botanical Museum, first named me to the Museum staff as a Research Fellow early in the 1960s, when I had just finished graduate school. The appointment immeasurably assisted my prosecution of this and other work in economic botany, including, in particular, research on hallucinogenic plants of the New World.

I am no less indebted to Professors Reed C. Rollins and Richard A. Howard, respective Directors of the Gray Herbarium and Arnold Arboretum of Harvard University. It was an interest in the history of botanical exploration in the Caribbean, held in common with Professor Howard, that originally drew me to graduate studies at Harvard. Subsequently, I wrote my doctoral thesis on the taxonomy and ethnobotany of the tropical American leguminous genus *Anadenanthera*, under the direction of Professor Rollins. At that time my first thoughts about this catalogue began to take shape and were discussed with Professor Rollins. Without the generous permission of both Professors Rollins and Howard to use the entire collections of the Gray Herbarium and Arnold Arboretum, the project could not have been undertaken.

Mrs. Edward H. Metzger and Dr. D. Doel Soejarto aided me in carefully and selectively collecting information from herbarium sheets. Drs. Shiu Ying Hu and Lily M. Perry especially, and on occasion other herbarium personnel, offered translations and interpretations of field notes and gave freely of their time and advice to help unravel knotty problems.

Professor Benjamin R. Hershenson, of the Massachusetts College of Pharmacy, is responsible almost exclusively for constructing the medical index to the catalogue, a painstaking task that required accuracy and judgment.

For their generous academic support, my special appreciation goes to: Professor Heber W. Youngken, Jr., Dean of the College of Pharmacy at the University of Rhode Island; Dr. Bryce Douglas, Director of Research at the Smith Kline and French Laboratories; Professor Robert F. Raffauf, of the College of Pharmacy at Northeastern University (formerly of Smith Kline and French); and Dr. Gordon H. Svoboda of Eli Lilly and Company. They worked in close cooperation with our team, enthusiastically receiving and analyzing portions of our data as they were gathered.

Perhaps more than any of my professional friends who were not intimately involved with the project, the late Professor Daniel H. Efron, of the National Institute of Mental Health and the University of Maryland, comprehended the broader implications of what we were trying to achieve. While the catalogue was in preparation, he suggested on a number of occasions appropriate forms for the presentation of portions of our data. He made recommendations of major importance to the publishing of the book.

We received financial aid from Eli Lilly Research Laboratories and from Smith Kline and French Laboratories, as well as from the National Institute of Mental Health, all to accomplish the gathering of data. The Massachusetts College of Pharmacy supported Professor Hershenson's portion of the work, while a grant from the National Library of Medicine made it possible to publish the manuscript through Harvard University Press.

S. v. R. A.

Foreword

In 1961 Dr. Siri von Reis Altschul, as an extension of her work in taxonomy and ethnobotany and her interest in several natural-products research programs then under way in American pharmaceutical circles, conceived an ambitious idea. This idea materialized, thanks to her persistence, and led to preparation of the present volume. I believe that the many kinds of specialists who undoubtedly will utilize the information in the following pages may appreciate a few words on the conception and development of Dr. Altschul's idea.

In science's search for novel biodynamic constituents and potential new therapeutic agents from the plant kingdom, several avenues of approach have been followed: a random sampling of the existing flora; direct study of plant uses among peoples still living in primitive cultures; and reliance upon information in the literature.

The last of these avenues apparently has been the one most commonly followed in pharmaceutical circles. Generally speaking, it has been unreliable. Scraps of information about the use of a plant may be gleaned easily from all kinds of literature—including even history, reports of missionaries and travelers, and other rather casual and diffuse writings. Investigators frequently begin by consulting regional floras, compendia prepared by botanists that usually include economic notes about the species growing naturally in the area covered by the particular flora. These notes on plant uses are seldom first-hand; more often they are second-, third- or fourth-hand repetitions, citing neither sources of the original reports nor voucher specimens from which the identifications were made. Consequently, errors and uncertainties are frequent even in the floristic literature, which should be—and undoubtedly is—the most reliable available. Recent pharmaceutical research more than once has become disoriented because of erroneous reports in floras.

It occurred to Dr. Altschul, as a direct outgrowth of her survey of literature reports about the use of hallucinogenic snuffs in the New World, that too much reliance had been placed on these data in floristic, historic and other works—data completely divorced from specimens or other means of checking identification. The natural corollary then occurred to her: Why not utilize the almost untapped wealth of ethnobotanical notes hidden away on labels in our herbaria?

The more she pondered the potentialities inherent in a search of herbaria and the longer she discussed it with her colleagues, the more Dr. Altschul was convinced that this source of information on folk medicine ought to be utilized.

A herbarium report itself, she reasoned, was usually first-hand, jotted down on the spot by the collector of the plant in the locality and at the time of collection. It was on a label, often in the collector's own handwriting, and physically attached to a specimen. Consequently, there could be no uncertainty about the identity of the plant. The season of collection, ecological notes, description of the plant, vernacular names, and other information would often be found on the label—all of these data occasionally of importance either for evaluation of the use or for possible re-collection of fresh material of the plant for chemical study. Some of the collections would be recent, others very old. The old specimens frequently preserved notes about uses that have disappeared with the cultures or peoples who employed the plant and therefore would be even more valuable to modern researchers.

The herbaria of the world comprise vast collections of plants from all parts of the world during the past two centuries. Thus they make it economically feasible to pursue significant ethnobotanical studies in deserts or in tropical rain forests, at sea level or on the highest mountains, in Asia, Australia, Africa, Europe, North or South America—all without leaving the herbaria.

Dr. Altschul eventually set about to search the estimated 2,500,000 specimens preserved in the combined collections of the Gray Herbarium and the Arnold Arboretum at Harvard University. It was a major task, but one which she carried through over a period of four and a half years, with the results that are set forth in the following pages. A combination of enthusiasm and energy, faith in the value of the work, and meticulous organization of the information retrieved mark her accomplishment. It is offered as an indicator and, at times, as a guide for the diverse kinds of specialists dedicated to an examination of the world's vegetation; it is also a potential model for what may prove to be in older and larger herbaria—as well as in newer and smaller ones—a key to advantageous use of the plant kingdom as a source of valuable new chemical constituents.

Richard Evans Schultes, Director
Botanical Museum of Harvard University

Drugs and Foods from Little-Known Plants

Introduction

The purpose of this book is to bring to light little-known data relating to unusual drug and food plants. It is my hope that the species in this catalogue of potential use to man may be studied and saved before they are extinguished. Some of these species have been partially domesticated by primitive societies that now are disappearing, along with their knowledge and traditions; many more are wild and likely to become extinct as civilized man disturbs their natural environments.

The concept of the book is based on trust in the elements of empiricism in primitive medicine and nutrition. Along with this goes faith that amassing folklore and presenting it against taxonomic patterns may yield insights into medical validities and, incidentally, into chemotaxonomic affinities of interest to the botanical systematist. The fact that the notes herein are firsthand and that the collections cited can be traced in most instances directly to their localities should provide the pharmacognocist with a head start in his search for drugs.

Species Included and Excluded

The data in this catalogue were gathered in four and a half years, starting in January 1962. A search was carried out through the estimated 2,500,000 sheets of dried, pressed specimens of angiosperms, or flowering plants, encompassed in the combined collections of the Gray Herbarium and the Arnold Arboretum of Harvard University. Approximately 5000 species of possible interest were found and are presented herewith.

The orchid family was not fully surveyed, as its representatives are housed separately at Harvard in the Oakes Ames Orchid Herbarium of the Botanical Museum; this large group was considered at the time to be relatively too well-known to warrant a special study. Furthermore, random search through several thousand sheets in the Ames Herbarium yielded no notes on utilization.

1

Also excluded are gymnosperms, pteridophytes, bryophytes, fungi, and algae, partly because it was believed that they would be accompanied by fewer notes of interest and partly because of the desirability of limiting an already sizable undertaking. It should be mentioned that an unreckoned number of sheets was out on loan to other herbaria and museums while the search was being conducted; to that extent, this catalogue does not reflect completely the riches of its sources. One may well wonder, too, how many potentially interesting species we passed over in the herbarium for lack of information on their uses. The percentage represented there could be fairly high, since field collectors are taxonomists more often than ethnobotanists, and this tends to be reflected in their notations.

Internationally recognized taxonomic types were omitted from this catalogue since they have been published previously in journals of systematics and are relatively well-known. Similarly, species whose Latin names alone suggested medicinal or nutritional properties, although obviously pertinent, were eliminated as having been published earlier. However, these were retained in the catalogue when they offered, in addition, corroborative or otherwise interesting label data. Still other species, where the writing on the labels was in exotic characters, were excluded because of potential difficulties of interpretation and reproduction.

The greatest number of species eliminated from the catalogue were those with uses well enough known to have been included in J. C. Th. Uphof's *Dictionary of Economic Plants* (Weinheim, Germany: J. Cramer, 1959). Each relevant species encountered in the herbarium was checked against Uphof's compendium. Species with uses enumerated therein were omitted from our lists, unless the herbarium specimen added new aspects to the uses given or extended the geographical distribution of a previously known use.

In addition, where one collection of a species offered a note of interest, notations otherwise of little interest from other collections of the same species sometimes were included if they bore notes which might enhance insights into the use of that species. Common names associated with species of interesting uses occasionally were found applied to other species. Where this occurred, an attempt was made to include all of the species bearing that common name on the chance that all might possess similar properties.

When a common name or a specific use was found in more than one collection of a given species within an "average-sized" country (if one may use such an expression), the repeats were not recorded, as this compilation was not intended to be a frequency study. The imbalance in geographical representation of species in this catalogue is unfortunately unavoidable. Herbarium collections tend to be specialized in particular geographical areas, and the collections of the Gray Herbarium and the Arnold Arboretum are richer in representatives from eastern Asia, South America, and the South Pacific than they are, for example, in those from Europe or Africa.

Of a total of 294 families examined in the course of searching, 178—or more than half—are represented here; these encompass about 1700 genera and 5100 species. The *Compositae* are the best represented, comprising appɪoximately 15 percent of the species; the *Euphorbiaceae*, 8 percent; *Leguminosae*, 7 percent; *Labiatae*, 6 percent; *Rubiaceae, Lauraceae,* and *Liliaceae,* 4 percent each; *Piperaceae,* 3 percent; *Rosaceae, Boraginaceae, Asclepiadaceae,* and *Rutaceae,* 2 percent each; *Verbenaceae, Apocynaceae, Fagaceae, Araceae,* and *Flacourtiaceae,* 1 percent each. The largest single category of plants represented is that of edible species (1255). Next are aromatic species (992), and then species referred to in the herbarium notes as "medicinal" (407). These are followed by poisonous species (284), species used against gastrointestinal disorders (241), species with reputed analgesic properties (190), species used to treat injuries (179), respiratory ailments (140), and skin diseases (116), species from which beverages are made (112), and so on, including all species with notes hinting at possible biodynamic properties of any kind.

Indexes to families and genera are presented to facilitate handling for readers interested in the folklore of particular plants or plant groups. For those who would use the book for locating plants associated with specific medicinal capacities—diaphoretics or febrifuges, for example—there is a medical index referring to diseases and therapeutic properties. The last indexes many of the notes in the catalogue but does not include common names, which are usually vague in application. It is recommended that readers interested in specific medicinal properties not confine themselves to plants cited in the index but thoroughly explore the text as well.

Editorial Principles Employed

The use of this catalogue should not require a background in systematic botany or herbarium techniques, but for the benefit of those familiar with the disciplines and for other readers interested in how certain problems were handled, the following remarks may be helpful.

The species and the notes belonging with them are arranged herein as they were found in the integrated collections of the Gray Herbarium and the Arnold Arboretum. This means that species are to be found under the genera which encompass them and that genera are to be found under their respective families. Families are arranged according to the Engler and Prantl system of phylogenetic sequence, progressing from the presumed primitive families, such as grasses, to the advanced composites. Because the system is linear, it probably does not reflect a truly evolutionary scheme, but it works well from a practical point of view.

Genera within families are arranged systematically according to the Dalla Torre and Harms system, which carries out the Engler and Prantl concept of phylogenetic sequence. Under this system the species within genera may be arranged in alphabetical order. But where genera are large, the species may be organized to reflect taxonomic (presumably evolutionary) relationships. Species within an extensive genus, of wide occurrence, at times may be arranged according to geographical distribution.

Following this system in our text serves to highlight groups of similar uses among related plants and helps to pinpoint genera, families, or clusters of families which may have similar biodynamic properties. The suggestion of common active constituents becomes especially strong where—as is evident in a number of instances in the text—within a single genus or species, plants widely separated geographically are reported to be used for similar purposes.

Family names throughout are spelled in accordance with G. H. M. Lawrence's *Taxonomy of Vascular Plants* (New York: Macmillan, 1955). Generic and specific names for the most part are spelled as they were in the herbarium, except that some obvious errors were corrected with the help of *Index Kewensis*.

Throughout the text specific and infraspecific epithets have been decapitalized. Since the *International Code of Botanical Nomenclature* (Utrecht, Netherlands: Kemink en Zoon, 1956) permits authors naming new species in some cases to capitalize such epithets, a portion of our entries no doubt is inexact. For the sake of expediency only, the correct forms were not ascertained.

In the interest of brevity, and because this book is not meant to be in any way a taxonomic work, the botanical authorities have been omitted from all Latin binomials.

It may be helpful to the reader to show the format of each entry in the catalogue and to explain the various elements in some detail:

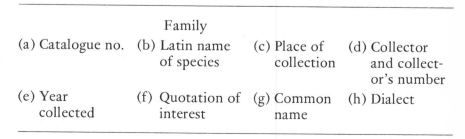

Family			
(a) Catalogue no.	(b) Latin name of species	(c) Place of collection	(d) Collector and collect-or's number
(e) Year collected	(f) Quotation of interest	(g) Common name	(h) Dialect

(a) Each species has been assigned a number (from 1 to 5178), for ease of handling and for indexing. The numbers serve no function outside the book, although closely related numbers may suggest relationships between the species, genera, or families involved. During the long preparation of this volume the author and editors decided to omit certain entries, to move misplaced materials, and to add some species. In view of the extent and complexity of the catalogue, we felt it was better sometimes to omit numbers—or to add them by the use of *A, B,* and so on—than to risk the errors that might occur if we were to alter the whole numbering system.

(b) In all instances identifications of the plants are recorded as they were indicated in the herbarium. Where there were discrepancies between the specific identification on a sheet and that on the folder enclosing it, the name on the folder was used. Even where the two names agreed, identifications sometimes were questionable, or the name used was not the currently accepted one. In order, however, for a researcher to be able to find specimens as desired among the extensive collections, the label names have been retained in the present work.

(c) Under each species, one or more collections are entered, each beginning with the name of a country or other political or geographical unit to indicate where, in general, the collection was made. These names are presented as they appeared on the labels attached to the herbarium specimens. In some instances the names are in archaic or foreign spellings and may refer to countries or political entities that no longer exist. Almost all these names have been retained herein because they can be useful in estimating the ages of undated collections, as well as in delimiting collection localities in countries where political events have resulted in border changes.

In other instances, especially among older collections, the country of collection could not be determined. Where this was the case, an approximation was substituted, where possible, by using the same broad, politically artificial (though floristically valid) system of geographical designation as is employed in the arrangement of specimens in the herbarium. The units of this system consistently are capitalized in their entirety, to distinguish them from countries, islands, and the like. These units are defined below and are used herein exactly as in the herbarium:

E. ASIA	China, Manchuria, Mongolia, Tibet, Siberia, Kamchatka, Japan, Korea, Formosa, Hainan
INDIA	India, Burma, Siam, Indo-China, Central Asia, Afghanistan, Ceylon, Andaman Islands
W. MALAYSIA	Malay Peninsula, Sumatra, Java, Borneo, Celebes, small islands in the area of Amboina

PAPUASIA   New Guinea, Solomon Islands, Admiralty Islands, New Britain and small islands in the area (Melanesia)

POLYNESIA   Polynesia, Micronesia, Mariana and Caroline Islands to Hawaii and the Marquesas Islands (Pacific Basin), Fiji, Sandwich, Guam, Samoa, New Hebrides

AUSTRALIA   Australia, New Zealand, Tasmania

EUROPE   including the Mediterranean district and eastward to Persia and North Central Asia, Canary Islands, Azores, Turkestan

(d) In each entry the country is followed by the name of the collector (or collectors) and his number. Where such numbers were absent, herbarium numbers are given if available, or some combination of names and numbers to identify the collection. The abbreviation "s.n." denotes the absence of a collector's number.

(e) The year of collection is given next, usually in whatever form or abbreviation occurred on the specimen label. Hence one may encounter in the text, for instance: 46, '46, "46, 194, 946, or 1946. This procedure was followed in part because it was not always clear in which century the collection was made. Furthermore, we have insisted on fidelity here because we are aware of the possibility that numbers assumed to be dates sometimes may have been intended to signify something entirely different.

(f) The year is followed by the note or notes of possible interest to the medical or health sciences. Here we have quoted verbatim from herbarium labels, including variations and errors in orthography and punctuation, following each label's capitalization or decapitalization, and the like.

(g) If there is a common name for the plant and this name is not included in the quotation, it will follow the quote, in a new quotation. Single quotes within double quotes mean that the single quotes were double quotes on the herbarium label from which the excerpt was taken. "N.v." appears often in this portion of the entries and designates the vernacular name. Many species have been included in this catalogue on the sole basis of the suggested symbolism of their common names. It was believed advisable to include more, instead of fewer, such species, rather than risk missing something useful. In line with this belief, notes indicating aromatic plants, edible plants, plants used in magical ceremonies, and so forth are included. Decisions to include or exclude materials in foreign tongues and local primitive dialects were difficult to make; some place names may have slipped in as common names, especially from very old and almost illegible labels.

(h) In each entry, along with the common or vernacular name(s), available information as to language or dialect or locality of the name may be given, sometimes in parentheses or following the name. Occasionally the language will appear after the quotation at the very end of the entry, often in abbreviated form. In all instances the data are as they appear on the herbarium labels themselves.

It is hoped that those who use this work will understand that the lack of uniformity in the entries arises mainly from variations in the styles of collectors' notations and from our desire to be faithful, as far as feasible, to the data as they appeared on the herbarium labels. Throughout, the aim has been to combine accuracy with facility of interpretation.

For data beyond those enumerated, the researcher may want to go to the specimen itself in any particular instance for details of locality, season of collection, and possible additional reports of uses not deemed relevant to our purposes here.

PANDANACEAE

1   *Pandanus conoideus* New Guinea / M. S. Clemens 445 / 1935 / "Fruit ...edible"

2   *P. tectorius* var. *suvaensis* Solomon Is. / S. F. Kajewski 2299 / 1930 / "When ripe the fruit is pressed out and drunk by the natives" / "Mataram"

3   *P. tectorius* Caroline Is. / C. C. Y. Wong 528 / 1948 / "The fruits are eaten" / "N.v. choi"
     Fiji Is. / A. C. Smith 1281 / 1934 / "Inflorescence bracts used for scenting coconut oil"

POTAMOGETONACEAE

4   *Potamogeton crispus* China / C. W. Wang 76896 / 1936 / "Edible when young"

APONOGETONACEAE

5   *Aponogeton eberhardtii* Annam / Poilane 968 / 1920 / "...tubercule ...comestible" / "Annte: Cây choi"

ALISMACEAE

6   *Caldesia* indet. Philippine Is. / A. L. Zwickey 777 / 1938 / "'Katabatuba' (Lan.); used as spinach"

7   *Echinodorus grandiflorus* var. *grandiflorus* Brazil / Y. Mexia 4179 / 1929 / "Infusion of dry leaves used for blood purifier" / "'Chapeo de couro' 'Cha Mineiro'"

8   *Echinodorus* indet. Colombia / R. E. Schultes 3599 / 1942 / "Med. for cholic"

GRAMINEAE

9   *Euchlaena mexicana* U.S.: Mass. (Cult. Original plants from Honduras) / P. C. Standley 27317 / 1950 / "'Maiz café'"

10   *Coix lachryma-jobi* China / S. K. Lau 550 / 1932 / "Fr. black, edible. Used to make wine" / "I Mai"
     Japan / E. Elliott 41 / '46 / "Food use: Grain eaten. . . . Medicinal use: Decocted seeds are good for stomach and are diuretic" / "Juzudamo"
     Philippine Is. / C. O. Frake 475 / 1958 / "Leaves burnt and applied to goiter" / "Gemuni / Sub."

11   *Imperata koenigii* Japan / E. Elliott 123 / '46 / "Food use: Young buds are eaten by children. Medicinal use: The root is well known diuretic" / "Chigaya"

12   *I. cylindrica* China / W. T. Tsang 23674 / 1934 / "...root edible"

13   *Miscanthus sinensis* Japan / E. Elliott 15 / '46 / "Medicinal use: Rootstock ...diuretic" / "Susuki"

14   *Elionurus* indet. Argentina / Rodriguez 38065 / 1916 / "N.v. 'Pasto amargo' o 'pasto puna'"

15   *Sorghum halepense* Philippine Is. / E. Fénix 128 / 1938 / "Medicinal"

Philippine Is. / Farinas & Abordo 11407 / 1950 / "Grain for human and animal food; stored for fodder" / "Batad"

16   *Themeda gigantea* Philippine Is. / H. C. Conklin 1138 / 1958 / "Leaves used as topical medicine for chest pains" / "baylod bantad"

17   *Arundinella deppeana* El Salvador / P. C. Standley 21754 / 1922 / "'Zacate amargo'"

18   *Thysanolaena maxima* Philippine Is. / C. O. Frake 745 / 1958 / "Leaves applied to pin worms" / "glingu betung Sub."

19   *Paspalum notatum* Bolivia / J. Steinbach 5273 / 1921 / "Es pasto excelente para postreros y el más usado en toda la región"

20   *Digitaria longiflora* Caroline Is. / C. C. Y. Wong 529 / 1948 / "The entire plant is used in the medicine for swelling veins (fely nigof) by pounding the plant, squeezing the sap into a bowl . . . adding water of a young coconut before mixture is drunk" / "N.v. cefer"

21   *Coridochloa cimicina* N. Sumatra / J. A. Lörzing 12881 / 1928 / "Gras, getrocknet stark nach Kumarin duftend"

22   *Echinochloa colonum* Mariana Is. / B. C. Stone 4079 / 1962 / "'chagua n agaga' (Chamorro). Jungle rice"

23   *Echinochloa* indet. Kenya / A. G. Curtis 516 / 1923 / "Natives eat seed and say better than rice"

24   *Lasiacis procerrima* Brit. Honduras / P. H. Gentle 2638 / 1938 / "'wild rice'"

25   *L. sorghoidea* Brit. Honduras / P. H. Gentle 2196 / 1938 / "'Rat rice'"

26   *Setaria pallide-fusca* Caroline Is. / C. C. Y. Wong 324 / 1948 / "Used as medicine" / "N.v. galewel"

27   *Cenchrus echinatus* Tonga Is. / T. G. Yuncker 15089 / 1953 / "Leaves sometimes used in treatment of wounds" / "Hefa"

28   *Pennisetum alopecuroides* Philippine Is. / C. O. Frake 512 / 1958 / "Roots applied to stomach to reduce size in pregnancy(!)" / "gakbas Sub."

29   *P. macrostachyum* Philippine Is. / C. O. Frake 385 / 1957 / "Roots used as yeast ingredient and as a remedy for general malaise" / "gulayulay Sub."

30   *Cyrtococcum accrescens* Philippine Is. / Edaño & Gutierrez 238 / 1957 / "Medicine" / "sabilaw"

Philippine Is. / C. Frake 525 / 1958 / "Antidote for snake bite—whole plant applied" / "Pangisul Sub."

31   *Leptaspis urceolata* Solomon Is. / S. F. Kajewski 2246 / 1930 / "The native superstition is that the leaves are heated and rubbed on the legs of the babies to ensure that they will walk quickly" / "Ortsikau"

32   *L. cumingii* Philippine Is. / G. E. Edaño 1912 / 1949 / "Medicine to stop falling of hair. The root is ground, soaked in water and the liquid is used to wash the hair" / "Bugluson Ma"

33   *Hierochloe redolens* Colombia / Killip & Smith 18295 / 1927 / "'Itámo real' / Valuable remedy, much used by aged people"
      Colombia / Killip & Smith 19678 / 1927 / "'Itámo real' / One of the most valuable of medicinal plants, used for many diseases"

34   *Stipa pekinensis* var. *planifolia* China / Cheo & Yen 337 / 1936 / "Spike for medicinal use"

35   *S. sibirica* Kashmir / R. R. Stewart 18120 / 1939 / "poisonous to animals particularly horses"

36   *Sporobolus indicus* Society Is. / Setchell & Parks 150 / 1922 / "'New Zealand grass' . . . pest, cattle will not eat"

37   *Calamagrostis eminens* Peru / O. Tovar 860 / 1952 / "N.v. 'sora-sora'"

38   *Cynodon dactylon* Honduras / P. C. Standley 20409 / 1949 / "'Zacate de gallina'"

39   *Gymnopogon spicatus* Brit. Honduras / W. A. Schipp 787 / 1931 / "Shunned by all kinds of animals"

40   *Eleusine indica* W.I.: Dominica / W. H. Hodge 2462 / 1940 / "driveway weed; alleged by natives to cause deflation of the bellies of corpses"

41   *Triodia irritans* Australia: Queensland / W. E. Everist 1929 / 1939 / "Local name porcupine grass. In times of drought, horses will eat this grass if tussocks are turned upside down. Infls. eaten in any season"

42   *Centotheca latifolia* New Hebrides Is. / S. F. Kajewski 338 / 1920 / "Chewed by natives and applied to burns" / "Now-now"

43   *Uniola paniculata* Bahama Is. / R. A. & E. S. Howard 10114 / 1948 / "'Sea oats'"
      Bahama Is. / C. B. Lewis s.n. / 1954 / "'Wild rice'"

44   *Secale africanum* S. Africa: Transvaal / C. P. Sutherland s.n. / '37 / "Wild Rye"

45   *Semiarundinaria scabriflora* China / W. T. Tsang 22097 / 1933 / " . . . fr. edible" / "Kam Chuk Tsai"

46   *Sinocalamus latiflorus* Indochina / W. T. Tsang 30192 / 1940 / "Cult. for edible shoots" / "Taai Ma Chuk: Big Jute Bamboo"

47   *Sinobambusa maculata* China / F. A. McClure 20573 / 1937 / "Shoots edible" / "Kuang chu"

48   *Phyllostachys nana* China / F. A. McClure 20591 / 1937 / "Shoots edible" / "Shui chu"

49   *Bambusa aurinuda* Indochina / W. T. Tsang 30198 / 1940 / "Semi-cult. for edible shoots" / "Taai Wong Chuk: Large Yellow Bamboo"

50   *Bambusa* indet. Indochina / W. T. Tsang 30587 / 1940 / "Semi-cult...shoots edible when pickled" / "Taai Lak Chuk: Big Thorny Bamboo"

51   *Dendrocalamus affinis* China / F. A. McClure 20555 / 1937 / "Shoots edible" / "T'ung chu"

52   *D. flagellifer* China / H. Fung 21151 / 1937 / "Shoot edible" / "Mao chu; Hwang chu"

53   *Cephalostachyum pergracile* Burma / J. Keenan et al. 1829 / 1961 / "The inside of the cane is eaten. Rice is baked inside canes. Local Karen name—Wa-Ciaw"

54   *Schizostachyum dumetorum* China / F. A. McClure L.U. 18546 / 1929 / "Used for drug plant (Rhizomes sold in drug stores)" / "Tang Chuk; Liu Chuk"

55   *S. lumampao* Philippine Is. / C. O. Frake 567 / 1958 / "...medicine for puerperium" / "glebuk Sub."

56   *Dinochloa scandens* Philippine Is. / Sulit & Conklin 5183 / 1953 / "Fr. edible cooked" / "Balkawi Mang."

CYPERACEAE

57   *Hypolytrum compactum* Philippine Is. / C. Frake 389 / 1957 / "Young stem chewed up with betel and applied for ulcers" / "Salag-salag Sub."

58   *Cyperus brevifolius* Philippine Is. / H. G. Gutierrez 61–357 / 1961 / "...decoction of whole plant febrifuge"
      New Hebrides Is. / S. F. Kajewski 276 / 1928 / "Common Name Polell. Polell and leaves of Denyung (Cane grass) are macerated together in cold water and drunk for spleen trouble. Polell with leaf of Naiwas tree and Tomi-rirri and Nesiv-nesip (shrub) macerated in cold water by women in state of pregnancy for good health"

59   *C. cyperoides* Philippine Is. / G. E. Edaño 1617 / 1949 / "An infusion of the seeds is given as a remedy for toothache" / "Uñgit Ma"

60   *C. diffusus* Philippine Is. / G. E. Edaño 1866 / 1949 / "The root is used as a medicine for diseased lips known as (singao)" / "Uñgit"

61   *C. compressus* Tonga Is. / T. G. Yuncker 15298 / 1935 / "Roots used for scenting oil" / "Tongan name: Pakopako"

62   *C. mollipes* Tanganyika Terr. / F. G. Carnochan 144 / 1928 / "'Muliwakamba' eaten by Kamba, a quail like bird"

63   *C. muricatus* Tanganyika Terr. / F. G. Carnochan 153 / 1928 / "'Mu solyu' Root used as food and as a scent"

64   *Fuirena umbellata* Neth. New Guinea / P. van Royen 4587 / 1954 / "Tubers edible"

65   *Fimbristylis globulosa* Philippine Is. / C. O. Frake 590 / 1958 / "Medicine for splenomegaly" / "tingel Sub."

66   *Bulbostylis capillaris* Paraguay / W. A. Archer 4931 / 1937 / "Sold by herb dealers in market at Asunción, . . . 'esparto mi.' Used as blood purifier in female disorders"

67   *Scleria mitis* Venezuela / M. Haman 8 / 1918 / "Eaten by cattle and horses" / "'Cortadera' (Cutting grass)"

68   *Scleria* indet. Philippine Is. / A. L. Zwickey 675 / 1938 / "'Tuor' (Lan.); lvs. of this and Kalawag applied to fracture"

69   *S. lithosperma* Caroline Is. / C. C. Y. Wong 411 / 1948 / "The plant is used in the treatment of coughing (fely colcol) the entire plant is pounded and squeezed into a bowl and the water of a young coconut is added before the mixture is drunk" / "N.v. beiby (Ruming); Yebey (Rul)"

70   *S. margaritifera* Caroline Is. / C. C. Y. Wong 250 / 1947 / "The flowers are used as medicine for säfein ik (medicine for fishing) the flowers are rubbed on the arm with a piece of coconut cloth in order to get more fish, this is done by both sexes" / "N.v. nikaasafeesaf"

PALMAE

71   *Phoenix hanceana* China / W. T. Tsang 136 / 1928 / " . . . used for making wine" / "Poh Tsung Shue"

      China / S. K. Lau 5772 / 1935 / " . . . fruit, edible"

72   *P. humilis* China / C. W. Wang 76630 / 1936 / "Fruit . . . edible"

73   *P. roebelinii* China / J. F. Rock 2531 / 1922 / " . . . the central young shoots are eaten by the natives"

74   *Phoenix* indet. Burma / F. G. Dickason 7470 / 1938 / "Fruit edible" / "'ku chuk' (Haka Chin) 'singh namaung' or 'thakyetsu'"

75   *Cocothrinax argentea* Bahama Is. / R. A. & E. S. Howard 10209 / 1948 / " . . . berries edible" / "'Thatch Palm'"

76   *Erythea aculeata* Mexico / H. S. Gentry 5190 / 1939 / "used for . . . food, etc." / "Palmilla"

77   *Sabal uresana* Mexico / H. S. Gentry 1210 / 1935 / "Warihos use . . . the terminal leaf, in bud, for petates" / "Palma, Mex. Tacu, W."

78   *S. mauritiiformis* Colombia / Killip & Smith 14291 / 1926 / "Palma amarga"

79   *Eugeissona tristis* Malaya / C. X. Furtado 33034 / 1937 / "Fruits edible"

80   *Calamus thysanolepsis* China / W. T. Tsang 28640 / 1938 / "...fr. ...edible" / "Wong Tang"

81   *Calamus* indet. Indochina / W. T. Tsang 20388 / 1939 / "...fr.... edible"

82   *Caryota mitis* China / C. I. Lei 1488 / 1936 / "Fruit...poisonous" / "Ka Long"

83   *C. monostachya* Annam / Poilane 4889 / 1922 / "...les graines sont chiquées avec le bétel, la pulpe provoque de vive démangeaisons plus que l'ortie et elles persistes plus longtemps" / "Annamite: Cây dông dinh"

84   *Geonoma binervis* Honduras / P. C. Standley 54689 / 1927–28 / "Young inflorescence cooked and eaten" / "'Pacuca'"

85   *Euterpe subruminata* Brit. Guiana / A. C. Smith 2551 / 1937 / "Heart edible"

86   *Oenocarpus bacaba* Brit. Guiana / A. C. Smith 2584 / 1937 / "...fruit used to prepare a cocoa-like drink"

87   *Ptychorhaphis* indet. Philippine Is. / M. D. Sulit 3019 / 1948 / "...fruit substitute for betel nut" / "Buringot Bis.?"

88   *Pinanga* indet. Philippine Is. / H. G. Gutierrez 61–138 / 1961 / "...substitute for A. catechu"

89   *Pinanga* indet. New Hebrides Is. / S. F. Kajewski 547 / 1928 / "Chewed by natives as betel nut but used only when other types are not available" / "Betel nut"

90   *Areca catechu* China / S. K. Lau 266 / 1932 / "...flesh of the fruit as medicine. Skin of the fruit may be eaten, boil the flower to be medicine" / "Pan Long"

91   *A. triandra* Indochina / Poilane 14748 / 1928 / "...les fruits sont chiqués comme l'aréquier"

92   *Orbignya cohune* Mexico / Y. Mexia 1032 / 1926 / "Very important as a source of oil. Natives break nut and sell almond" / "'Coquitos de aceite'"

      Guatemala / J. A. Steyermark 39210 / 1940 / "Fruit edible, like coconut" / "'Coroz'"

93   *Astrocaryum munbaca* Colombia / R. E. Schultes 3885 / 1942 / "Fr. orange, edible" / "Nombre Huitoto: rui-rĕ-gö"

94   *Bactris major* Honduras / P. C. Standley 54714 / 1927–28 / "Seeds eaten" / "Biscoyol"

ARACEAE

95 *Pothos rumphii* Philippine Is. / A. L. Zwickey 171 / 1938 /
" . . . leaves mixed with hard tobacco to make a softer blend" /
"'Ganalum' (Lan.); 'Takoling' (Bis.)"

96 *Pothos* indet. Philippine Is. / C. O. Frake 513 / 1958 / "Leaves applied for headache" / "'pusan' (Sub.)"

97 *Pothoidium lobbianum* Philippine Is. / Sulit & Conklin 5094 / 1953 /
"Roots smashed and applied as plaster to part bitten by centipede" /
"'Ayasun'" Mang. (Hanunoo)

98 *Heteropsis* indet. Ecuador / R. C. Gill 3 / 1941 / " . . . bark has medicinal value" / "'Mimbre'"

99 *Anthurium scolopendrinum* Colombia / O. Haught 1375 / 1934 /
"Root mass infested by fierce ants, invariably, so far as I have seen"

100 *Anthurium* indet. Colombia / R. E. Schultes 3609 / 1942 / "Medicinal" / "'Faḿ-bĕ'"

101 *Acorus pusillus* China / W. P. Fang 12176 / 1938 / "Cultivated near a temple . . . "

102 *Raphidophora* indet. New Guinea: Papua / Womersley & Simmonds 5047 / 55 / "Stems faintly resinous, aromatic"

103 *R. korthalsii* Philippine Is. / A. L. Zwickey 796 / 1938 / " . . . inner bark chewed; it stains saliva red and (in time) the teeth black" /
"'Palipay' (Lan.)"

104 *R. merrillii* Philippine Is. / H. C. Conklin 435 / 1953 / "Bark of aerial roots chewed to blacken teeth (lukúb gamút; gínan sa tungúd)" /
"'Amlúng kamūrang'?"

105 *Raphidophora* indet. Philippine Is. / C. O. Frake 370 / 1957 /
"Whole leaf 1 M.—plant cut off at attachment used in agric. ritual" /
"'Pipay'" Sub.

106 *Monstera friedrichsthalii* El Salvador / S. Calderón 1358 / 1922 /
"'Piña anona'"

107 *Epipremnum pinnatum* Fiji Is. / O. Degener (& E. Ordonez) 14180 /
1941 / "Exported formerly to America for medicine. Fijians drink the scraped bark with water for dysentery" / "'Yalu'"

108 *Scindapsus* indet. Caroline Is. / C. C. Y. Wong 488 / 1948 / "The stems are chewed for tooth trouble (*fely na ngol*), about a piece of 2½ inches long, and then the contents after chewing is spit out" /
"'Gumoi'"

109 *Spathiphyllum* indet. El Salvador / P. C. Standley 22318 / 1922 /
"Fls. eaten in soup and as a vegetable" / "'Gusnáy'"

110   *Spathiphyllum* indet. El Salvador / P. C. Standley 20166 / 1922 / "Spadix eaten, cooked with eggs" / "'Huisnay'"

111   *Lysichiton americanum* U.S.: Wash. / F. G. Meyer 807 / 1937 / "...flowers musk scented"

112   *Calla palustris* U.S.: Mass. / Gray 550 / no year / "'Water Arum'"

113   *Cyrtosperma chamissonis* Caroline Is. / C. C. Y. Wong 158 / 1947 / "The corms are eaten. Children's medicine is made from the fruit and flower stalk, *ettin,* by pounding and mixing with coconut water and coconut milk" / "'Puna'"

Caroline Is. / C. C. Y. Wong 367 / 1948 / "The corms are eaten after they are cooked by boiling"

114   *Lasia spinosa* China / C. W. Wang 77843 / 1936 / "Root stock introduced to Lushan"

Siam / A. Kostermans (Exped.) 56 / 1946 / "...fruit is eaten after putting in brine" / "'Rong kui kuo'" Karieng

Burma / P. Khant D. R. (Herb.?) 270 / 48 / "Leaves edible" / "'Za-yit'"

115   *Amorphophallus rivieri* China / H. C. Chow 8507 / 1938 / "...poisonous when raw"

Philippine Is. / H. C. Conklin 745 / 1953 / "...sap of flower very itchy" / "'Burúy'"

116   *Amorphophallus* indet. Siam / Wichian (Exped.) 200 / 1946 / "...young shoots edible after roasting and crushing out the juice" / "'Chou mei'" Karieng

117   *Amorphophallus* indet. Siam / Wichian (Exped.) 322 / 1946 / "...eatable after stuffing" / "'Tsjo dao'" Karieng

118   *Amorphophallus* indet. Sumatra / W. N. & C. M. Bangham 788 / 1932 / "Fruits orange red. Natives use them for medicine for skin, for itches. Native name 'Panam'"

119   *A. campanulatus* Fiji Is. / O. Degener 13992 / 1940 / "Considered poisonous by natives"

120   *Montrichardia arborescens* Brit. Honduras / P. H. Gentle 3582 / 1941 / "Fruits boiled or roasted and eaten " / "'Camotillo'"

122   *Homalomena aromatica* Indo-China / A. Pételot 7592 / 1940 / "*Stem very aromatic*"

123   *Homalomena* indet. Philippine Is. / M. D. Sulit 3942 / 1950 / "Juice of roots or stems cures pain caused by water snake, by applying it to the affected parts or wound" / "'Rampano'" Tagb.

124   *Schismatoglottis calyptra* Philippine Is. / M. D. Sulit 1873 / 1947 / "...young leaves are cooked and eaten"

125 *Schismatoglottis* indet. Philippine Is. / A. L. Zwickey 120 / 1938 / " . . . frt. and young lvs. cooked and eaten" / " 'Posau' "

127 *Aglaonema philippinensis?* Philippine Is. / P. Añonuevo 45 / 1950 / "Edible. The trunk is roasted for a few minutes before serving" / " 'Palante'."

128 *Dieffenbachia oerstedii* Honduras / Yuncker, Koepper & Wagner 8395 / 1938 / " . . . crushed leaves with disagreeable odor"

129 *Dieffenbachia* indet. Panama / I. M. Johnston 1165 / 1946 / " . . . juice blistering skin"

130 *D. seguine* W.I.: Dominica / J. S. Beard 656 / 1954 / " . . . stem containing a very caustic juice"

W.I.: Dominica / W. H. & B. T. Hodge 2929 / 1940 / " . . . juice supposedly irritating—'scratches' "

Brazil / Krukoff's 6th Exped. 7637 / 1936 / "Used in Curare of Tecuna Indians"

131 *D. paludicola* Brit. Guiana / A. S. Hitchcock 17030 / 1919 / " . . . odor of skunk when fresh. Raphides produced irritation of hands in handling split portions"

132 *Dieffenbachia* indet. Brazil / Krukoff's 6th Exped. 7672 / 1935 / "Used in Curare"

133 *Alocasia* indet. Solomon Is. / S. F. Kajewski 1923 / 1930 / "The sap of these plants have a severe stinging effect on the bare skin. Some of the Taros are used to put in food to poison natives. The tongue swells up, also the mucous membrane of the throat. Possibly the effect of the calcium oxalate crystals would work off after a time, but the native naturally believes he is poisoned and resigns himself to death" / " 'Utakau' "

134 *A. wenzelii* Philippine Is. / A. L. Zwickey 791 / 1938 / " . . . juice from petiole for itch from 'Sagay' or centipede and scorpion bites" / " 'Bugiang' " Lan.

135 *Schizocasia portei?* New Guinea: Papua / Womersley & Simmonds 5088 / 55 / "Juice irritant"

136 *Colocasia maclurei* China / F. A. McClure 20587 / 1937 / "Erect herbaceous drug plant. . . . Corms used in treatment of scrofula"

137 *Caladium puberulum* Colombia / O. Haught 1922 / 1936 / "Inflorescence has peculiar resinous odor—not fetid. Latex abundant, cut ends turn orange within a few minutes. When collected swollen basal parts of spathes were crowded with beetles"

138 *C. colocasia* Hawaiian Is. / O. Degener 8940 / 1930 / "Cultivated by Hawaiians"

139 *Xanthosoma helleboriaefolium* Panama / Woodson, Jr., Allen & Seibert 1606 / 1938 / " ... cooked with other herbs and tea drunk for snake bites"

140 *X. hoffmannii* El Salvador / S. Calderón 860 / 1922 / " 'Hoja de culebra' "

141 *X. roseum* Guatemala / J. D. S. s.n. / 1899? / "Tubers edible. 'Osch' Indian name"

142 *X. brasiliense* W.I.: Dominica / W. H. Hodge 2466 / 1946 / "Young leaves a favorite as 'calalou' (calalon?) for soups; leaves must be scalded else irritating"

143 *Syngonium podophyllum* Mexico / Y. Mexia 1200 / 1926 / "Fr. edible" / " 'Huevo de toro' "

144 *S. ternatum* Brit. Guiana / J. S. de la Cruz 1469 / 1922 / "Very poisonous. Juice milky"

145 *Syngonium* indet. Colombia / A. E. Lawrance 688 / 1933 / "The white sap is exceedingly mucilaginous. Perhaps could be used in glue trade"

146 *Taccarum weddellianum* Bolivia / Cárdenas 3388 / 1935 / "Root poison"

147 *Spathantheum orbignyanum* Bolivia / Cárdenas 3650 / 1944 / " 'Katari papa' = 'Snake's potato' "

148 *Typhonium divaricatum* China / W. T. Tsang 25445 / 1935 / " ... poisonous"

China / H. C. Chow 8555 / 1938 / "Cultivated. Yielding an edible tuber"

Philippine Is. / E. Quisumbing 45–13 / 1945 / "Medicinal plant"

149 *T. trilobatum* Siam / Wichian 317 / 1946 / " ... leaves heated on fire to soften them, (as poultice for boils)"

150 *Arisaema* indet. China / Y. W. Taam 820 / 1938 / "fr. poisonous"

151 *A. consanguineum* China / Sun & Chang 170 / 1939 / " 'Snake Maize' "

152 *A. brevistipitatum* China / W. T. Tsang 20149 / 1932 / "Medicinal value"

153 *A. jacquemontii* Nepal / D. G. Lowndes 985 / 1950 / "The tubers are eaten"

154 *Cryptocoryne cruddasiana* Burma / F. Kingdon Ward 260 / 1939 / "Pollination is probably effected by minute larvae found in the spathe"

155 *Pistia stratiotes* El Salvador / P. C. Standley 19147 / 1921–22 / "Cultivated" / " 'Lechuga de sapo' "

Philippine Is. / C. O. Frake 768 / 1958 / "Leaves applied to ulcers and headache" / " 'Dunsul' " Sub.

156 *Aracea* indet. Mexico / G. B. Hinton (Herb.) 8798 / 35 / "Poisonous"

157 *Aracea* indet. Colombia / R. E. Schultes 3615 / 1942 / "Medicinal"

158 *Aracea* indet. Colombia / R. E. Schultes 3662 / 1942 / "Medicinal"

159 *Aracea* indet. Colombia / R. E. Schultes 3512 / 1942 / " 'Culebrilla huasca' "

160 *Aracea* indet. Colombia / R. E. Schultes 3424 / 1942 / "Used as vermifuge" / " 'Tzen-den-day' "

161 *Aracea* indet. Colombia / R. E. Schultes 3437 / 1942 / "Sap placed in ear for curing aches"

162 *Aracea* indet. Brit. New Guinea / L. J. Brass 6623 / 1936 / " . . . all parts anise-scented"

163 *Aracea* indet. Brit. New Guinea / L. J. Brass 7361 / 1936 / "Pl. when crushed emits sickening sweetish odor"

FLAGELLARIACEAE

164 *Joinvillea elegans* Fiji Is. / O. Degener 15139 / 1941 / "Fijians squash root and tie onto wound ("nail poke") in foot to promote healing" / "Vavara Serua"

165 *Hanguana malayana* var. *anthelmintica* Brit. New Guinea / L. J. Brass 7925 / 1936 / "Pithy stems and runner eaten raw by natives"

ERIOCAULACEAE

166 *Eriocaulon sinicum* China / H. H. Chung 2711 / 1924 / " . . . bought from the shop selling fresh medicinal plants . . . used in Chinese medicine"

BROMELIACEAE

167 *Puya flocosa* Venezuela / J. Steyermark 68623 / 1944 / "Leaves used as purgative when boiled" / " 'arua-yek' "

168 *P. gummifera* Ecuador / W. H. Camp E2202 / 1945 / " 'Achupalla'— base of plant, especially the expanded leaf-base, eaten by the common people and said to be 'good for the kidneys'; also fed to cattle, pigs, etc."

169 *P. pyramidata* Peru / T. H. Goodspeed 8177 / 1938 / " . . . the leaves are used to fatten guinea pigs" / " 'Achupalla chica' "

170 *P. hamata* Ecuador / J. N. Rose 22778 / 1918 / "Tender base of leaves used for salad also made into flour for the Indians" / "Arua-rongo"

171 *Pitcairnia pulchella* Ecuador / J. Steyermark 52770 / 1943 / "Mules eat leaves" / " 'shagraquihua' "

172 *Tillandsia benthamiana* Mexico / W. P. Hewitt 38 / 1945 / " . . . cooked to a soup in water, mixed with 50% alcohol as preservative, then taken

1 teaspoonful each morning as a cure for anaemia and kidney trouble"
/ "Herba de pajaro Lichen de Encino Mescalito"

173   *T. circinnata* Dominican Republic / R. A. & E. S. Howard 8367 /
1946 / "Local name: 'guajaca'"

174   *T. lorentziana* Brazil / J. Eugenio 2211 / 1941 / "'Sapucaia'"

175   *T. complanata* Ecuador / J. A. Steyermark 53250 / 1943 / "Indians
use the dry basal leaves to wrap tamales of borlios (farina de maiz
with sugar)"

176   *T. streptocarpa* Brazil / M. & R. Foster 1058 / 1940 / " . . . has exqui-
site fragrance"

177   *Bromelia karatas* Venezuela / J. A. Steyermark 61026 / 1945 /
" . . . fruit edible crude, with sweet delicate flavor, . . . brownish hairs
at base of fruit and on fruit used to put on burns; dry immediately
after placing on burns" / "'kurucujurro'"

COMMELINACEAE

178   *Pollia* indet. New Britain / A. Floyd 3499 / 54 / "The fruit is used as
medicine" / "La kamamuta (W. Nakanai)"

179   *P. thyrsifolia* Philippine Is. / C. Frake 491 / 1958 / "Leaves applied
for amenorrhea" / "galupu? ulangan Sub."

180   *Commelina diffusa* Bolivia / J. Steinbach 5292 / 1921 / "Se usa el
agua que junta en la hoja guardaflor contra el mal del ojos" / "Santa
Lucia"

Peru / G. Klug 1638 / 1930 / "Aromatic tea"

Caroline Is. / C. C. Y. Wong 329 / 1948 / "The entire plant is used in
the medicine for general rundown with the loss of the use of the legs
and hands, general swelling (mocunuk) and general weakness usually
born with this feeling, (mogotugut); the entire plant is pounded,
placed in a coconut sheath and squeezed, an old coconut is grated
and mixed with the sap of the plant, then squeezed and the mixture
is drunk" / "fanu"

Fiji Is. / O. Degener 14997 / 1941 / "Juice said to be cure for wounds,
also used to remove particles from eyes"

181   *C. erecta* Bolivia / J. Steinbach 6732 / 1924 / "La hoja enveltoria . . .
encierra un liquido transparente . . . y es usado para curar el mal de
ojos (inflamación epidemica)"

182   *C. benghalensis* China / W. T. Tsang 581 / 1923 / " . . . medicine" /
"Paak Yat Sai Ts'a"

183   *C. chinensis* var. *ciliata* China / Sun & Chang 727 / 1939 / " . . . leaves
edible"

184   *C. communis* Japan / E. Elliott 2 / '46 / "Young leaves boiled, washed
with water, eaten with salt and Miso. Medicinal use: The whole plant

is decocted and taken to cure heart disease, diarrhoea, throat swelling and impurity of blood" / "Tsuyukusa"

185 *C. obliqua* China / L. Y. Tai T1402 / 1942 / "Young stem and leaf edible"

186 *Commelina* indet. Philippine Is. / R. B. Fox 7 / 1949 / " ... medicinal" / "Kuwán-Kuwán Neg."

187 *Aneilema bracteatum* China / H. H. Chung 2710 / 1924 / " ... used in medicine, Chinese name 'Cocks tongue'"

188 *Aneilema* indet. China / L. Y. Tai T1397 / 1942 / "Young stem and leaf edible"

189 *Aneilema* indet. New Britain / A. Floyd 6487 / 54 / "The cooked leaves are applied to skin rashes" / "La pohouma (W. Nak.)"

190 *Campelia zanonia* El Salvador / P. C. Standley 19320 / 1921 / "Remedy for gonorrhoea" / " 'Cana de Cristo' "

PONTEDERIACEAE

191 *Monochoria vaginalis* Philippine Is. / C. O. Frake 508 / 1958 / "Corms pounded and applied for ulcers" / "Sesela Sub."

192 *Pontederia cordata* Colombia / G. Klug 1877 / 1930 / "Remedy for 'caracha' "

STEMONACEAE

193 *Stemona tuberosa* China / W. T. Tsang 21441 / 1932 / " ... roots cooked with pig's feet can be used as medicine to heal rheumatism" / "Taai Chun Kan Keuk"

China / W. T. Tsang 22391 / 1933 / "Fr. edible" / "Chuk Ko Shue"

LILIACEAE

194 *Tofieldia gracilis* Japan / M. Furuse s.n. / 1948 / "(Chyabo-zekishoo)"

195 *T. japonica* Japan / M. Furuse s.n. / 1956 / "(Iwa-syoobu)"

196 *T. nuda* Japan / M. Furuse s.n. / 1957 / "(Yasyuu—hanazekisyoo)"

197 *T. nutans* Japan / T. Yamazaki 3127 / 1952 / "Nom. Jap.: Tisima—zekisyo"

198 *T. nutans* var. *kondoi* Japan / M. Furuse s.n. / 1957 / "(Apoi-zeki-syoo)"

199 *T. okuboi* Japan / M. Furuse s.n. / 1958 / "(Hime-iwa-syoobu)"

200 *Narthecium asiaticum* Japan / M. Furuse s.n. / 1958 / "(Kimkooka)"

201 *Chamaelirium luteum* U.S.: Va. / Fernald & Long 11804 / 1940 / "odor heavy"

202 *Heloniopsis japonica* Japan / S. Suzuki 91 / 1947 / "Nom. Jap.: Shôjobakama"

Japan / K. Uno 5878 / 1931 / "shirobana-syōjobakama"

203   *H. orientalis* Japan / M. Furuse s.n. / 1958 / "(Syoojoo-bakama)"

204   *Metanarthecium foliatum* Japan / K. Watanabe s.n. / 1891 / "Nebari-nogi-ran"

205   *Amianthium muscaetoxicum* U.S.: Md. / C. C. Plitt 676 / 1901 / "'Fly-poison'"

       U.S.: Ga. / Harper & Mizell 60 / 1930 / "'Puppy tails'"

206   *Schoenocaulon coulteri* Mexico / J. Gregg s.n. / 1848 / "root used for snuff"

207   *S. jaliscense* Mexico / H. S. Gentry 6555 / 1941 / "Cebadilla. Roots employed for killing maggots in wounds; powdered and applied or decocted to make wash"

208   *S. officinale* Venezuela / Curran & Haman 1196 / 1917 / "'Cebadilla.' Used by the Germans in making poisonous gases"

209   *Stenanthium gramineum* U.S.: Va. / H. A. Allard 3236 / 1937 / "Very stinking"

210   *Stenanthella frigida* Mexico / C. V. Morton 15572 / 40 / "Flower & root smell like fungus. Said to poison cattle"

211   *Zygadenus densus* U.S.: S.C. / B. L. Robinson 177 / 1912 / "Sold by negroes in market under name 'Black Snakeroot'"

212   *Z. elegans* U.S.: N.M. / O. St.John 2 / 1896 / "Flower ill scented"

213   *Z. micranthus* U.S.: Calif. / H. N. Bolander 137 / 1864 / "Called hog's potato by the farmers of that section because the hogs root them out"

214   *Zygadenus* indet. Mexico / G. B. Hinton et al. 8747 / 35 / "Vernac. Lirio"

215   *Veratrum virginicum* U.S.: Iowa / Pammel & Zimmerman 285 / 1925 / "Poisonous to stock"

216   *V. tenuipetalum* U.S.: N.M. / O. St.John s.n. / 1895 / "called Elk-weed"

217   *V. anticleoides* Japan / K. Uno 21284 / 1937 / "karafuto-syurosō"

218   *V. grandiflorum* Japan / M. Furuse s.n. / 1958 / "(Baikei-soo)"

219   *V. maximowiczii* Japan / M. Furuse s.n. / 1958 / "(Awoyagi-soo)"

220   *V. nigrum* China / T. P. Gordeev 270 / 1937 / "Medicinal plant"

221   *V. nigrum* var. *japonicum* Japan / M. Furuse s.n. / 1948 / "(Shūro-soo)'

222   *V. stamineum* Japan / M. Furuse s.n. / 1958 / "(Ko-baikei-soo)"

223   *V. maackii* Korea / R. K. Smith s.n. / 1937 / "Shurosō Packsai Yuro"

224   *V. maackii* var. *parviflorum* Japan / M. Furuse s.n. / 1958 / "(Awoyagi-soo)"

225 *Gloriosa superba* Katanga / Quarré 1443 / 1928 / "toxique pour le betail"

226 *Tricyrtis affinis* Japan / M. Furuse s.n. / 1954 / "(Yamaji-no-hototo-gisu)"

227 *T. flava* Japan / T. Komeda 22337 / 1937 / "cyabo-hototogisu"

228 *T. latifolia* Japan / M. Furuse s.n. / 1935 / "(Tamagawa-hototogisu)"

229 *T. macropoda* Japan / M. Furuse s.n. / 1958 / "(Yama-hototogisu)"

230 *T. japonica* Japan / K. Uno E37 / 1938 / "yamojino-hototogisu"

231 *Asphodelus fistulosus* Mexico / Shreve 9328 / 1939 / "'Estrella del norte'"

232 *A. tenuifolius* Arabia / H. Dickson 15 / 61 / "Vern. name Bur wug. Eaten only in summer when dry by animals"

233 *Anthericum consanguineum* Mexico / Y. Mexia 8822 / 1937 / "'Amol'"

234 *A. eleutherandrum* Guatemala / J. A. Steyermark 50515 / 1942 / "'chichuisa'"

235 *A. torreyi* Mexico / A. López 65 / 1948 / "'Clabellina'"

236 *Pasithea coerulea* Chile / J. West 3963 / 1935 / "'Pajarito'"
Chile / E. E. Gigoux s.n. / 1887 / "'Varilla de San José'"

237 *Chlorogalum pomeridianum* U.S. or Canada / Fremont's 2nd Exped. 81 / 1884 / "'Ammolé'"

238 *Stypandra caespitosa* Tasmania / O. Rodway H972 / "Vern. Name 'Tufted Lily'"

239 *Excremis coarctata* Colombia / Killip & Smith 18721 / 1927 / "Gramalote"

240 *Dianella ensifolia* China / W. T. Tsang 16469 / 1927 / "Name reported: Fung Ch'e Hop Ts'o"
Japan / K. Uno 8742 / 1931 / "Nipp. kikyō-ran"
Fiji Is. / L. Reay 33 / 1941 / "Common name: 'Me meirakalavu'"
New Hebrides Is. / S. F. Kajewski 346 / 28 / "Common name: Did-and-did"

241 *D. nemorosa* Sumatra / R. S. Boeea 8546 / 1935 / "'si tanggis'"
New Guinea: Papua / R. D. Hoogland 3918 / 1953 / "Local Name: Sienda (Orokaiva language, Mumuni)"
Philippine Is. / G. E. Edaño 11169 / 1949 / "Common name: Bugna. Dialect Ma"

242 *Hosta coerulea* Japan / K. Watanabe s.n. / 1891 / "Gibōshi"

243 *H. japonica* Japan / K. Uno 9295 / 1935 / "Nipp. nagaba-mizu-gibōshi"

244 *H. longipes* Japan / M. Furuse s.n. / 1952 / "(Iwa-gibooshi)"

245 *H. longissima* Japan / M. Furuse s.n. / 1958 / "(Mizu-gibooshi)"

246 *H. minor* Korea / K. Uno 23574 / 1938 / "Nipp. ko-giboshi"

247 *H. rhodeifolia* Japan / K. Uno 24092 / 1939 / "Nipp. to-gibashi"

248 *H. ventricosa* China / S. K. Lau 4668 / 1934 / "Used as medicine, Kau Tsit Lin"

249 *Hemerocallis citrina* China / W. T. Tsang 27710 / 1937 / "fragrant. Ching Chan Ts'oi"

250 *H. esculenta* Japan / S. Suzuki 339007 / 1947 / "Jap. name: Nikkô-kisuge"

251 *H. flava* China / Steward & Cheo 916 / 1933 / "Used as vegetable"
China / W. T. Tsang 25857 / 1935 / "edible. Kam Cham Ts'oi"
China / C. Y. Chiao 2880 / 1930 / "fls. used as a vegetable"

252 *Lomandra leucocephala* Australia / P. Ambrose s.n. / 1940 / "Iron Grass"

253 *Xanthorrhoea preissii* Australia / C. T. White 5174 / 1927 / "'Black-boy' or 'Grass Tree'"

254 *Allium schoenoprasum* var. *laurentianum* Newfoundland / Fernald, Long & Fogg 1520 / 1929 / "Local name: 'Chibbles'"

255 *A. cernuum* U.S.: Ill. / H. H. Smith 5678 / 1913 / "Indian name 'Chicago'"

256 *A. anceps* U.S.: Nev. / C. L. Anderson 23 / 1865 / "These onions come up very early in the spring . . . The Indians find them very useful being the first of green things appearing after the long winter"

257 *A. perdulce* U.S.: Neb. / H. Hopeman 882 / 1899 / "Com. 'Nuttall's Wild Onion'"

258 *A. victorialis* Kodiak Is. / W. J. Eyerdam 3106 / 1939 / "brittle stem edible but it has a very strong onion or garlic flavor"

259 *A. kunthii* Mexico / G. B. Hinton et al. 16522 / 44 / "Vernac. Name 'Ajo' Uses: to flavor food; to eat"
Mexico / R. M. Stewart 858 / 1941 / "'Cebolla Cimarron'"

260 *A. grayi* Japan / M. Furuse s.n. / 1958 / "(No-biru)"

261 *A. hookeri* China / C. W. Wang 77337 / 1936 / "Edible"

262 *A. japonicum* Japan / M. Furuse s.n. / 1954 / "(Yama-rakkyo)"

263 *A. macrostemon* China / H. T. Feng 180 / 1925 / "Uses: eating"

264 *A. nipponicum* Japan / E. Elliot 96 / 46 / "Food use: Leaves & bulbs are eaten together after boiled and seasoned with Misō. Medicinal use: Dried whole plant (decocted) cures blood diseases and diarrhoea & is good for nerves. It opens menses"

265   *A. schoenoprasum* var. *orientale* Japan / M. Furuse s.n. / 1958 / "(Shirouma-asatsuki)"

266   *A. schoenoprasum* var. *yezomonticola* Japan / M. Furuse s.n. / 1957 / "(Hime-ezo-negi)"

267   *A. splendens* Japan / K. Uno 20829 / 1937 / "Nipp. Chishima-rakkyo"

268   *A. taquetii* Korea / K. Uno 23262 / 1938 / "Nipp. saisu-yamarakkyō"

269   *A. thunbergii* Japan / M. Furuse s.n. / 1957 / "(Yama-rakkyoo)"

270   *A. tuberosum* China / C. I. Lei 978 / 1933 / "Oil from seeds is used for food"

Philippine Is. / A. L. Zwickey 654 / 1938 / "'Sakorab' (Lan.); for flavor & odor in fish"

271   *A. victorialis* var. *platyphyllum* Japan / M. Furuse s.n. / 1958 / "(Gyoojya-nimniku)"

Japan / S. Suzuki 93 / 1946 / "Nom. Jap. Gŷoja-nim'niku"

272   *Allium* indet. China / W. T. Tsang 26024 / 1935 / "fr. edible"

273   *A. wallichii* E. Nepal / Herb. Banerji 1243 / 60 / "local name 'Jimbur'; leaves used as a condiment"

274   *A. uliginosum* Sumatra / R. S. Boeea 8291 / 1935 / "si maroeli oele"

275   *Nothoscordum fragrans* Jamaica / W. Harris 12662 / 1917 / "'Wild Onion' flowers fragrant"

276   *N. andicola* Peru / C. Vargas 1623 / 1939 / "N. vulgar. chullcu"

277   *Lilium longiflorum* El Salvador / P. C. Standley 20487 / 1922 / "Azucena"

Japan / no collector s.n. / 1892 / "Jap. Tametomo-yuri"

278   *L. auratum* Japan / E. Elliot 90 / 1946 / "Uses: Bulbs are slightly bitter but are eaten cooked. It can be powdered and used to make delicious dumpling ball"

Japan / K. Watanabe s.n. / 1891 / "Ryōri-yuri"

Japan / no collector s.n. / 1894 / "Jap. Kuruma-yuri"

279   *L. brownii* China / W. T. Tsang 27945 / 1937 / "Pak Hop Fa. fr. edible"

280   *L. cordatum* Japan / M. Furuse s.n. / 1958 / "Uba-yuri"

281   *L. concolor* Japan / no collector s.n. / 1892 / "Jap. Aka-himeyuri"

China / Cheo & Yen 52 / 1936 / "Bulb edible"

282   *L. japonicum* var. *albomarginatum* Japan / no collector s.n. / 1892 / "Jap. Fukurin-sasayuri"

283   *L. lansium* Korea / E. H. Wilson 8509 / 1917 / "Takashima-yuri"

284   *L. maximowiczii* Japan / no collector s.n. / 1891 / "Ko-oni-yuri"

285   *L. makinoi* Japan / K. Uno 5805 / 1931 / "Nipp. sasa-yuri"

286   *L. medeoloides* Japan / M. Furuse s.n. / 1958 / "(Kuruma-yuri)"

287   *Fritillaria agrestis* U.S.: Calif. / C. B. Wolf 8324 / 1937 / "foliage glaucous;...odor like bad cheese"

288   *F. camschatcensis* Kodiak Is. / W. J. Eyerdam 3105 / 1939 / "Flower ...rather unpleasant odor, edible roots"

289   *Tulipa edulis* Japan / M. Furuse s.n. / 1958 / "(Amana)"

290   *Erythronium japonicum* Japan / M. Furuse s.n. / 1958 / "(Kata-kuri)"

291   *Lloydia alpina* Japan / K. Uno 20724 / 1937 / "Nipp. chishima-amana"

292   *L. triflora* Japan / M. Furuse s.n. / 1958 / "(Hosobano-amana)"

293   *Scilla chinensis* Japan / no collector s.n. / 1893 / "Jap. Tsurubo"

294   *Ornithogalum biflorum* var. *chlorolinea* Chile / D. Bertero 481 / 1828 / "Vulgo. Lagrima dela Virgen"

295   *Yucca elata* Mexico / M. Stewart 2655 / 1942 / "'Sollate'"

296   *Y. schottii* Mexico / S. S. White 4217 / 1941 / "'Dátil'"

297   *Y. torreyi* Mexico / W. P. Hewitt 179 / 1947 / "Common name: Palma angosta. Uses: flowers dried, boiled as tea and drunk for cough remedy"

298   *Hesperaloë funifera* Mexico / A. H. Schroeder 167 / 1941 / "'Saban-doque'"

299   *Nolina recurvata* Mexico / R. L. Dressler 2038 / 1957 / "'soyate'"

300   *Dasylirion inermes* Mexico / E. Palmer 644 / 1905 / "'Palma culona.' 'Zoyate'"

301   *D. leiophyllum* Mexico / W. P. Hewitt 246 / 1947 / "Uses: trunk distilled to produce alcoholic drink 'Sotohl' which has substance of an oil"

302   *Cordyline terminalis* Java / H. Hallier 85 / 1910 / "Inl. naam: handjoewang. Gebruik: obat sakit oeloe hati (= Mittel gegen Leberleiden)"

303   *C. roxburghiana* Philippine Is. / H. C. Conklin 127 / 1953 / "Common name: Di ulay. Dialect: Mang."

Philippine Is. / H. C. Conklin 847 / 1953 / "Common name: Maguey? Magé"

304   *C. fruticosa* Pitcairn I. / A. M. Christian s.n. / no year? / "'Tee leaves'"

Fiji Is. / A. C. Smith 5471 / 1947 / "Ngolo"

Fiji Is. / A. C. Smith 48 / 1933 / "Kokoto"

Fiji Is. / A. C. Smith 6478 / 1947 / "'Nggai.' Roots used for sweetening"

Caroline Is. / C. Y. Wong 267 / 1947 / "Native name: Tiinen cuuk. The name means tiin of the mountain. The unopened leaves are used to treat pain in the stomach: the leaves are pounded slightly and wrapped in a bundle of coconut cloth; this bundle is then tied with a string and worn around the neck so that the bundle lies next to the stomach; it is worn for 10 or 20 days and cannot be thrown away, or the sickness will come back. The unopen leaves are called macangan tiin. The underground portion of the trunk can be used as medicine for säfein mwöcamu (medicine for constipation); a coconut that has fallen to the ground but not yet sprouted is obtained & cut in half; one half is slightly roasted in a fire, both halves are grated in a wooden bowl, then a portion of the tiin bark is scraped into a bowl; the whole mixture is expressed with a piece of coconut cloth and the resulting liquid is drunk"

Caroline Is. / C. Y. Wong 344 / 1948 / "Native name: ric. Uses: The plant is used for medicine in bringing an erection of the penis (tu). The young unrolled leaves are taken and pounded then placed in a coconut sheath and squeezed; a young coconut is taken and the water is used to mix with the juice of the plant and drunk"

305   *C. australis* New Zealand / A. W. Anderson 144 / 1934–35 / "('Cabbage tree.' 'Ti')"

306   *Dracaena elliptica* Cochinchina / L. Pierre 1849 / 1877 / "Cay băè gŭe"

S. Sumatra / W. J. Lütjeharms 4869 / 1930 / "Inl. naam: Lindjoeang oetan"

307   *Dracaena* indet. N. Borneo / L. Apostel B.N.B. For. Dept. 2435/5 / 1926 / "Leaves edible when cooked"

308   *Dracaena* indet. Solomon Is. / S. F. Kajewski 2536 / 1931 / "Common name: Chil-in chili. When a man has a swelling of the joints the leaves are taken and macerated with water and rubbed on the affected parts"

309   *Pleomele* indet. Sumatra / R. S. Boeea 5936 / 1933 / "pagar sari"

310   *P. angustifolia* Philippine Is. / Sulit & Conklin 5098 / 1953 / "Common name: Bayibayi ambô. Uses: Decoction of roots used as a cure for stomach trouble"

Philippine Is. / P. Añonuevo 279 / 1950 / "Common name: Alibhid. Dialect Bis."

311   *Pleomele* indet. Philippine Is. / H. C. Conklin 275 / 1953 / "Common name: Mirawnūn"

312   *Astelia malayana* Sumatra / R. S. Boeea 6034 / 1933 / "si hilap tombak"

313 *A. alpina* Tasmania / E. Rodway H953 / 1930 / "Vern. name 'Artichoke.' 'Pineapple Grass'"

314 *Collospermum montanum* Fiji Is. / A. C. Smith 1978 / 1934 / "'Malatava ni vekau'"

Fiji Is. / A. C. Smith 5156 / 1947 / "Mbevu"

315 *Asparagus cochinchinensis* China / W. T. Tsang 28701 / 1938 / "Tin Tung.... Fl. white, fragrant; fr. edible"

China / W. T. Tsang 20474 / 1932 / "Tien Tung, for medicine"

China / R. C. Ching 5293 / 1928 / "the natives use this as medicine"

China / S. K. Lau 36 / 1932 / "Mai Lun (Lois)"

Japan / no collector s.n. / 1889 / "Kusa-sugi-kazura"

316 *A. schoberioides* Japan / no collector s.n. / 1894 / "Jap. Kiji-kakushi"

317 *Clintonia udensis* Japan / K. Uno 19887 / 1937 / "Nipp. tsubame-omoto"

318 *Smilacina hondoensis* Japan / M. Furuse s.n. / 1958 / "(Ohba-yuki-gasa)"

319 *S. japonica* Japan / K. Uno 21512 / 1937 / "Nipp. yuki-zasa"

320 *S. trifolia* Japan / K. Uno 20126 / no year? / "Nipp. tonakaiso"

321 *Smilacina* indet. China / T. T. Yü 19822 / 1938 / "Edible"

322 *Maianthemum bifolium* Japan / K. Ishida 5843 / 1913 / "Nipp. mai-zuru-so"

323 *Disporum smilacinum* Japan / M. Furuse s.n. / 1958 / "(Chigo-yuri)"

Japan / no collector s.n. / 1894 / "Jap. Tigo-yuri"

324 *D. viridescens* Korea / J. H. Rho (gift) Sungkyunkwang Univ. s.n. / 1954 / "K. N. peun aigi nari"

Korea / K. Uno 23113 / 1938 / "Nipp. oh-chigo-yuri"

325 *D. sessile* Japan / M. Furuse s.n. / 1958 / "(Hoochaku-soo)"

326 *Disporum* indet. Burma / F. G. Dickason 8535 / 1939 / "Taw-pi"

327 *D. chinense* Sumatra / R. S. Boeea 10731 / 1936 / "Kajoe soma-soma"

328 *Polygonatum* indet. Burma / F. G. Dickason 9705 / 1940 / "Young sprouts eaten in soup"

329 *Reineckea carnea* Japan / M. Furuse s.n. / 1958 / "(Kichijyoo-soo)"

330 *Paris chinensis* China / W. T. Tsang 20295 / 1932 / "For making medicine"

China / W. T. Tsang 22915 / 1933 / "Tsat Ip Yat Chi Fa. Used as medicine"

331 *P. hainanense* China / W. T. Tsang 21439 / 1932 / "Tsat Ip Yat Chi Fa ... root used as medicine to remove toxin"

332   *P. japonica* Japan / M. Furuse s.n. / 1958 / "(Kinugasa-soo)"

333   *P. quadrifolia* Japan / no collector s.n. / 1891 / "Kurumaba-tsuku-bane-soo"

334   *Trillium erectum* U.S.: Mass. / D. W. Rogers s.n. / 1886 / "Birthroot"

335   *T. apetalon* Japan / M. Furuse s.n. / 1946 / "(Enrei-sō)"

336   *T. kamtschaticum* Japan / K. Uno 20653 / 1937 / "ohbana-enreiso"

337   *T. smallii* Japan / M. Furuse s.n. / 1958 / "(Emrei-soo)"

338   *T. tschonoskii* Japan / M. Furuse s.n. / 1958 / "(Shirobana-emrei-soo)"

339   *Liriope koreana* Korea / K. Uno 23042 / 1938 / "Nipp. chosen-yaburan"

340   *L. minor* Japan / K. Uno 5840 / 1929 / "Nipp. hime-yaburan"

341   *L. graminifolia* China / W. T. Tsang 15891 / 1927 / "Ka Kam Cham Ts'o"

342   *Ophiopogon japonicus* Japan / M. Furuse s.n. / 1958 / "(Jyano-hige)"
Korea / R. K. Smith s.n. / 1934 / "Jano hige"

343   *O. japonicus* var. *wallichianus* Japan / no collector s.n. / 1891 / "Ōba-ja-no-hige"

344   *Peliosanthes stenophylla* China / W. T. Tsang 20366 / 1932 / "Shek Laan Fa"

345   *P. teta* China / Y. W. Taam 53 / 1937 / "Tu-mao-ken"

346   *Peliosanthes* indet. Sumatra / R. S. Boeea 5775 / 1933 / "doekoet poetar-poetar"

347   *Aletris farinosa* U.S.: Va. / S. Kouvacs 87 / 1933 / "Common name: Colic-root"

348   *A. foliata* Japan / M. Furuse s.n. / 1957 / "(Nebari nogiran)"

349   *A. spicata* Japan / K. Uno 5662 / 1914 / "Nipp. sokushin-ran"

350   *Geitonoplesium cymosum* Fiji Is. / A. C. Smith 4005 / 1947 / "'Wa mbitumbitu'"
Fiji Is. / A. C. Smith 4848 / 1947 / "'Wa ula'"
Fiji Is. / L. Reay 13 / 1941 / "'Wadakua.' Young shoots edible when roasted"
Fiji Is. / O. Degener 15169 / 1941 / "'Wambitumbitu' (Serua). Leaves used medicinally for some disease 'like smallpox.' Leaves are mashed, fresh water added all put in bamboo or bottle. Liquid applied to blisters night and morning"
Fiji Is. / O. Degener 15027 / 1941 / "'Waula' (Sabatu)"

351   *Luzuriaga polyphylla* Chile / J. West 4671 / 1935 / "'Quilineja, azahar'"

352   *Luzuriaga* indet. Solomon Is. / S. F. Kajewski 2641 / 1931 / "Common name 'Halon-vava-lise'"

353 **Rhipogonum** indet. Solomon Is. / S. F. Kajewski 2382 / 1930 / "Common name: 'Mar-mala-fari'"

354 **Smilax luculenta** Honduras / P. C. Standley 55107 / 1928 / "'Corona de Cristo'"

Brit. Honduras / P. Gentle 3261 / 1940 / "'Sarsaparilla'"

355 **S. moranensis** Mexico / G. B. Hinton 3478 / 33 / "Uses: Concoction of the root taken for kidneys" / "'Vern. name Palo de la Vida'"?

356 **S. pringlei** Mexico / G. B. Hinton 3543 / 33 / "Vernac. name: Nitamo"

357 **S. spinosa** El Salvador / P. C. Standley 22647 / 1922 / "'Bejuco de corona'"

El Salvador / P. C. Standley 21833 / 1922 / "'Espuela de gallo'"

358 **S. populnea** Dominican Republic / Pater Fuertes 444 / 1910 / "Vernac.: 'Longuey'"

359 **S. brasiliensis** Uruguay / G. Herter 359 / 1927 / "N.v. Zarzaparilla blanca"

360 **S. campestris** Brazil / Y. Mexia 5143 / 1930 / "'Pitanga Liso'"

361 **S. cumanensis** Brazil / R. Froes 11631 / 1939 / "'Japecunga'"

362 **S. ramiflora** Brazil / Y. Mexia 5271 / 1930 / "'Pitanga sem Espinha'"

363 **Smilax** indet. Brazil / C. Mendes 1800 / 1932 / "'Japecanga.' Roots used as a treatment for syphilis"

364 **S. arisanensis** China / W. T. Tsang 20434 / 1932 / "Kai Kung Tang"

365 **Smilax** indet. China / Y. W. Taam 372 / 1938 / "Shue Leueng Kung"

366 **S. glabra** China / W. T. Tsang 21180 / 1932 / "Fo Taan Tse T'ang"

China / W. T. Tsang 21575 / 1932 / "Min Pei Tse T'ang"

China / W. T. Tsang 22920 / 1933 / "Ngan Fan Tau ... fruit ... edible"

China / W. T. Tsang 20832 / 1932 / "Tsim Mi Ip Ma K'ap Lat"

367 **Smilax** indet. China / W. T. Tsang 24168 / 1934 / "Kam Kwong Lak"

368 **S. stenopetala** Formosa / Simada-Hidetarô 949 / 1936 / "'Satuma-sankirai'"

369 **S. china** China / J. Hers 1049 / 1919 / "name: shan cha 'mountain tea'"

China / W. T. Tsang 21750 / 1932 / "Uen Ip Ma Cap ... fruit ... edible"

Japan / M. Furuse s.n. / 1957 / "(Sarutori-ibara)"

Burma / F. G. Dickason 7407 / 1938 / "Common name Haka Chin.: 'artitsum tsuk'"

370 **S. davidiana** China / W. T. Tsang 28916 / 1938 / "Sai Yeung Ma Kap Tang"

371   *S. discotis* var. *concolor* China / W. T. Tsang 20868 / 1932 / "Ma Kap T'ang . . . fruit . . . edible"

372   *S. corbularia* China / W. T. Tsang 21343 / 1932 / "Chin Ip Ma Mei T'ang"

       Siam / native collector Royal For. Dept. S479 / 1949 / "Local name: Hua Khaw-yen Nuea. Tuberous root used for venereal disease remedy"

373   *S. hypoglauca* China / Tsang, Tang & Fung 17642 / 1929 / "Chuk Ip Kap"

       China / W. T. Tsang 25794 / 1935 / "Shue Leung Kung"

374   *S. lanceaefolia* China / Y. W. Taam 690 / 1938 / "Shu Leung Kown"

       China / W. T. Tsang 24634 / 1934 / "Chim Ip Ma Kap Tang"

       China / W. T. Tsang 25926 / 1935 / "Ma Kap Tang Fr. edible"

375   *S. opaca* China / C. I. Lei 227 / 1932 / "Pun Muk Tang"

376   *S. parvifolia* India: Assam / L. F. Ruse 119 / 1923 / "Vern. soh. krot. rit. Economic use: extract from leaves & roots is used for medicinal purposes by natives"

377   *S. megacarpa* China / S. K. Lau 398 / 1932 / "Ngai Mongdoe (Lai)"

       Indochina / Poilane 1469 / 1920 / "Annte.: Däy mau ch ou day cam ićh. Les fruits noir à maturité son comestible"

378   *S. megalantha* var. *maclurei* China / W. T. Tsang 25560 / 1935 / "Shue Leung Kung"

       China / W. T. Tsang 20816 / 1932 / "Tai Ip Mah Kaap Tun"

379   *S. nipponica* Japan / K. Uno 2611 / 1951 / "Nipp. Shiode"

380   *S. oldhami* Korea / R. K. Smith s.n. / 1933 / "Tachi shiode"

381   *S. ovalifolia* China / F. A. McClure 8446 / 1921 / "name reported— Ma kap"

382   *S. perfoliata* China / W. T. Tsang 16063 / 1927 / "Tai Ip Ma Kap Tang"

383   *S. riparia* China / W. T. Tsang 23228 / 1933 / " . . . fruit . . . edible. Ma Mi Kit"

384   *S. herbacea* China / C. I. Lei 633 / 1933 / "Ngou Chue Tang"

385   *S. sarumame* Japan / M. Furuse s.n. / 1958 / "(Saru-mame)"

386   *S. stans* Japan / H. Muroi 468 / 1950 / "Jap. name: Muruba-sankirai"

387   *S. scobinicaulis* China / J. Hers 1821 / 1921 / "'tao kow tze'"

       China / J. Hers 1257 / 1919 / "Chinese name: la tuan kin"

       China / E. H. Wilson 27 / 1907 / "'Chin-Pa-Ton'"

388   *S. sieboldi* Japan / H. Muroi 234 / 1952 / "Jap. name: Yamaga syu"

389 *S. taiheiensis* Formosa / K. Uno 10637 / 1934 / "Nipp. taikei-sankirai"

390 *S. trachyclada* Formosa / Simada-Hidetarô 1130 / 1937 / "'Agagata-osankiru'"

391 *S. thomsoniana* China / W. T. Tsang 20023 / 1932 / "Yuen Yip Ma Kaap Tang"

China / W. T. Tsang 21729 / 1932 / "Kau Shi Shue"

392 *S. bauhinioides* Indochina / Poilane 926 / 1920 / "Annte.: Cay cu'ôm lăng.... fruits... seraient comestibles"

393 *S. ferox* Burma / F. G. Dickason 8538 / 1939 / "(Chin) 'yum ghat.' Bur. name: 'Sein-da-baw'"

India: Assam / L. F. Ruse 54 / 1933 / "Vern.: Soh krot. Economic uses: the root is ground and the extract in water is used for stomach complaints"

394 *S. myrtillus* India: Assam / L. F. Ruse 117 / 1923 / "Vern. shi-ja-hroit. Economic uses: extract from the leaves and roots is used for medicinal purposes by natives"

395 *S. zeylanica* India / J. Fernandes 588 / 1949 / "'Gotisero'"

396 *S. calophylla* Sumatra / R. S. Boeea 8590 / 1935 / "'kajoe badja ra-poton'"

397 *S. leucophylla* Sumatra / W. J. Lütjeharms 3976 / 1936 / "Inl. naam: Akar banau   Taal: Mal. Palemb."

Philippine Is. / M. D. Sulit 3799 / 1950 / "Common name: Banag. Dialect Tagb. Tagbunua uses: roots boiled and decoction taken by women having irregular menses. Tops cooked as vegetables"

398 *Smilax* indet. Solomon Is. / S. F. Kajewski 2275 / 1930 / "The bark at the base of the stem is macerated and applied to an aching tooth to stop toothache"

399 *Smilax* indet. New Guinea / A. Floyd 3486 / 54 / "Native name: La tuoga balu (W. Nakanai). The juice of the stem is squeezed into the eyes as a cure for conjunctivitis"

400 *S. vitiensis* Fiji Is. / A. C. Smith 597 / 1933 / "'Takataka'"

Fiji Is. / A. C. Smith 1542 / 1934 / "'Kundrangi'"

Fiji Is. / E. H. Bryan, Jr. 303 / 1924 / "odor of washing soap"

Fiji Is. / A. C. Smith 4783 / 1947 / "'Wa mbitumbitu'"

Fiji Is. / no collector s.n. / no year / "'Wa rusi'"

Fiji Is. / O. Degener 15302 / 1941 / "'Wambitumbitu' (Serua). If a Fiji woman has a baby & within two months is pregnant again, the first baby gets sick, so Fijians say. The baby is then given drink from *Smilax* stem in water"

401 *Heterosmilax gaudichaudiana* China / W. T. Tsang 23830 / 1934 / "Yuen Ip Ma Kap Tang.... fr.... edible"

Indochina / Poilane 98 / 1920 / "Plante medicinale"

402 *H. japonica* China / W. T. Tsang 16802 / 1928 / "Siu Ma Kaap Tang"

AMARYLLIDACEAE

403 *Crinum asiaticum* Philippine Is. / C. O. Frake 509 / 1958 / "Corms mixed with lime for wounds" / "Buteli? Sub."

404 *Hymenocallis declinata* Bahama Is. / R. A. & E. S. Howard 10005 / 1948 / "Reported as very irritating to the skin" / " 'White Lily' "

405 *Agave rubescens* Mexico / H. S. Gentry 5215 / 1939 / "Fruits and fls. cooked and eaten" / "Mescal"

406 *A. wocomahi* Mexico / H. S. Gentry 6340 / 1941 / "Makes good mescal. Fls. eaten"

407 *Curculigo conoc* Indochina / Poilane 1502 / 1920 / " ... les feuilles sont comestibles" / "Annamite: Cay cō no'e"

408 *C. disticha* Indochina / Poilane 10240 / 24 / "Les racines et la base de la tige sont employé contre les maux de gorge   les bouillir et boire" / "Moi: dor oto"

409 *C. brevipedunculata* Philippine Is. / M. D. Sulit 3946 / 1950 / "Juice of stem is rubbed against part of the body bitten by centipede to alleviate pain (Tagb.)" / "Lamba Tagb."

TACCACEAE

410 *Tacca* indet. Philippine Is. / Sulit & Conklin 4638 / 1952 / "Underground rootstock is medicinal by Mangyan" / "Tangisang-danin"

411 *T. leontopetaloides* Caroline Is. / C. C. Y. Wong 368 / 1948 / "Tubers used as food" / "N.v. cobcob"

412 *T. pinnatifida* Fiji Is. / Degener & Ordonez 13995 / 1940 / "Corm used for arrowroot starch. 'If chicken eat the scrapings of root, they die'"

DIOSCOREACEAE

413 *Dioscorea convolvulacea* var. *grandifolia* Mexico / P. C. Standley 2328 / 1936 / " ... fleshy root eaten raw or roasted by the Warihios and Mexicans" / "Chichiwo, W."

414 *D. remotiflora* Mexico / G. B. Hinton 6694 / 34 / " ... root edible sold in markets"

415 *D. remotiflora* var. *maculata* Mexico / Y. Mexia 8716 / 1937 / "Long thickened roots crushed and thrown in water to poison fish" / " 'Bejuco Costilludo' "

416   *D. cayennensis* W.I.: Dominica / W. H. & B. T. Hodge 3377 / 1940 / "(yam jaune la fevre) used for fever; tea made from leaves"

417   *D. polygonoides* Jamaica / G. R. Proctor 8648 / 1954 / "'Bitter yam'"

418   *D. pozucoensis* Peru / F. Woytkowski 5982 / 1960 / "... medicinal roots" / "'Zarzaparilla'"

419   *Dioscorea* indet. Brazil / R. Froes 2000 / 1932 / "Root reported as poisonous" / "'Cura de porco'"

420   *D. hainanensis* China / S. K. Lau 492 / 1932 / "The root is edible from April to Sept. After that it has to be boiled"

421   *D. bulbifera* Solomon Is. / S. F. Kajewski 2648 / 1931 / "... yam. It is poisonous ... resorted to in times of stress for food. The poisonous principle ... is got rid of by repeated washings in water" / "Bitsor"
Philippine Is. / C. O. Frake 714 / 1958 / "Leaves applied for pinworm" / "Penubulen Sub."

422   *Dioscorea* indet. New Britain / A. Floyd 6473 / 54 / "Inedible. Juice used for sores" / "La Poo (W. Nakanai)"

423   *D. nummularia* Philippine Is. / C. O. Frake 869 / 1958 / "Fruits applied for internal pain"

424   *D. alata* Fiji Is. / O. Degener 15405 / 1941 / "Starch edible when cooked, juice poisonous" / "'Ngelemila' (Ra)"

425   *Rajania cordata* W.I.: Dominica / W. H. Hodge 3216 / 1940 / "'wawa': roots edible and eaten by Caribs"

IRIDACEAE

426   *Eleutherine bulbosa* W.I.: Dominica / D. Taylor 30 / 1941 / "'chalotte caraïbe' ... used as a cure for irregular menstruation and menopause"

427   *E. plicata* Peru / F. Woytkowski 5744 / 1950 / "... bulbs color medicinal: bulbs cooked and taken (decoction) for women hemorrhagia" / "'Yabar-piri-piri'"

428   *E. palmifolia* Philippine Is. / C. O. Frake 880 / 1958 / "Corms applied for Charlie horse" / "Tambang"
Philippine Is. / A. L. Zwickey 36 / 1938 / "'Lasun a mariga' (Lan.); root in application to draw thorn from foot, and for insect stings, wounds and boils"
Philippine Is. / M. D. Sulit 4964 / 1952 / "Bulb crushed as medicine for sting of poisonous fishes" / "Ceballas del Monte Bis."

429   *Sisyrinchium scabrum* Mexico / C. H. Mueller 2142 / 1935 / "... said to intoxicate and kill grazing animals"
Mexico / C. H. Mueller 2178 / 1935 / "... said to kill livestock during April and May"

430 *Gynandriris sisyrinchium* Kuwait / H. Dickson 31 / "61 / "Not eaten by animals when green"

MUSACEAE

431 *Musa paradisiaca* Fiji Is. / O. Degener 15096 / 1941 / "Eaten raw... ripe; cooked unripe.... Dry leaf used to roll tabacco.... 'Drala' leaves squashed, mixed with coconut oil is put in banana leaf and then heated near fire. This is then used to rub part where bone is broken" / "Vudi Serua"

432 *Heliconia subulata* Panama / W. L. Stern et al. 249 / 1959 / "Rhizome used in conjunction with Stern et al. #253 for treatment of cancer of skin. Shredded and combined with Stern et al. #253 and cooking oil and used as a hot plaster" / "'Platavillo morado'"

433 *H. vitiensis* Fiji Is. / O. Degener 14352 / 1941 / "Fijians eat seed cooked in water" / "'Paka' (Sabatu)"

ZINGIBERACEAE

434 *Zingiber cassumunar* Siam / A. Kostermans 64 / 1946 / "...flowers, fruit, shoots edible" / "pu pên"

435 *Z. zerumbet* Fiji Is. / A. C. Smith 6777 / 1947 / "...roots used as medicine for infants' coughs" / "'Ndrove'"

Tonga Is. / T. G. Yuncker 15221 / 1953 / "...roots used as medicine" / "Angoango"

Fiji Is. / O. Degener 14926 / 1941 / "Rootstock used for stomach trouble" / "'Naimbita' (Sabatu)"

Fiji Is. / A. C. Smith 4010 / 1947 / "...medicine for coughs prepared from rhizome" / "'Ndalasika'"

436 *Alpinia speciosa* W.I.: Dominica / W. H. Hodge 3211 / 1940 / "...leaves and flowers boiled—brew used by Caribs to cure stomach gas or indigestion"

China / W. T. Tsang 26691 / 1936 / "Used as medicine"

437 *A. chinensis* China / W. T. Tsang 24739 / 1934 / "Used for medicine" / "Ma lat"

438 *A. japonica* China / W. T. Tsang 42113 / 1934 / "...fruit...edible; used for medicine" / "Ma Lat"

439 *Alpinia* indet. China / W. T. Tsang 25932 / 1935 / "Seed used as medicine" / "Ma Lat"

440 *Alpinia* indet. Solomon Is. / S. F. Kajewski 1820 / 1930 / "Roots eaten by natives as an aid to chewing betel nut, to bring out the narcotic"

441 *A. brevilabris* Philippine Is. / P. Añonuevo 212 / 1950 / "Leaves used instead of banana leaves to make rice soft"

442 *A. elegans* Philippine Is. / Edaño & Gutierrez 149 / 1957 / "Fruit yellow (ripe)...sweet—a little bit sour.... Edible" / "tagbac"

443 *Alpinia* indet. Philippine Is. / M. D. Sulit 2726 / 1948 / "...rhizomes chewed for stomach trouble" / "Togus Bis."

444 *Alpinia* indet. Philippine Is. / Sulit & Conklin 5141 / no year? / "Decoction of underground roots given to person to stop excessive vomiting" / "Banglay Mang."

445 *A. purpurata* Fiji Is. / A. C. Smith 8 / 1933 / "Used for medicine" / "'Thevunga'"

446 *Alpinia* indet. Caroline Is. / C. C. Y. Wong 556 / 1948 / "The plant is used for medicine for sore ear (fely lantali). The young stems and leaves are pounded and the sap is squeezed into the ear" / "cifif"

447 *Renealmia cernua* Panama / W. L. Stern et al. 87 / 1959 / "Used for fevers as a bath, also for typhoid; ½ lb. in 10 liters of water, boil down to 8 liters"

448 *R. pedicellaris* Brit. Guiana / A. C. Smith 2835 / 1937 / "Herb; entire plant boiled and used externally as bath to reduce fever by Waiwais"

449 *Amomum speciosum* Sumatra / W. N. & C. M. Bangham 819 / 1932 / "Natives eat flower for vegetable" / "'Kintjaeng'"

450 *Amomum* indet. Philippine Is. / P. Añonuevo 68 / 1950 / "Fruit is edible. Root is used in the preparation of cough medicine" / "Tekala Bukidnon"

451 *Hedichium coccineum* India: Khomi / N. L. Bor 16033 / 42 / "The Agamos believe that if a flower is warn in the ear it is a powerful repellant of evil spirits and disease"

452 *H. philippinense* Philippine Is. / C. O. Frake 610 / 1958 / "Corms applied to ulcers" / "gilengileng Sub."

453 *Hornstaedtia lycostoma* New Britain / A. Floyd 3495 / 54 / "The raw seeds are eaten by children" / "Ekiai (W. Nakanai)"

454 *Hornstaedtia* indet. Solomon Is. / S. F. Kajewski 2192 / 1930 / "The plant is of more than passing interest, as the leaves are heated and applied to the skin. Native babies are light coloured and a superstition is held that if this not applied all sorts of things may happen. The skin will not go black and it will stink.... Also the leaves are heated and the sap wrung out so it drops on the sores of the natives" / "Artsi-alu"

455 *Afromomum granum-paradisi* W.I.: Dominica / W. H. Hodge 3204 / 1940 / "'poivre guinée'; leaves used to make tea, seeds 'to make dogs hunt'"

456 *Curcuma* indet. Solomon Is. / S. F. Kajewski 1959 / 1930 / " . . . plant . . . subject of sorcery . . . roots supposed to kill people. . . . The plant to my knowledge is not poisonous"

457 *Globba nutans* W.I.: Martinique / H. & M. Stehlé 5729 / 1945 / "Medicinal" / "Nom vulgaire 'A tough mouse'"

458 *Costus* indet. Burma / F. G. Dickason 6519 / 1937 / " . . . young shoots eaten" / "Hplan taung mwe"

459 *C. sericeus* Caroline Is. / C. C. Y. Wong 407 / 1948 / "The plant is used for the treatment of sickness of a chill (fely bathir), the entire plant is pounded, squeezed and mixed with water before it is drunk. The plant is used in the treatment of boils (fely-lot), the young plant is used, 6 inches of the plant, by pounding and squeezing the sap onto the boil, then spreading it over the boil" / "thowel (Rumung); wänim (Rul)"

CANNACEAE

460 *Canna* indet. Colombia / R. E. Schultes 3403 / 1942 / "Leaves ground up and used for treatment of skin infection" / " 'Khang-gu-pa'-chu' "

461 *C. indica* China / W. T. Tsang 23344 / 1933 / " . . . fruit . . . edible" / "Sz Kwai Chiu"

MARANTACEAE

462 *Actoplanes* indet. Solomon Is. / S. F. Kajewski 2193 / 1930 / "The juice of this plant is put on sores of small boys . . . " / " 'Mollita' "

463 *Phrynium placentarium* China / S. Y. Lau 20211 / 1932 / "Cultivated"

464 *Calathea klugii* Peru / J. M. Schunke 204 / 1935 / " 'Bijahuia chica' "

465 *C. peruviana* Peru / J. M. Schunke 9 / 1935 / " 'Bijan' "

466 *C. ursina* Peru / J. M. Schunke 294 / 1935 / " 'Bijahuio' "

467 *Calathea* indet. Colombia / R. E. Schultes 3837 / 1942 / "Medicinal"

468 *Ischnosiphon obliquus* Peru / J. M. Schunke 314 / 1935 / " 'Bijaii' "

469 *I. ornatus* Peru / J. M. Schunke 7 / 1935 / " 'Bijahuia' "

470 *Monotagma anathronum* Peru / Y. Mexia 6241 / 1931 / " 'Bijao' " / " . . . leaves used for wrapping"

471 *Thalia trichocalyx* Nicaragua / C. F. Baker 2383 / 1903 / " 'Iril la fruta se toma en la sopa' "

472 *Marantacea* indet. Solomon Is. / S. F. Kajewski 1879 / 1930 / "The leaves are used by the natives, to roll up pig and other foods for cooking" / " 'En-i-wus' "

SAURURACEAE

473 **Saururus chinensis** China / W. T. Tsang 25423 / 1935 / "...used for medicine" / "Tong Pin Kui"

474 **Houttuynia cordata** China / W. T. Tsang 25311 / 1935 / "...fr. edible" / "Kau Tap I"

475 **Anemopsis californica** Mexico / H. S. Gentry 4241 / 1939 / "Roots boiled as potion for catarrh. Bark of root roasted, ground and mixed with lard as ointment for sores" / "Yerba del Mansa"

PIPERACEAE

476 **Piper obtusifolium** U.S.: Fla. / L. J. Brass 15798 / 1945 / "...slightly aromatic"

477 **P. alleni** Panama / P. H. Allen 270 / 1937 / "...roots used by Indians to deaden pain; leaves used as snake bite remedy"

478 **P. auritilaminum** Honduras / Yuncker, Koepper & Wagner 6263 / 1938 / "Crushed leaves very aromatic"

479 **P. auritum** Mexico / A. Dugès s.n. / 1897 / "Les fueilles froissées est mâchées sentent fortement l'anis et la cannelle—ou les emploie comme condiment"

El Salvador / P. C. Standley 20550 / 1922 / "Herb...very fragrant when crushed. Juice removes ticks" / "'Santa María'"

El Salvador / P. C. Standley 21236 / 1922 / "lvs. fragrant" / "'Candela de ixote'"

Honduras / P. C. Standley 53402 / 1927–28 / "Crushed leaves have aromatic odor. Young leaves cooked and eaten" / "'Matarro'"

Panama / R. J. Seibert 623 / 1935 / "Odor of sarsparilla"

Colombia / O. Haught 1806 / 1935 / "Plant very fragrant"

480 **P. auritum** var. **grandifolium** Mexico / Y. Mexia 9131 / 1938 / "...Plant has a pleasant, spicy odor and is used to flavor food" / "'Hierba Santa'"

481 **P. discolor** Brit. Honduras / P. H. Gentle 4179 / 1942 / "'Spanish Elder'"

482 **P. jaliscanum** Mexico / Y. Mexia 1177 / 1926 / "Used medicinally" / "'Tanqua'" "'Cordoncillo'"

483 **P. lucigaudens** var. **alleni** Panama / P. H. Allen 270 / 1937 / "...roots used by Indians to deaden pain; leaves used as snake bite remedy"

484 **P. marginatum** Panama / I. M. Johnston 472 / 1944 / "...very faint licorice odor"

Panama / I. M. Johnston 319 / 1944 / "...with strong odor and taste of licorice"

Panama / I. M. Johnston 1376 / 1946 / " . . . with licorice odor"

Dominican Republic / E. J. Valeur 418 / 1929 / "'Aniseto'"

485 *P. middlesexense* Brit. Honduras / P. H. Gentle 3180 / 1940 / "'Spanish Elder'"

486 *P. multinervium* Brit. Honduras / W. A. Schipp 4 / 1929 / "'Caribs' use the leaves for medicinal purpose"

487 *P. patulum* El Salvador / P. C. Standley 19954 / 1922 / "'Santa María'"

488 *P. san-joseanum* Panama / P. C. Standley 28207 / 1923–24 / "Used as tea" / "'Hinojo' . . . 'Cow-foot' (W. Indian)"

489 *P. san-joseanum* var. *panamanum* Panama / Woodson, Jr., Allen & Seibert 1608 / 1938 / " . . . odor of sarsaparilla"

491 *P. triquetrum* El Salvador / P. C. Standley 20972 / 1922 / "'Santa María'"

El Salvador / S. Calderón 2565 / 1930 / "Plant very fragrant" / "'Cordoncillo'"

492 *P. tuberculatum* El Salvador / P. C. Standley 19106 / 1921–22 / " . . . Fr. used like pepper" / "'Cordoncillo blanco'"

Brit. Honduras / P. H. Gentle 3448 / 1940 / "'Spanish Elder,' 'Cordoncillo'"

Brazil / Y. Mexia 6065 / 1931 / "'Pimienta Longa'"

493 *P. uncatum* El Salvador / P. C. Standley 21840 / 1922 / "'Santa María'"

494 *P. yzabalanum* El Salvador / P. C. Standley 19953 / 1922 / "'Santa María'"

495 *P. dilatatum* W.I.: Grenada / G. R. Proctor 16822 / 1957 / "'mal-estomach'"

496 *P. hispidum* Cuba / W. L. White 698 / 1941 / "Collected as host for Corticium-like fungus on dead branches"

Colombia / R. E. Schultes 3438 / 1942 / "Fruit dried, powdered and used to delouse dogs. 'Cordoncillo.' Leaves smashed and mixed with Phyllanthus for fish poison"

Colombia / Schultes & Smith 2038 / 1942 / "Tea of leaves used by Ingas for malarial fevers" / "'Cordoncillo'"

Ecuador / Y. Mexia 8407 / 1936 / "Said to be used crushed in water to kill head lice" / "'Pipilongo'"

497 *P. marginatum* var. *catalpaefolium* W.I.: Trinidad / W. E. Broadway s.n. / 1932 / "Freshly cut strong smelling" / "'Bois l'anée'"

Venezuela / H. Pittier 13086 / 1929 / "N.v. 'Anisillo'"

498   *Piper* indet. W.I.: Dominica / W. H. & B. T. Hodge 2062 / 1940 /
       " . . . used as soap for bathing" / " 'mal estomach' "

499   *Piper* indet. W.I.: Dominica / W. H. & B. T. Hodge 1293 / 1940 /
       " . . . infusion of dry leaves supposedly good for stomache"? /
       " 'malestomach' "

500   *Piper* indet. W.I.: Dominica / W. H. & B. T. Hodge 3252 / 1940 /
       " 'mal bouck' "?

501   *Piper* indet. W.I.: Dominica / W. H. & B. T. Hodge 3226 / 1940 /
       " 'Dr. Bush' "

502   *Piper* indet. / W.I.: Dominica / R. G. Fennah 25 / 1940 / " 'feuille
       mal estomac' "

503   *Piper* indet. W.I.: Dominica / W. H. & B. T. Hodge 1812 / 1940 /
       " 'mal estomac' "

504   *Piper* indet. W.I.: Dominica / P. Beard 1457 / 1946 / " 'Doctor Bush' "

505   *P. poiteanum* Surinam / Geykes s.n. / 1939 / "Used in making oerali
       poison. Roots taste peppery. These can be boiled" / " 'Arakumpane'
       (Way.) . . . "

506   *P. bartlingianum* Surinam / Geykes s.n. / 1939 / "(likely conspecific
       with A. Smith 2826 & 2827 from Brit. Guiana and reported as a com-
       ponent of 'Balauitu' (Waiwai arrow poison)" / "Used in preparation
       of oerali-poison by the Wayana Indians . . . "
       Brit. Guiana / A. C. Smith 2826 / 1937 / "Component of 'Balauitu'
       (Waiwai arrow-poison). Stems used"

507   *Piper?* indet. Brit. Guiana / A. C. Smith 2827 / 1937 / " . . . com-
       ponent of 'Balauitu' (Waiwai arrow poison); stems used"

508   *P. aequale* Venezuela / L. Williams 11395 / 1939 / " 'Anicillo' "
       Colombia / F. W. Pennell 12150 / 1922 / "Leaves used in decoction
       to cure rheumatism"

509   *P. pseudoglabrascens* Venezuela / L. Williams 11701 / 1939 / " 'Ani-
       cillo' "

510   *P. demerarum* Venezuela / L. Williams 11628 / 39 / " 'Anicillo' "

511   *P. mollicum* Venezuela / L. Williams 11551 / 1929 / " 'Anicillo' "

512   *P. bogotense* var. *ovalilimbum* Colombia / Killip & Smith 19662 /
       1927 / "Put in baths as cure for ulcers"

513   *P. aduncum* Colombia / Killip & Smith 15516 / 1926 / "Remedy for
       infection & inflammation" / " 'Cordoncilla' "

514   *P. dichroöstachyum* Colombia / Schultes & Villarreal 5187 / 1943 /
       "fragrant"

515   *P. futuri* Colombia / R. E. Schultes 3626 / 1942 / "*Very* aromatic.
       Medicinal"

516 *P. lenticellosum* Ecuador / W. H. Camp E-3525 / 1945 / "Lvs. pungent-aromatic..."

517 *P. barbatum* Ecuador / Y. Mexia 7688 / 1935 / "'Luto'"

518 *P. nubigenum* Ecuador / Y. Mexia 7659 / 1935 / "Used by natives as disinfectant in wounds, in infusion" / "'Luto'"

519 *P. aduncum* var. *exotum* Peru / Y. Mexia 8094 / 1936 / "Leaves used in infusion to wash wounds" / "'Matico'"

520 *P. antirheumaticum* Peru / Y. Mexia 8034 / 1936 / "An infusion of leaves and aments is used to bathe rheumatic joints" / "'Matico'"

521 *P. callosum* Peru / Y. Mexia 6157 / 1931 / "... fruit and foliage have a strong odor of wintergreen; tea is made from them" / "'Huausa' 'Anisa del Monte'"

522 *P. tingens* Peru / Y. Mexia 6072 / 1931 / "Said to be used by Indians to dye teeth black... 'Cordoncillo'"

523 *Piper* indet. Peru / Y. Mexia 04105 / 1935 / "Infusion of plant used to wash wounds" / "'Mogo-mogo'"

524 *P. colubrinum* Bolivia / J. Steinbach 6131 / 1924 / "Fruta comible" / "'Ambaivillo'"

525 *Piper* indet. Bolivia / I. Steinbach 7290 / 1925 / "fruta comible cuelga debajo las ramas" / "'Ambaibillo'"

526 *P. alegreanum* Brazil / Y. Mexia 4005 / 1929 / "Roots, stems and leaves used in decoction for toothache" / "'João Borandi'"

527 *P. aquilibaccum* Brazil / Y. Mexia 4465 / 1930 / "Plant 5 m. high with green flower and fruit edible when ripe" / "'João Borandy'"

528 *P. aquilibaccum* var. *stoloniferum* Brazil / Y. Mexia 4351 / 1930 / "Eaten by natives"

529 *P. obliquum* Brazil / Krukoff's 6th Exped. to Braz. Amaz. 7634 / 1936 / "Used in Curare of Tecuna Indians"

530 *P. vellozianum* Brazil / Y. Mexia 4485 / 1930 / "Catkins eaten by natives" / "'João Barandy'"

531 *Piper* indet. Brazil / Ducke 1649 / 1944 / "'Pimenta longa' vel 'Pimenta da macaco'"

532 *Piper* indet. Brazil / Ducke 1640 / 1944 / "'Pimenta longa'"

533 *Piper* indet. Brazil / R. Froes for B. A. Krukoff 2021 / 1933 / "Roots of medicinal value, used as a sudorific" / "'Jubaraudy'"

534 *Piper* indet. Brazil / A. Ducke 1649 / 1944 / "'pimenta longa' ou 'pimenta de macaco'"

535 *Piper* indet. Brazil / Krukoff's 6th Exped. to Braz. Amaz. 7546 / 1935 / "Component of Curare of Tecuna Indians"

536 *Piper* indet. Brazil / Krukoff's 6th Exped. to Braz. Amaz. 6972 /

1934 / "Vine; stem contains a substance producing local anaesthesia, used by Indians to cure tooth ache" / "'Cipo de dor dente'"

537 *P. hancei* China / Steward & Cheo 439 / 1933 / "Plant medicinal" / "'P'a Yai Shiang'"

538 *P. laetispicum* Hainan / C. I. Lei 377 / 1933 / "...fruit green poisonous"

539 *P. sarmentosum* Hainan / S. K. Lau 401 / 1932 / "Use to cook with rice as medicine to decrease fever & aid digestion"

Indochina / F. Evrard 527 / 1921 / "Très odorant.... Employé comme bétel par les moïs"

540 *Piper* indet. China / C. W. Wang 73248 / 1936 / "fruit edible"

541 *Piper* indet. China / C. W. Wang 74894 / 1936 / "medicine"

542 *Piper* indet. China / C. W. Wang 74927 / 1936 / "medicine"

543 *Piper* indet. China / C. W. Wang 76178 / 1936 / "medical"

544 *Piper* indet. China / H. T. Tsai 61372 / 1934 / "fls. edible"

545 *Piper* indet. China / C. W. Wang 74762 / 1936 / "medical"

546 *Piper* indet. Siam / A. Kostermans 745 / 1946 / "...leaves very strongly taste of P. Betle"

547 *Piper* indet. Siam / A. Kostermans 714 / 1946 / "leaves tasting of P. Betle, ..."

548 *Piper* indet. India / J. Fernandes 520 / 1949 / "The pepper vine. Vernac. 'Mirivel'"

549 *Piper* indet. Siam / Wichian 321 / 1946 / "...plants piperaceous smelling; eatable"

550 *Piper* indet. Siam / native collector S 436 (Flora of Siam, Royal Forest Dept. 4542) / 49 / "Latex used in remedy skin diseases" / "'Mowb.'"?

551 *Piper* indet. Siam / T. Smitinand 915 / 51 / "Bark whitish with brown lenticels, spicy odour" / "'Yarn Prik Nok'"

552 *Piper* indet. Siam / T. Smitinand 781 / 51 / "...betel-like scented"

553 *P. chlorocarpum* Solomon Is. / S. F. Kajewski 1980 / 30 / "The fruit is chewed at the same time as betel nut and lime to bring out the stimulating effect of betel" / "'Koluga'"

554 *P. exorcista* Solomon Is. / S. F. Kajewski 2185 / 30 / "This piper is used to drive or banish a poison given to the kanakas by another person. Also to drive out a devil the man possesses. A woman takes a leaf in either hand, and draws both across the abdomen until they meet, this is supposed to banish the poison or devil as the case may be" / "'Urugu'"

555   *P. kava-quua* Solomon Is. / L. J. Brass 3319 / 32 / "Juice of leaves & soft young shoots, taken as remedy for stomach pains" / "'Kava-qwua'"

556   *P. oleosum* Solomon Is. / S. F. Kajewski 2133 / 1930 / "Never have I seen a plant with leaves of such high oil content, as they fairly exude it. The drying paper surrounding these specimens was similar to grease paper when the specimens were dry. The oil has a strong aromatic perfume, faintly like Eucalyptus. It should have strong antiseptic and medicinal virtues...." / "'Kuta-vuta'"

557   *P. qua-seek* Solomon Is. / S. F. Kajewski 2646 / 1931 / "The leaves are heated and applied to sores" / "'Qua-seek'"

558   *P. surrogatus* Solomon Is. / S. F. Kajewski 1982 / 1930 / "This is used for a substitute for another piper, for chewing with betel nut" / "'Kulomai'"

559   *P. waimamura* Solomon Is. / L. J. Brass 2583 / 1932 / "Whole plant slightly aromatic"

560   *Piper* indet. New Guinea / M. S. Clemens 2031 / 1936 / "Hepatics"

561   *Piper* indet. Solomon Is. / S. F. Kajewski 1585 / 1930 / "This is a wonderful aromatic plant, and you will notice that the plant specimens have a quite strong odour between aniseed and liquorice even when dry. This plant is worth cultivating for the fragrant oil the leaves contain"

562   *Piper* indet. Solomon Is. / S. F. Kajewski 2077 / 1930 / "This fruit has rather an interesting touch among natives as the fruit is burned to frighten the devil away. Also the leaves are scattered round to keep his Satanic majesty away" / "'Malacogie'"

563   *Piper* indet. Solomon Is. / S. F. Kajewski 2047 / 1930 / "The leaves are taken by the natives and rubbed on the forehead to alleviate headaches" / "'Cumi-ku'"

564   *Piper* indet. Brit. New Guinea / L. J. Brass 5662 / 1933 / "...lvs. aromatic..."

565   *Piper* indet. Dutch New Guinea / L. J. Brass 10299 / 1938 / "...aromatic"

566   *Piper* indet. Neth. New Guinea / L. J. Brass 13710 / 1939 / "...leaves aromatic"

567   *Piper* indet. Dutch New Guinea / L. J. Brass 11670 / 1938 / "...aromatic"

568   *Piper* indet. Dutch New Guinea / L. J. Brass 11671 / 1938 / "...aromatic"

569   *Piper* indet. Neth. New Guinea / L. J. Brass 12931 / 1939 / "...aromatic"

570   *Piper* indet. Neth. New Guinea / L. J. Brass 14042 / 1939 / "...aromatic"

571   *Piper* indet. Dutch New Guinea / L. J. Brass 11054 / 1938 / "...aromatic"

572   *Piper* indet. Dutch New Guinea / L. J. Brass 9134 / 1938 / "...aromatic"

573   *Piper* indet. Brit. New Guinea / L. J. Brass 8128 / 1936 / "...lvs. aromatic..."

574   *Piper* indet. Brit. New Guinea / L. J. Brass 7281 / 1936 / "All parts aromatic..."

575   *Piper* indet. Brit. New Guinea / L. J. Brass 7027 / 1936 / "Pl. aromatic..."

576   *Piper* indet.New Guinea / J. S. Womersley 4301 / 1951 / "The fruits of this 'Daka' are chewed by natives with betle nut, Cultivated"

577   *Piper* indet. New Guinea / L. J. Brass 28371 / 1956 / "...plant aromatic"

578   *P. abbreviatum* Philippine Is. / C. O. Frake 726 / 1958 / "Leaves applied for spleenomegaly" / "'malaneb-ulangan'" Sub.

579   *P. albidirameum* Philippine Is. / C. O. Frake 514 / 1958 / "Leaves applied to deep ulcers" / "'Gapin manamed'"

580   *P. caninum* Philippine Is. / C. Frake 365 / 1957 / "Economic Uses: In a religious myth" / "'gapinapit'" Sub.

581   *P. penninerve* Philippine Is. / C. O. Frake 155 / 1958 / "Leaves applied for ulcers" / "'deling'"

582   *P. retrofractum* Philippine Is. / Sulit & Conklin 5112 / 1953 / "Roots boiled—decoction given to patient to relieve stomach trouble" / "'Sarimarâ-mabirú'" Man. (Hanunoo)

583   *P. subpeltatum* Philippine Is. / A. L. Zwickey 82 / 1938 / "Lvs. cooked with rice"

584   *Piper* indet. Philippine Is. / E. Maliwanag 327 / 1941 / "Leaves are chewed with betel nut" / "'sangilo'" Mang.

585   *P. excelsum* var. *tahitianum* Rapa Is. / F. R. Fosberg 11512 / 1934 / "...herbage has odor of licorice"
       Rapa Is. / St.John & Maireau 15582 / 1934 / "...foliage with slight resinous odor"

586   *P. latifolium* New Hebrides Is. / S. F. Kajewski 718 / 1928 / "This is the wild Kava and is not used for drinking purposes" / "'Wild Kava'"
       Austral Is. / H. St.John 16715 / 1934 / "Tea, ..."
       Austral Is. / A. M. Stokes 114 / 1921 / "Uses: leaves pounded, cooked in water or coconut milk, taken internally for cramps" / "'avaavairai'"

587   *P. puberulum* Fiji Is. / O. Degener 15042 / 1941 / "Take bark, cook and drink liquid for fever. Cook leaf and drink tea for bloody stool" / " 'Gakawa' or 'Yangonagona' " Serua

Fiji Is. / O. Degener 15005 / 1941 / "Squash leaf in water and give liquid to drink to women sick after childbirth" / " 'Yangonangona' " Serua

588   *P. puberulum* var. *glabrum* Tonga Is. / T. G. Yuncker 15323 / 1953 / "Leaves used for preparing medicine" / " 'kavakava'uli' " Tongan

589   *P. tristachyon* Society Is. / E. H. Quayle 214 / 1921 / "Juice from roots formerly fermented for intoxicant" / " 'ava ava' "

590   *Piper* indet. Caroline Is. / C. Y. C. Wong 170 / 1947 / "Elbert says the plant is used for healing wounds" / " 'Enes' "

591   *P. novae-hollandiae* Australia / M. S. Clemens 43579 / 1944 / " . . . red fruit, sought for birds"

Australia / L. J. Brass 2334 / 1932 / "Berries green, aromatic & pungently flavoured"

592   *P. capense* Nyasaland / L. J. Brass 17041 / 1946 / " . . . pungently aromatic"

593   *P. guineense* Liberia / D. H. Linder 672 / 1926 / " . . . green aromatic leaves"

594   *P. umbellatum* Nigeria / W. D. MacGregor 483 / 1930 / " 'Yaw' (Yoruba)"

595   *Pothomorphe umbellata* El Salvador / P. C. Standley 19326 / 1921 / " . . . strong odor" / " 'Santa María' "

El Salvador / P. C. Standley 19197 / 1921–22 / "Lvs. used with hen fat to reduce swellings" / " 'Santa María grande' 'Cordoncillo' "

El Salvador / P. C. Standley 22653 / 1922 / " . . . lvs. fragrant. Juice destroys ticks" / " 'Santa María' "

Mexico? C. & E. Seler 4998 / 1907 / " 'Yerva Santa monter' "?

Colombia / Schultes & Smith 2081 / 1942 / " 'yerba santa maría' "

Colombia / E. Dryander 2414 / 1939 / " 'Santa María' "

596   *P. iquitosensis* Peru / J. M. Schunke 105 / 1935 / "Medicinal" / " 'Santamaria' "

597   *P. peltata* Brazil / Y. Mexia 6067-a / 1931 / "Leaves are placed on a swelling as in toothache" / " 'Tapeba' "

Bolivia / J. Steinbach 5536 / 1921? / "Se usa contra llagas sipiliticas" / " 'Matico' "

Colombia / Pennell, Killip & Hazen 8566 / 1922 / "para ulceras y conos" / " 'Santa Maria' "

Colombia / R. E. Schultes 3457 / 1942 / "Medicinal"

598   *Peperomia maculosa* Honduras / Yuncker, Dawson & Youse 5878 / 1936 / "Cut stems have a pungent odor"

599   *P. major* Guatemala / J. A. Steyermark 44663 / 1942 / "Fresh leaves applied on knees for granos" / "'ye-pa-setas'"

600   *P. pellucida* El Salvador / P. C. Standley 22657 / 1922 / "'Herba del sapo'"

        Fiji Is. / Degener & Ordonez 13670 / 1940 / "Fijians use as poultice"

601   *P. quadrifolia* Mexico / A. Dugès 323 A&B / no year / "Piperacée nommée verdolaguilla à Morelia, et employée comme antilithique"

        El Salvador / P. C. Standley 19426 / 1921 / "'Cuartillo'"

        El Salvador / P. C. Standley 20097 / 1922 / "Remedy for skin diseases" / "'Cuartillo'"

602   *P. distachya* Haiti / W. J. Eyerdam 502 / 1927 / "strong acrid smell when crushed"

603   *Peperomia* indet. W.I.: Dominica / W. H. & B. T. Hodge 2005 / 1940 / " . . . everywhere on branches of Theobroma; lves. boiled as tea for colds"

604   *P. elongata* var. *guianensis* Brit. Guiana / J. S. de la Cruz 1365 / 1922 / "Myrmecophilous"

605   *P. elongata* Venezuela / B. Maguire 33137 / 1952 / " . . . fls. with slight aromatic fragrance"

606   *P. glabella* var. *melanostigma* Colombia / Schultes & Smith 2059 / 1942 / "Supposed remedy for conjunctivitis" / "'tre-gwen' 'gwee-nan'" Ingano

607   *P. serpens* Colombia / R. E. Schultes 3589 / 1942 / "Remedy for sting of conga ant"

608   *P. emarginella* Colombia / Schultes & Smith 3028 / 1942 / "Pounded, mixed with tobacco & urine to poultice bites of cungamanda ant" / "'cungamanda-ambe'"

609   *P. reflexa* China / W. P. Fang 3457 / 1928 / "Rare, one kind of Chinese medicine"

610   *P. contractispica* Solomon Is. / L. J. Brass 3015 / 1932 / "Leaves have a pungent, peppery flavour"

611   *P. leptostachya* New Hebrides Is. / S. F. Kajewski 359 / 28 / "Sap of leaf used to make faces of the natives glossy" / "'Nimtoro-orah'"

612   *P. marutoi* Solomon Is. / L. J. Brass 3392 / 1932 / "Leaves Aniseed scented"

613   *Peperomia* indet. Dutch New Guinea / L. J. Brass 11227 / 1938 / "slightly aromatic"

614   *Peperomia* indet. Neth. New Guinea / L. J. Brass 13032 / 1939 / "plant aromatic"

615   *Peperomia abyssinica* Nyasaland / L. J. Brass 16962 / 1946 /
"... slightly pungent taste, not aromatic"

CHLORANTHACEAE

616   *Chloranthus glabra* Philippine Is. / C. O. Frake 725 / 1958 / "Leaves applied for internal pain" / "puti?—selimbangun Sub."

Philippine Is. / K. I. Pelzer 55 / 1952 / "Root used for the preparation of cough medicine or mixture" / "Kumok Bukidnon"

617   *C. officinalis* Philippine Is. / C. O. Frake 376 / 1957 / "Leaves pounded and applied to ulcers" / "timug Sub."

Philippine Is. / C. O. Frake 649 / 1958 / "Leaves applied to burns" / "belekbut Sub."

Philippine Is. / R. B. Fox 20 / 1948 / "A medicine for aching in the joints of the leg, PAMILAJ. The leaves are heated on the fire, formed into a tapol, and placed on the joints. This medicine is usually employed for aching in the knees; sometimes ankles" / "NIMÚNMO Egóngot"

618   *Sarcandra glabra* China / W. T. Tsang 24737 / 1934 / "For Cure Rheumatism" / "Chau Tait Cha"

China / W. T. Tsang 21387 / 1932 / "... roots used for medicine" / "Tsau Chit Ch'a"

India: Assam / L. F. Ruse 11a / 1923 / "... leaves and shoots are used as a poultice for rheumatism" / "Ia-khai"

Philippine Is. / A. L. Zwickey 805 / 1938 / "'Makadudug' (Lan.); frts. used to prevent boils"

Philippine Is. / R. S. Williams 1362 / 1904 / "Root with ginger-like fragrance"

619   *Ascarina irvingbaileyana* Solomon Is. / S. F. Kajewski 2014 / 1930 / "The leaves of this tree are pounded together with the spittle from chewing betel nut and applied to relieve pain in the head" / "Tubulai"

SALICACEAE

620   *Populus laurifolia* China / J. Hers 1128 / 1919 / "Chinese name: ku yang 'bitter poplar' so called because, though the leaves resemble those of P. suaveolens, they are not edible"

621   *Salix lasiandra* U.S.: Nev. / L. A. Smith 2880 / 29 / "Forage value—not eaten here but is taken by sheep elsewhere where found"

622   *S. babylonica* China / H. T. Feng 64 / 1925 / "For making furniture and medicine"

623   *S. oxycarpa* India: Punjab / W. Koelz 8443 / 1936 / "Feed twigs to cattle in Kolung in winter"

MYRICACEAE

624 *Myrica cerifera* Brit. Honduras / H. O'Neill 8589 / 1936 / "'Tea bark'"

625 *M. esculenta* China / W. T. Tsang 24314 / 1934 / "... fruit edible" / "Sai Yeung Mui Shue"
China / Steward & Cheo 304 / 1933 / "Edible"

626 *M. rubra* China / Steward, Chiao & Cheo 16 / 1931 / "Fr. edible seen in market" / "Liang Feng Yah"

627 *M. sapida* India: Assam / L. F. Ruse 10 / 1923 / "... fruit eaten as dessert also pickled and used as a condiment, extract from bark is used as aid for catching fish" / "Soh-lia"

628 *Myrica* indet. Burma / F. G. Dickason 8354 / 1939 / "Fruit very tasty very sour. Should make very fine drink from the juice" / "'Mahhmon' Shan 'Ta Sang' Taungthu"

BETULACEAE

629 *Ostryopsis davidiana* China / J. Hers 2703 / 1923 / "leaves used as substitute for tobacco" / "hu chen tze (the tiger's nut)"

630 *Corylus sieboldiana* China / J. Hers 2011 / 1922 / "... the bristles are very irritating and easily cause inflammation of the skin; to cure this, the natives use the leaves of the same species" / "mao chen tze"

631 *C. jacquemontii* Kashmir / Keshavanand 175 / 1906 / "Lopped for fodder" / "Krin Kash; Urni Gujur"

632 *Betula corylifolia* Japan / E. H. Wilson 6847 / 1914 / "... shoots with taste of B. leuta"

633 *Alnus firma* Japan / E. H. Wilson 8371 / 1917 / "... young branches used for feeding cattle"

634 *A. formosana* Formosa / E. H. Wilson 10080 / 1918 / "Commonly planted by savages to fertilize field"

FAGACEAE

635 *Castanea henryi* China / R.-C. Ching 1490 / 1924 / "... edible fruit"

636 *C. seguinii* China / H. H. Hu 266 / 1920 / "Edible"

637 *Castanopsis delavayi* China / Pater Siméon Ten 343 / 1917 / "... fructus edulis"

638 *C. fordii* China / W. T. Tsang 28666 / 1938 / "... fr. edible" / "Mo Chui Shua"

639 *C. hystrix* China / W. T. Tsang 23622 / 1934 / "... flower fragrant, edible" / "Hung Yuen Shue"

640 *C. formosana* China / S. K. Lau 336 / 1932 / "... fruit edible.... To burn the fruit for food" / "Wong Mei Tsz"

641   *C. indica* China / W. T. Tsang 26768 / 1936 / "...fr....edible"

642   *C. lamontii* China / W. T. Tsang 28533 / 1938 / "...fr. edible" / "Pak Nai Chu Shue"

643   *C. namdinhensis* China / W. T. Tsang 22800 / 1933 / "...fruit smoky, edible" / "Suen Sum Yuen Shue"

644   *C. fabri* Indochina / W. T. Tsang 27086 / 1936 / "...fr....edible"

645   *C. pyriformis* Indochina / Poilane 7928 / 23 / "...fruits sont commestibles"

646   *C. tribuloides* India: Assam / L. F. Ruse 14 / 1923 / "Fruit sold in bazar as dessert"

647   *C. philippensis* Philippine Is. / C. O. Frake 612 / 1958 / "...medicine for asthma, headache (bark)" / "gemulaun Sub."

648   *Lithocarpus hungmoshanensis* China / F. A. McClure 739 / 1929 / "Fruits eaten by women to recover weight after giving birth..." / "Ch'a Ngut"

649   *L. litchi* China / Steward & Cheo 828 / 1933 / "Ripe fruit when roasted is good to eat"

650   *L. tsangii* China / W. T. Tsang 21612 / 1932 / "...fruit...edible" / "Pui kwoh Shue"

651   *L. thomsonii* Indochina / W. T. Tsang 27354 / 1936 / "...fr. edible"

652   *Lithocarpus* indet. Philippine Is. / C. O. Frake 769 / 1958 / "Roots boiled and used for stomach ache" / "gasusu Sub."

653   *Lithocarpus* indet. Philippine Is. / M. D. Sulit 1244 / 1946 / "Eaten by wild pigs" / "Ulayan Tag."

654   *Quercus garryana* U.S.: Calif. / W. A. Dayton 549 / 1913 / "Forage value....Good browse"

655   *Q. kelloggii* U.S.: Calif. / W. A. Dayton 347 / 1913 / "With Ceanothus the most valuable cattle browse of this range"

656   *Q. albocincta* Mexico / H. S. Gentry 1460 / 1935 / "The acorns are sweet and edible" / "Kusi, Mex. Hachuká, W."

657   *Q. rhodophlebia* Mexico / G. B. Hinton 6807 / 34 / "...acorns edible roasted also as an adulterant for coffee"

658   *Q. acutissima* China / W. T. Tsang 27902 / 1937 / "...fr. edible" / "Chui Tsoi Shu"

659   *Q. liaotungensis* China / J. F. Rock 13482 / 1925 / "...acorns sweet, edible"

660   *Q. mongolica* China / J. Hers 2015 / 1922 / "...the leaves of this species are used for packing meat" / "tsai shu"

661   *Quercus* indet. Philippine Is. / C. O. Frake 863 / 1958 / "Bark applied to fungous infections" / "Kalingan Sub."

ULMACEAE

662   *Celtis pallida* Mexico / A. Dugès 4894A / 1896 / "...baie...que mangent les enfants" / "Granjeno"

663   *C. boninensis* China / E. H. Wilson 8302 / 1917 / "...edible shoots used to feed cattle"

664   *C. bungeana* China / W. T. Tsang 21862 / 1933 / "Fr. edible"

665   *Trema orientalis* Brit. N. Borneo / Damai B.N.B. For. Dept. 4769 / 1935 / "Medicinal plant; used for bathing when woman gives birth" / "Balek balek angin jantan (Brunei)"

666   *T. cannabina* Solomon Is. / S. F. Kajewski 2478 / 1931 / "The leaves are heated with lime and applied to boils" / "Hai-art-se"

Solomon Is. / S. F. Kajewski 2430 / 1931 / "When a man or woman has pain in the body, the leaves are chewed with betel nut and spat over the body" / "Guyarci"

Solomon Is. / S. F. Kajewski 2325 / 1930 / "The leaves are eaten by the natives for sickness of the stomach" / "Bless-ci"

667   *Parasponia orientalis* Fiji Is. / O. Degener 15471 / 1941 / "Take bark and put in water and drink as remedy for 'blood in stomach after lifting something heavy'"

668   *Aphananthe aspera* China / W. T. Tsang 27816 / 1937 / "...fr....edible" / "Wai Yeung Song Shue"

669   *Gironniera cuspidata* Indonesia / Kostermans 18580 / 1961 / "Fruit ...sweet, edible" / "kaju belikat"

MORACEAE

670   *Fatoua villosa* Japan / E. Elliott 146 / '46 / "Food use: Young plant well cooked and eaten in time of scarcety" / "'Kuwa kusa'"

671   *F. pilosa* Philippine Is. / A. L. Zwickey 398 / 1938 / "...causes mild itching" / "'Sagay a manok' (Lan.)"

672   *Paratrophis* indet. Solomon Is. / S. F. Kajewski 2396 / 1930 / "The young leaves are cooked and eaten by the natives. This seems to be common over the Pacific...." / "Torga-bagi"

673   *P. tahitensis* New Hebrides Is. / S. F. Kajewski 818 / 1929 / "Young leaves cooked and eaten by the natives as...food"

674   *Morus rubra* China / W. T. Tsang 25919 / 1935 / "...fr. dark, edible" / "Song Che Shue"

675   *Trophis mexicana* El Salvador / P. C. Standley 20183 / 1922 / "...fr. edible" / "'Raspa-lengua'"

676   *Chlorophora tinctoria* Mexico / H. S. Gentry 6130 / 1941 / "Fruit edible but caustic to lips" / "Palo Moro"

Costa Rica / H. E. Stork 3389 / 1932 / "Fruits.... Sweet even before ripening"

Brazil / Y. Mexia 5329 / 1930 / "Milky juice, used in toothache; vermifuge" / "'Tajuba'"

677 *Malaisia scandens* China / H. Fung 20243 / 1932 / "...fr. edible" / "Poh Lak Pam"

678 *Broussonetia papyrifera* China / H. T. Feng 18 / 1924 / "...fruits can be eaten"

China / Steward & Cheo 88 / 1933 / "Leaves used to feed pigs"

679 *Taxotrophis ilicifolia* China / S. K. Lau 1596 / 1933 / "...fruit edible" / "Ka Tuk"

Indochina / Eberhardt 4535 / no year / "fruit comestible" / "Nom indigène: Cay quit gui   nom tho: may tēo"

Philippine Is. / Sulit & Conklin 5067 / 1953 / "Young leaves and tops used as vegetable by Mangyan" / "ulus buladlad Mang."

680 *Streblus asper* China / H. Fung 20069 / 1932 / "...odor sweet; fruit edible" / "Sing Kwan Tsz"

681 *Phyllochlamys taxoides* China / S. K. Lau 187 / 1932 / "Fr....use to boil soup"

Philippine Is. / Sulit & Conklin 4735 / 1952 / "Sliced roots infused in water obtained from stems of some vines is medicine against bad effect caused by 'anito'"

682 *Dorstenia contrajerva* Mexico / H. S. Gentry 1026 / 1934 / "Medic. 'bueno por muelo y por corimiento de la cabeza'" / "Barboria, Mex."

683 *D. contrajerva* var. *houstonii* El Salvador / P. C. Standley 19204 / 1921–22 / "Lvs. used to flavor cigarettes, also for stomach ache and snake bites" / "'Contrahierba'"

684 *D. drakena* Mexico / H. S. Gentry 1551 / 1935 / "Roots boiled for fevers" / "Barboria, Mex."

685 *Helianthostylis paraensis* Brazil / Ducke 1537 / 1944 / "...pulpa dulci"

686 *Poulsenia armata* Peru / H. Youngken 114 / no year / "Cascara Medic. Bark used in medicine" / "'Yanchana'"

687 *Parartocarpus venenosus* New Britain / A. Floyd 6448 / 54 / "Fruit eaten by natives. The latex is applied to sores" / "la Geo"

New Guinea / P. van Royen 4737 / 1954 / "The ripe fruits are pulverised and then used for healing wounds. Edible" / "kebötrek (Je dialect)"

Solomon Is. / F. S. Walker B.S.I.P. 238 / 1946 / "...fruit...highly esteemed for its flavour"

Solomon Is. / S. F. Kajewski 2199 / 1930 / "The bark is scraped and put into the food of dogs to poison them, . . . probably it is only a superstitious poison" / "Targatz"

688 *Artocarpus hypargyreus* China / W. T. Tsang 21182 / 1932 / " . . . fruit yellow, edible" / "Cheung Kwan Tse Shue"

689 *A. nitidus* China / S. K. Lau 566 / 1932 / "Fr. . . . edible" / "Kiam Wo (Lois)"

690 *A. styracifolius* China / W. T. Tsang 23259 / 1933 / " . . . fruit . . . edible"

691 *A. hirsutus* India: Bombay / J. Fernandes 1585 / 1950 / " . . . small fruits attractive to monkeys. . . . Seeds and echinous skins of both wild and cultivated fruit are preserved" / "'Haeblesa' Kanarese"

692 *A. lakoocha* Burma / J. F. Smith 44 / 1914 / "Fruit fleshy, . . . much relished by monkeys"

693 *A. melinoxylus* Indochina / J & M. S. Clemens 3510 / 1927 / " . . . fruit edible"

694 *A. vrieseanus* var. *papillosus* Solomon Is. / S. F. Kajewski 2501 / 1931 / "The bark is macerated with water and drunk to relieve a pain in the stomach" / "Ku-kupe"

Solomon Is. / S. F. Kajewski 1920 / 1930 / "Young fruit chewed as a substitute for betel nut" / "Pumbai"

695 *A. vrieseanus* var. *subsessilis* New Guinea: Papua / R. D. Hoogland 4813 / 1954 / "Wood straw, with penetrating unpleasant smell"

696 *A. blancoi* Philippine Is. / M. D. Sulit 4519 / 1952 / " . . . fruit cooked as vegetable"

697 *A. rubrovenius* Philippine Is. / M. D. Sulit 4993 / 1952 / "Bark used by Mangyan as substitute for betel nut"

698 *A. sericicarpus* Philippine Is. / M. D. Sulit 2834 / 1948 / "Said to be edible fruit"

699 *Helicostylis asperifolia* Brazil / Ducke 1534 / 1944 / " . . . fructibus edulibus" / "'Mao de gato'"

700 *Ogcodeia ternstroemiiflora* Bolivia / B. A. Krukoff 11111 / 1939 / "Frts. edible" / "'Coca-coca'"

701 *Pseudolmedia spuria* Brit. Honduras / W. A. Schipp 1271 / 1934 / " . . . fruits . . . edible of fine flavour" / "'wild cherry'"

702 *Brosimum sapiifolium* Costa Rica / P. H. Allen 5877 / 1951 / "Fruits reported to be edible"

703 *B. terrabanum* El Salvador / P. C. Standley 2233 / 1922 / " . . . lvs . . . . good forage . . . " / "'Ojushte'"

Honduras / P. C. Standley 54672 / 1927–28 / "Leaves eaten by stock. Seeds eaten boiled, also made into tortillas" / "'masica'"

704   *Ficus cotinifolia* Mexico / H. S. Gentry 1043 / 1934 / "Fruit eaten by natives. Medic. milky sap used for bowel ailments" / "Nacopooli, Mex. Wohtoli, W."

Mexico / J. Bequaert 62 / 1929 / "...leaves used as horse fodder"

705   *F. maxima* Mexico / M. Martinez 64 / '62 / "Latex utilizado como purgante" / "'higuera prieta'"

706   *F. pringlei* Mexico / R. L. Dressler 2027 / 1957 / "...fruit...eaten by parrots and small boys" / "Higuerón"

707   *F. pertusa* Mexico / Y. Mexia 640 / 1926 / "Fr.... edible" / "'Camichin'"

708   *F. segoviae* Mexico / Y. Mexia 8833 / 1937 / "Fruit eaten by domestic animals" / "'Higuera'"

709   *F. subscabrida* Cuba / J. G. Jack 6962 / 1929 / "Fruit greedily eaten by swine"

710   *F. palmata* India: Punjab / W. Koelz 8268 / 1936 / "Natives eat young leaves as vegetable"

711   *F. hypogaea* Sumatra / W. N. & C. M. Bangham 983 / 1932 / "Natives say they use the leaves with opium"

712   *F. ramantacea* Fed. Malay States / G. anak Umbai for A. H. Millard KL 1635 / 59 / "Poisonous"

713   *F. adenosperma* var. *glabra* Solomon Is. / S. F. Kajewski 2480 / 1931 / "The bark is macerated with water and drunk to cure colds" / "Arngea"

714   *F. chrysochaete* Solomon Is. / S. F. Kajewski 1983 / 1930 / "The leaves of this tree are boiled with pig or opossum to give it a flavour" / "Labutsie"

Solomon Is. / S. F. Kajewski 2148 / 1930 / "The young leaves are eaten after being boiled" / "Karsikerie"

715   *F. copiosa* Solomon Is. / S. F. Kajewski 1836 / 1930 / "Young leaves are boiled and eaten by the natives" / "Tu-nan-ni"

716   *F. edelfeldtii* Solomon Is. / S. F. Kajewski 2419 / 1931 / "When the natives have no betel nut, they eat this leaf with lime and pepper leaves" / "Gwyembure"

717   *F. erythrosperma* Solomon Is. / S. F. Kajewski 2550 / 1931 / "The bark is macerated and drunk to cure gonorrhoea" / "Tangia"

718   *F. gul* New Britain / A. Floyd 3490 / 54 / "Leaves inedible, but fed to dogs to enhance their pig hunting ability" / "La vovali (W. Nakanai)"

719   *F. indigofera* Solomon Is. / S. F. Kajewski 2321 / 1930 / "The sap of this tree is applied by natives to sores" / "Esch-armbora"

720   *F. odoardi* Solomon Is. / S. F. Kajewski 2618 / 1931 / "The leaves are mixed with lime and piper leaf and chewed as a substitute for betel nut" / "Alafasu"

721   *F. septica* Solomon Is. / S. F. Kajewski 2412 / 1931 / "When a man has a bruise the leaves of this tree are heated and applied to it" / "Nure"

Solomon Is. / S. F. Kajewski 2597 / 31 / "The leaves are rubbed on the skin to cure Buckra (Tinea)" / "Trure"

722   *F. zylosycia* Solomon Is. / S. F. Kajewski 2528 / 1931 / "The natives have a superstition that if bits of this tree are given with the food of pigs, they will fatten quickly" / "Gatutoo"

723   *Ficus* indet. Solomon Is. / S. F. Kajewski 2642 / 1931 / "The bark is macerated and drunk for stomach ache" / "Cove Cove"

724   *F. longipedunculata* Philippine Is. / Sulit & Conklin 5140 / 1953 / "Pounded bark used as plaster to sprained part of the body" / "Balete-buladlad Mang."

725   *F. minahassae* Philippine Is. / Sulit & Conklin 5154 / 1953 / "Water exuded from stem is medicine for cough" / "Ayimit Mang."

726   *F. retusa* Philippine Is. / C. O. Frake 706 / 1958 / "Bark boiled and drunk for puerperium" / "mulawan Sub."

727   *Ficus* indet. Philippine Is. / C. O. Frake 783 / 1958 / "Leaves applied to internal pains" / "busyuz-usa Sub."

728   *F. barclayana* Fiji Is. / H. St.John 18096 / 1937 / " . . . ripe fruit put in sore cavities in teeth" / "'masimasi'"

729   *F. kajewski* Fiji Is. / H. St.John 18213 / 1937 / "Leaves used for greens" / "'leweto'"

Fiji Is. / H. St.John 18264 / 1937 / " . . . fr. . . . eaten by birds" / "'nambuluwai'"

730   *F. masonii* Fiji Is. / Degener & Ordonez 13541 / 1940 / " . . . leaves edible when cooked" / "Kau"

731   *F. obliqua* Fiji Is. / H. St.John 18037 / 1937 / "Used as medicine for sore joints" / "'bak'"

Fiji Is. / H. St.John 18271 / 1937 / " . . . sap milky, used for bird lime" / "baka"

732   *F. pritchardii* Fiji Is. / H. St.John 18013 / 1937 / "Juice used for cuts" / "'kamba'"

Fiji Is. / H. St.John 18268 / 1937 / " . . . fr. . . . eaten by people, pigs and birds"

733   *F. senfftiana* Caroline Is. / C. Y. Wong 332 / 1948 / "The fruits are eaten; the process of preparing the fruits is by placing . . . in a pot . . . to boil . . . until the fruits are soft, then they are pounded . . . coconut . . . milk is squeezed into the pounded material . . . " / "N.v. woce"

734   *F. storkii* Fiji Is. / O. Degener 14984 / 1941 / "Leaf is sweet and eaten raw by Fijians" / "'Losilosi' (Serua)"

735   *F. tinctoria* Caroline Is. / C. Y. Wong 227 / 1947 / "The fruits are eaten when green . . . cooked and pounded into poi, treating it as tapioca or breadfruit. It can be made into mwatyn by mixing poi with expressed coconut milk" / "N.v. öwöön"

Caroline Is. / C. Y. Wong 313 / 1948 / "The inner bark is used in making fish lure by scraping (ketiker) the outer bark before the inner bark is stripped from the limb. . . . This bark is used because 'the fish like the smell of the bark' . . . birds eat the fruit" / "N.v. wooagey"

736   *Ficus* indet. Caroline Is. / C. Y. Wong 151 / 1947 / "The scraped bark is used as medicine to guard against soome (soul of the dead) by pounding and mixing with water; the medicine is drunk by men and women. The fruit and leaves are used as itang's (shaman) medicine. The roots are mixed with grated coconut meat squeezing it over kon (poi) . . . adding a special taste" / "ääy"

737   *Myrianthus holstii* Tanganyika Terr. / H. J. Schlieben 2769 / 1932 / " . . . Früchte essbar"

738   *Pourouma aspera* Peru / Y. Mexia 8172 / 1936 / "'Papaya del Monte'"

739   *Cecropia mexicana* Mexico / Y. Mexia 9225 / 1938 / "Fruit favorite food of toucans" / "'Chancarro' 'Huagadeug' (Zapotecans)"

740   *C. latiloba* Bolivia / J. Steinbach 7264 / 1925 / "Fruta comible, muy dulce y sana" / "ambaico dulce o blanca"

741   *C. tolimensis* Colombia / Schultes & Villarreal 5128 / 1942 / "Diffusion of fruits for fevers" / "'guarumo'"

URTICACEAE

742   *Urtica echinata* Argentina / Eyerdam & Beetle 22372 / 1938 / "Vile smell, stings viciously"

743   *Pilea hyalina* El Salvador / P. C. Standley 22206 / 1922 / "Hierba de masamora"

744   *P. mongolica* China / W. T. Tsang 20869 / 1932 / "Food for pigs" / "Fi Yuk Ts'o"

745   *Elatostema griffiana* China / W. T. Tsang 20178 / 1932 / "Food for pigs" / "Shek Beg Ngai"

746   *Elatostema* indet. New Guinea / R. Pullen 1083 / 1958 / " . . . eaten as a cooked vegetable"

747   *E. banahaense* Philippine Is. / W. Beyer 6854 / 1948 / "Mashed leaves and stems and fruits . . . used as medicine for bodily itch" / "Piw-wut" "Ifugau"

748   *Elatostema* indet. Philippine Is. / C. O. Frake 815 / 1958 / "Leaves used to poison fresh water crabs" / "tamesi Sub."

749   *Pipturus* indet. Solomon Is. / S. F. Kajewski 2546 / 1931 / "The bark is macerated and drunk for colds" / "Sortora"

750   *Poikilospermum suaveolens* Philippine Is. / Sulit & Conklin 5058 / 1953 / "Water exuding from cut stem is medicine for pink eyes" / "Anopal Mang."

       Philippine Is. / G. E. Edaño 2017 / 1950 / ". . . cut the trunk and stem and apply to numbed (pasma) parts of the body" / "Litid Vis."

       Philippine Is. / K. I. Pelzer 74 / 1950 / "Sap used to treat cuts, water of vine used as substitute for water"

       Philippine Is. / M. D. Sulit 727 / 1945 / "The stem is reservoir for drinking water" / "Hanopal Tag."

       Philippine Is. / A. L. Zwickey 711 / 1938 / ". . . frts. . . . with odor of banana oil"

751   *Leucosyke capitellata* Philippine Is. / E. Maliwanag 179 / 1941 / "Leaves used as poultice on swollen face" / "'anagosi' (Mang.)"

752   *L. corymbosa* Fiji Is. / A. C. Smith 6727 / 1947 / ". . . tea made from leaves" / "'Tavitanggai'"

       Fiji Is. / Degener & Ordonez 13809 / 1940 / "Not eaten by Fijian. Fruit to me, however, tastes like that of Pipturus of Haw."

753   *Parietaria floridana* Bermuda Is. / E. Manuel 468 / 64 / ". . . apparently grown for tea previously"

PROTEACEAE

754   *Brabejum stellatifolium* Algeria / E. H. Wilson s.n. / 1922 / "'Wild Almond'"

755   *Franklandia fucifolia* Australia: W. Australia / C. T. White 5404 / 1927 / "Flower . . . strongly vanilla-scented"

756   *Grevillea elaeocarpifolia* New Hebrides Is. / S. F. Kajewski 350 / 1928 / "Nuts eaten by natives"

757   *Helicia artocarpoides* Brit. N. Borneo / H. G. Keith 1626 / 1932 / "Medicine—sore eyes, ulcers, jaws—root used" / "Ambwitil (Murut)"

758   *Finschia chloroxantha* Solomon Is. / Walker & White 183 / 1945 / ". . . edible, pleasantly tasting kernel"

759   *Dryandra floribunda* Australia: W. Australia / E. H. Wilson s.n. / 1920 / "'Parrot Bush'"

LORANTHACEAE

760   *Dactyliophora salomonia* Solomon Is. / S. F. Kajewski 2497 / 1931 / "The leaves are heated and rubbed on sore legs" / "Bitorchi"

761   *Gaiadendron punctatum* Ecuador / W. H. Camp E2078 / 1945 / "'Violeta'—an infusion of the flowers taken for a cough. Giler swears that the gentle call this 'violeta' and use it for a cough; in Cuenca during an epidemic of cough the dried flowers of *Viola* sp. were sold as a cure. It may be . . . some sort of confusion"

762 *Macrosolen tricolor* China / S. K. Lau 424 / 1932 / "Used by native people to cure stomachache by boiling water with root" / "Lio-tei"

763 *M. cochinchinensis* Indochina / Poilane 1319 / 1920 / "Les feuilles sont employées pour faire une boisson analogue au thé" / "Annte.: Cay coi trang   mois aloang to leeo"

764 *Loranthus confusus* Philippine Is. / C. O. Frake 575 / 1958 / "Tree roots boiled and drunk for dysentery" / "Gendis ulangan Sub."

765 *Loranthus* indet. Philippine Is. / C. O. Frake 357 / 1957 / "Love charm" / "(1) Delakep (2) Doribun Sub."

766 *Loranthus* indet. Philippine Is. / C. O. Frake 875 / 1958 / "Leaves applied for various skin diseases" / "Kalambuway Sub."

767 *Loranthus* indet. Philippine Is. / C. Frake 369 / 1959 / "Leaves pounded and applied to ulcers" / "Gimara ulangan"

768 *Loranthus* indet. Philippine Is. / G. E. Edaño 1999 / 1950 / "Pound the leaves in mortar and apply to stomach before giving birth" / "Managuimpol Vis."

769 *Loranthus* indet. Philippine Is. / G. E. Edaño 1700 / 1949 / "The stem of the plant is pounded and is used as a medicine for cuts" / "Pawawot Ma"

770 *Helixanthera parasitica* Indochina / Poilane 1421 / 1920 / "Employé pour faire la boisson" / "Annte.: Cay edi"

771 *Taxillus caloreas* var. *fargesii* China / T. T. Yü 1571 / 1932 / "medicinal material"

772 *Oryctanthus cordifolius* Honduras / P. C. Standley 54749 / 1927–28 / "Suelda con Suelda"

773 *Struthanthus haenkii* Mexico / P. D. Standley 1301 / 1935 / "Medic. Cook herbage in water and wash wound of bite or sting" / "Tohi, W."

774 *S. flexicaulis* Brazil / Y. Mexia 5522 / 1931 / "'Herva de passaro'"

775 *S. syringifolius* Brit. Guiana / A. C. Smith 2435 / 1937 / "Tea made from leaves is used to cure toothache" / "'Pipi-dik' (Wapisiana)"

776 *Psittacanthus collum-cygni* Brazil / R. Froes 1740 / 1932 / "Juice used for hemorrhages" / "'Tonton'"

777 *P. cordatus* Paraguay / T. M. Pedersen 3145 / 1955 / "Locally used as a remedy against Malaria" / "'Caavó-firey'"

778 *P. falcifrons* Brazil / R. Froes 1718 / 1932 / "'Ervo de passarinho'"

779 *P. cupulifer* Peru / J. M. Schunke 332 / 1935 / "Suelda con suelda"

780 *Phoradendron randiae* Dominican Republic / R. A. & E. S. Howard 9676 / 1946 / "Fruit yellow, eaten and enjoyed by the kids"

781 *P. trinervium* W.I.: Dominica / W. H. & B. T. Hodge 2524 / 1940 / "...tea from lves. used for colds"

782   *P. crassifolium* Brazil / Y. Mexia 5123 / 1930 / "'Herva de Passarin'"

783   *P. platycaulon* Brazil / B. A. Krukoff 1248 / 1931 / "'Herva de passerinho'"

784   *P. piperoides* Brazil / Y. Mexia 6011 / 1931 / "'Herva do passaro'"

SANTALACEAE

785   *Exocarpus latifolius* Neth. New Guinea / P. van Royen 5231 / 1955 / "The leaves are used together with betel and causes the teeth to become red" / "kalanje (Waifoi dialect)"

786   *E. aphylla* Australia: Queensland / S. L. Everist 746 / 34 / "'Currant Bush'"

787   *E. cupressiformis* Australia: S. Australia / C. M. Eardley s.n. / 1938 / "'Native Cherry'"

788   *Dufrenoya platyphylla* Burma / F. G. Dickason 7430 / 1938 / "Gurkhas use for tea" / "Ser Bawn (Haka Chin)"

789   *Scleropyrum wallichiana* N. Siam / native collector Royal For. Dept. 4516 / 1949 / "Fruits edible" / "Kaw Nam"

790   *Santalum yasi* POLYNESIA: Tongatapu / T. G. Yuncker 15161 / 1953 / "...root used in preparing Tongan perfume" / "Ahi"

OPILIACEAE

791   *Champereia manillanc* Philippine Is. / C. O. Frake 756 / 1958 / "Young leaves applied for splenomegaly" / "gelemuntay Sub."
Philippine Is. / C. O. Frake 574 / 1958 / "Leaves pounded and applied for headache and stomachache" / "Gelunub Sub."
Philippine Is. / M. D. Sulit 1386 / 1947 / "Young leaves said to be edible" / "Panalayapin Ilak"

792   *Agonandra racemosa* Mexico / G. B. Hinton 3124 / 33 / "Triturated with sugar taken as physic" / "Selva"

793   *A. brasiliensis* Venezuela / L. Williams 12695 / 1940 / "...fruto... con pulpa comible" / "Aceituno"

OLACACEAE

794   *Ptychopetalum olacoides* Surinam / G. Stahel 269 / '44 / "Aphrodisiacum"

795   *Olax laxiflora* China / W. T. Tsang 23863 / 1934 / "fr. red, edible"

796   *Ximenia americana* New Hebrides Is. / S. F. Kajewski 797 / 1929 / "Fruit has an almond taste"

797   *Scorodocarpus borneensis* Brit. N. Borneo / Abunawas B.N.B. For. Dept. A832 / no year? / "Fruit...edible" / "Bawang Otau (Malay)"

798   *Anacolosa griffithii* Indochina / Poilane 14687 / 1928 / "Employé contre les maux de nez fumer les feuilles, raper l'écorce et mélanger au tabac" / "Cambodgien: Dông pra"

799   *Heisteria chippiana* Brit. Honduras / P. H. Gentle 2897 / 1939 / "'copalche macho' 'wild cinnamon'"

800   *Heisteria* indet. Colombia / R. E. Schultes 3481 / 1942 / "Bark infusion to treat for blows or fracture of head in form of hot bath" / "Nombre kofán: 'a-bwa'-ko-pee'"

BALANOPHORACEAE

801   *Ombrophytum zamioides* Bolivia / R. S. Shepard 246 / 1921 / "Root edible" / "'Mantioca'"

802   *Balanophora fungosa* Fiji Is. / O. Degener 15418 / 1941 / "Fijians at the time of planting yam put this plant in ground with idea of stimulating yam to grow" / "Tumbutumbu Ra"

ARISTOLOCHIACEAE

803   *Asarum arifolia* var. *arifolia* U.S.: Va. / Fernald & Griscom 4388 / 1935 / "'Rabbit's Tobacco'"
      U.S.: Miss. / R. McVaugh 8538 / 1947 / " . . . roots smelling of ginger"

804   *A. arifolia* var. *callifolia* U.S.: Fla. / Herb. Chap. s.n. / no year / "'Heart leaf'"

806   *A. insigne* China / W. T. Tsang 24772 / 1934 / "Kam I Wan. Used for medicine"

807   *A. nipponicum* Japan / M. Furuse 20192 / 1951 / "Kama woi"

808   *A. sieboldi* China / Prof. Dyson 99 / 1925 / "For Chinese medicine. Flowers very fragrant"
      Korea / R. K. Smith s.n. / 1925 / "Usuba saishin Churi pul Sai sin"

809   *A. variegatum* Japan / M. Furuse 18231 / 1951 / "Kobano-kamawoi"

810   *Apama corymbosa* Sumatra / R. S. Toroes 3825 / 1933 / "andor lasi-lasi"

811   *A. macrantha* Sumatra / R. S. Boeea 6557 / 1934–35 / "ambolas tombak"
      Sumatra / R. S. Boeea 7113 / 1934–35 / "kajoe attoelmak"

812   *Aristolochia tomentosa* U.S.: Tex. / C. L. Lundell 8437 / 1940 / "'Dutchman's pipe'"

813   *A. anguicida* El Salvador / S. Calderón 148 / 1921 / "'Champipito'"
      El Salvador / P. C. Standley 19441 / 1921 / "'Guaco.' Plants used to cleanse clothes in washing"

El Salvador / P. C. Standley 21262 / 1922 / "'Guaco.' Remedy for stomach ache"

814 *A. arborea* El Salvador / S. Calderón 287 / 1921 / "'Guaquito de la tierra'"

815 *A. cordiflora* Brit. Honduras / P. H. Gentle 3131 / 1939 / "'bastard contrayerba'"

El Salvador / P. C. Standley 23197 / 1922 / "'Güegüecho,' 'Güegüechito.' Lvs. remedy for female venereal diseases and infantile dysentery"

W.I.: Trinidad / W. E. Broadway s.n. / 1932 / "The large flower has a most objectionable odour"

816 *A. glandulosa* Cuba / R. A. Howard 6583 / 1941 / "'Zapatita de la reina'"

817 *A. grandiflora* El Salvador / P. C. Standley 20545 / 1922 / "'Champipe,' 'Güegüecho'"

Jamaica / W. T. Stearn 782 / 1956 / "flowers & fruits pendulous; corolla purple mottled, evil smelling"

818 *A. taliscana* Mexico / G. B. Hinton et al. 11605 / 37 / "Bejuco de guaco"

Mexico / Y. Mexia 1057 / 1926 / "Called 'Zapatilla' because of shape of flower"

819 *A. lindeniana* Cuba / Marie-Victorin, Clément & Alain 21501 / 1943 / "Nourrit la larve du *Papilio* sp. le plus beau papillon . . . de Cuba"

820 *A. mycteria* Mexico / Y. Mexia 8790 / 1937 / "'Hierba del Huaco.' Roots and stalk boiled and concoction drunk for scorpion sting"

821 *A. orbicularis* Mexico / G. B. Hinton 8467 / 1935 / "'Cuajo'"

822 *A. ovifolia* Brit. Honduras / W. A. Schipp 384 / 1929 / "'Guaco' or 'snakeroot'"

823 *A. quercetorum* Mexico / H. S. Gentry 5285 / 1939 / "Yerba del Indio. Roots decocted for stomach ailments"

824 *A. schippii* Brit. Honduras / W. A. Schipp 75 / 1929 / "'Snake root'"

825 *A. sylvicola* Panama / W. L. Stern et al. 204 / 1959 / "'Zaragosa'"

826 *A. trilobata* Brit. Honduras / P. H. Gentle 3604 / 1941 / "'contrayerba'"

827 *A. wrightii* Mexico / Stewart & Johnston 2107 / 1941 / "'Yerba del Indio'"

828 *A. chilensis* Chile / R. Wagenknecht 18538 / 1940 / "'oreja de zorro'"

829 *Aristolochia* indet. Colombia / G. Klug 1849 / 1930 / "'Oreja de tigre'"

830   *Aristolochia* indet. Brit. Guiana / A. C. Smith 2831 / 1937 / "'Mame-tala' (Waiuai). Entire plant boiled and used externally to reduce fever"

831   *Aristolochia* indet. Peru / F. Woytkowski 5432 / 1959 / "Medicinal"

832   *Aristolochia* indet. Philippine Is. / M. D. Sulit 4546 / 1952 / "Ubi-ubihan. Dialect Tag."

833   *A. arimensis* Japan / H. Muroi 203 / 1953 / "Japanese name: Arima uma no suzukusa"

834   *A. bracteata* E.I. / Wight 88 / before 1870 / "it is used for the cure of cutaneous diseases"

835   *A. elegans* Philippine Is. / E. Canicosa 424 / 1949 / "Tunbaigor"

836   *A. tagala* Solomon Is. / S. Kajewski 2225 / 30 / "Mai Mai. When a man wants a wife, he hides behind a tree and picks a small piece and blows along it towards the woman, which is supposed to act as a kind of love charm"

837   *Thottea* indet. Sumatra / R. S. Boeea 9917 / 1936 / "'si marete-ate'"

838   *Thottea* indet. Sumatra / R. S. Boeea 9534 / 1936 / "Kajoe si mar-saping"

839   *Thottea* indet. Sumatra / R. S. Boeea 7438 / 1934 / "kajoe pinggoe batoe"

POLYGONACEAE

840   *Eriogonum atrorubens* Mexico / W. P. Hewitt 44 / 1945 / "...root astringent and chewed for gums" / "Yerba Colorada"
Mexico / H. S. Gentry 8484 / 1948 / "...eaten by cattle"

841   *E. tenellum* Mexico / W. P. Hewitt 206 / 1947 / "...boiled and tea taken as medicine" / "Suchiaca"

842   *Emex spinosa* Kuwait / H. Dickson 124 / 62 / "The root can be eaten by Badu.... Grazed on by animals" / "Ah Hambizan"

843   *Rumex crispus* Brazil / W. A. Archer 3426 / 1936 / "Calves sometimes poisoned by eating this" / "'lengua de vaca'"
China / W. T. Tsang 21942 / 1933 / "...food for pig"

844   *R. vesicarius* Kuwait / H. Dickson 7 / 61 / "Leaves eaten by Badu" / "Hamaith"

845   *Polygonum aubertii* China / J. Hers 813 / 1919 / "Chinese name: suan ki ki 'leaves sour'"

846   *P. capitatum* China / W. T. Tsang 22936 / 1933 / "...used as medicine" / "Sha Tan Tsz"

847   *P. fagopyrum* China / W. T. Tsang 21959 / 1933 / "...fr. edible" / "Sam Kok Mak"

848   *P. hydropiper* China / W. T. Tsang 21762 / 1932 / "...root as medicine to wound" / "Sai Yeung Lat Lind"

       INDIA / L. Pierre s.n. / 1866 / "...feuilles mangées en salade"

849   *P. longiseta* Japan / E. Elliott 149 / 1946 / "Food use: Leaves boiled in ash water, soaked in fresh water; then eaten with salt and miso (in time of scarcety). Medicinal use: Whole plant good for gastric cancer" / "Inutade"

850   *P. nepalense* China / W. T. Tsang 27576 / 1937 / "Used for feeding pigs" / "Ye Sam Kok Mak Tsoi"

851   *P. palmatum* China / W. T. Tsang 23124 / 1933 / "...used as medicine" / "Wat Tso"

852   *P. sphaerocephalum* China / W. T. Tsang 22990 / 1933 / "...used as medicine" / "Fo Tan Mo"

853   *P. amplexicaule* Kashmir / W. Koelz 8920 / 1936 / "Roots said to be used for kidney medicine"

854   *Polygonum* indet. New Guinea / R. Pullen 1638 / 1959 / "Used as tobacco when Nicotiana not available" / "Pirum-pirum-man (Timbunke = Sepik)"

855   *P. apoense* Philippine Is. / P. Añonuevo 191 / 1950 / "The leaves are pounded and the juice is dropped in the ear to cure deafness" / "Cabanaybanay Dialect Manobo"

856   *P. minus* New Hebrides Is. / S. F. Kajewski 339 / 28 / "mixed with Nunpor-lell Ney-wass, Ne-cit-ersif, for medicine for sickness on left side of stomach, for to drink" / "Neta-pea"

857   *Fagopyrum* indet. Thailand / T. Smitinand 7032 / 1960 / "Young shoot is edible; leaves as poultice used for broken limbs by Musaor people" / "Chu pe ne (musaor)"

858   *Muehlenbeckia tamnifolia* Argentina / Eyerdam & Beetle 22245 / 1938 / "'Zarzaparrilla'"

859   *M. rhyticarya* Australia: Queensland / C. T. White 10672 / 1936 / "...said to be liked by stock"

860   *Coccoloba barbadensis* Mexico / H. S. Gentry 6804 / 1943 / "Fruit reported edible when black or dark purple" / "Roble"

861   *C. belizensis* Honduras / P. C. Standley 53555 / 1927–28 / "Fruit... much eaten by birds" / "'Uva'"

862   *C. dussii* W.I.: St. Vincent / R. A. Howard 11185 / 1950 / "...fruit black, very astringent"

863   *C. pubescens* W.I.: Lesser Antilles (La Désirade) / G. R. Proctor 21232 / 1960 / "'Ti Raisin'"

864   *Triplaris americana* El Salvador / P. C. Standley 22216 / 1922 / "'Palo mulato, Mulato, Canilla de mula'"

865   *T. noli-tangere* Brazil / B. A. Krukoff 1126 / 1931 / " ... resin used as perfume" / " 'Brew branco' "

866   *Ruprechtia splendens* Mexico / G. B. Hinton 5376 / 33 / "Sangre de Toro"

CHENOPODIACEAE

867   *Rhagodia hastata* Australia: Queensland / L. S. Smith 516 / 1936 / "Readily eaten by both sheep and cattle which it is said to keep in good condition"

868   *R. parabolica* Australia: Queensland / C. T. White 11666 / '61 / " ... eaten by stock only as a last resort"

869   *Chenopodium album* Mexico / A. Dugès 8 / 1907 / "Sous le nom de Cuanzontle ou Cuanzoncle on mange les extrémités fleuries frites enveloppées d'oef; c'est un bien pauvre legume"

870   *C. berlandieri* Mexico / J. Gregg 549 / 1848 / " 'Quelito,' used as salad"

871   *C. ambrosioides* Mexico / G. B. Hinton 3857 / no year / "Concoction taken to cure fright. Also to flavor beans" / "Epasote"

Mexico / H. S. Gentry 1475 / 1935 / "Medic. herbage boiled and eaten for colic" / "Lipasote, Mex. Pasote, W."

Chile / A. A. Beetle 26181 / 1939 / "Used as tea for stomach trouble by natives" / " 'Pyco' "

Colombia / Killip & Smith 19665 / 1927 / "Remedy for tropical anemia" / " 'Baíco' "

Bolivia / J. Steinbach 5135 / 1920 / "De odor fuerte como especies Planta medicinal; se usa en untos calientes" / "Karé"

872   *C. graveolens* Mexico / G. B. Hinton 6513 / 34 / " ... medicinal" / "Epasote de Perro"

873   *C. quinoa* Honduras / P. C. Standley 28135 / 1951 / " ... leaves were cooked and eaten as quelites"

874   *C. incisum* Argentina / J. West 6308 / 1936 / "Used medicinally by people of region"

875   *C. hircinum* Paraguay / W. A. Archer 4933 / 1937 / "Used as tea for internal injuries" / " 'quino quino' "

876   *C. quinoa* var. *glomerulatum* f. *kcella* Bolivia / M. Cárdenas 3502 / 1944 / "Edible seeds" / " 'Kcella quinoa' "

877   *C. auricomum* Australia: Queensland / C. T. White 11900 / 1941 / "Said to be an excellent vegetable and good fodder for stock" / " 'Blue bush' "

878   *C. polygonoides* Australia: Queensland / L. S. Smith 514 / 1938 / "Picked out by cattle in preference to other herbage"

879 *Atriplex canescens* Mexico / R. M. Stewart 842 / 1941 / "'Costilla de Vaca'"

880 *A. semibaccata* Chile / A. Garaventa 1463 / 1929 / "utilizada como forraje"

Australia: Queensland / C. T. White 12234 / 1934 / "...freely eaten by stock"

881 *A. muelleri* Australia: Queensland / C. T. White 9528 / 1933 / "...not...eaten by stock except when other feed not available"

882 *Eurotia lanata* U.S.: Mont. / E. J. Woolfolk s.n. / 36 / "Grazed by C. S. H."

U.S.: Wyo. / J. G. Jack 1003 / 1918 / "'Winter Fat'"

883 *Bassia eriophora* Kuwait / H. Dickson 65 / 61 / "...not eaten by animals" / "Ah Guttainah"

884 *B. muricata* Kuwait / H. Dickson 123 / 62 / "...not eaten by animals"

885 *Kochia scoparia* China / Cheo & Yen 274 / 1936 / "...lvs. edible"

AMARANTHACEAE

886 *Deeringia amaranthoides* Solomon Is. / S. F. Kajewski 2164 / 1930 / "The vine is cut and the sap is used to wash the skin of pigs suffering with skin diseases" / "Tugiama"

887 *D. polysperma* Philippine Is. / C. O. Frake 564 / 1958 / "Roots used to prevent drunkeness"

Philippine Is. / C. O. Frake 643 / 1958 / "...young leaves applied during puerperium" / "Libuatan"

888 *Celosia cristata* Japan / E. Elliott 72 / '46 / "Young plant cooked or put in Miso soup.... Flower comb, dried and decocted, is good for dysentery and is diuretic" / "Keito"

889 *C. argentea* Siam / den Hoed & Kostermans 700 / 1946 / "...taste of boiled leaves like spinach somewhat bitter;...was eaten in large quantities as a green vegetable against diseases like beriberi and pellagra by prisoners of war at Death-railway in Siam.... good results"

890 *Amaranthus palmeri* Mexico / G. B. Hinton 5532 / 34 / "Ash for soap making" / "Quelite de marrano"

891 *A. spinosus* Mexico / G. B. Hinton 2663 / 32 / "Food for pigs"

W.I.: Barbados / G. R. Cooley 8639 / 1962 / "'Wild spinach, Baggy'"

W.I.: St. Kitts / G. R. Cooley 8844 / 1962 / "When young used as salad"

Siam / den Hoed & Kostermans 694 / 1946 / "...eaten in large quantities by P.O.W.... at 'death railway' in Siam; good vegetable when boiled like spinach"

Siam / Wichian 312 / 1946 / " . . . roots rubbed with salt, to be ripped; for relieving malnutrition"

892 *A. hybridus* Colombia / Killip & Smith 19661 / 1927 / "Remedy for indigestion" / "'Bledo'"

893 *A. blitum* China / W. T. Tsang 22731 / 1933 / " . . . fl. dark red edible; it is eaten to remove poison from system" / "Ka In Tsoi"

894 *A. tricolor* New Britain / J. H. L. Waterhouse 943 / 1935 / "Excellent spinach"

895 *Amaranthus* indet. Philippine Is. / A. L. Zwickey 722 / 1938 / " . . . edible; lvs. and roots in treatment for boils" / "'Marodu' (Lan.)"

896 *Cyathula prostrata* Philippine Is. / C. Frake 467 / 1958 / "Leaves burnt and applied to ringworm"

897 *Pupalia lappacea* Philippine Is. / L. E. Ebalo 1010 / 1941 / " . . . leaves used as medicine for boils" / "Name in Subano: 'mulohamad'"

898 *Aerva* indet. Philippine Is. / E. Maliwanag 190 / 1941 / " . . . fruits used as medicine for toothache" / "'dama' (Mang.)"

899 *Ptilotus clementi* Australia: Queensland / S. F. Pearson 112 / 1941 / " . . . eaten by horses and kangaroos" / "Tassel Top"

900 *Achyranthes indica* W.I.: St. Lucia / J. Clarke s.n. / 62 / "Known as Marie pourrie and used for treatment of venereal diseases"

901 *A. japonica* Japan / E. Elliott 130 / '46 / "Boiled young leaves may be eaten. Medicinal use: subterranean stem is used for diuretic" / "Inokusuchi"

902 *A. bidentata* China / H. T. Feng 36 / 1924 / "Some of it . . . used as medicine for perspiration"

903 *Alternanthera sessilis* China / W. T. Tsang 22909 / 1933 / " . . . flower grey edible" / "Chuk Tsit Tsoi"

China / W. T. Tsang 21108 / 1932 / " . . . flower, white, edible, fruit" / "Ngo Cheung Tsoi"

New Guinea / K. G. Heider 11 / 1962 / "Smoked as medicine" / "Vern. name: asukdlek (Dani lang.)"

904 *Gomphrena dispensa* Panama / Macbride & Featherstone 2 / 1922 / "Has peculiar odor suggesting fresh fish"

905 *G. decumbens* Mexico / A. López 82 / 1948 / "'Retama' buena par enfermidades interiores"

906 *G. nitida* Mexico / A. López 53 / 1948 / "'Rosetilla' 'buena para los bados bajinales'"

907 *Iresine herrerae* Mexico / M. Martínez 93 / '63 / "'carne de gallina'"

908 *I. interrupta* Mexico / P. C. Standley 1260 / 1935 / "Has a strong feral odor"

909 *Philoxerus vermicularis* Venezuela / Curran & Haman 443 / 1917 / "Yerba paloma"

NYCTAGINACEAE

910 *Acleisanthes longiflora* Mexico / J. Gregg 88 / 1848 / "'Yerba de la rabia' 'Yerba Santa'"

911 *Mirabilis jalapa* Mexico / G. B. Hinton 631 / 32 / "'Maravilla' Ground up and applied as a poultice relieves bruises and strains; so the Indians say"

912 *M. viscosa* Peru / Y. Mexia 04108 / 1935 / "'Tabaquilla del Campo'"

913 *Mirabilis* indet. Hawaiian Is. / O. Degener 17268 / 1926 / "(fls. bright yellow. This clump is said by Hawaiians to be better medicine than other colored fls. Belief founded on chiefs color being yellow?)"

914 *Allionia incarnata* Mexico / A. Lopez 83 / 1948 / "'Yerba Mora' buena para los baños a quite la calentura"

915 *Boerhaavia erecta* Mexico / P. C. Standley 1045 / 1934 / "Medic. By boiling the plant in water a wash is obtained for sores" / "Mochi, Mex., Peneywa, W."

916 *B. intermedia* Mexico / A. Lopez 31 / 1948 / "'La mosa' serba par los granos"

917 *B. coccinea* Brazil / R. Froes 1861 / 1932 / "Roots used as an emetic" / "'Teja Pinto'"

Brazil / Reitz 1724 / 46 / "Medicinal" / "Herva tostae"

918 *Pisonia capitata* Mexico / P. C. Standley 1273 / 1935 / "Man and animals eat orange colored berries" / "Gumbro, Garumbuyo, Bynora, Mex. Koloka, W."

919 *P. irregularis* Bolivia / J. Steinbach 6464 / 1924 / "La madera se quema para ceniza que se ocupar en el beneficio del azucar y para hacer jabon" / "Ajillo"

920 *P. umbellifera* Philippine Is. / C. O. Frake 361 / 1957 / "Roots used for soap" / "Melikapuk Sub."

Philippine Is. / C. O. Frake 609 / 1958 / "Bark applied for stomachache" / "bintulung Sub."

921 *Neea parviflora* Colombia / G. Klug 1955 / 1931 / "'Yana muco' The leaves are chewed by the natives, making their teeth black but preserving them. These Indians have very sound and strong teeth. 'Yana' is black, 'muco' is chew"

922 *Salpianthus purpurascens* Nicaragua / A. Guarnier 1763 / 1935 / "'Pala de paloma'"

PHYTOLACCACEAE

923 *Achatocarpus nigricans* El Salvador / P. C. Standley 21046 / 1922 / "...fr....reported poisonous" / "'Limoncillo'"

924   *Rivina humilis* Mexico / R. M. Stewart 1563 / 1941 / "'Yerba de Cancer'"

Bahama Is. / R. A. & E. S. Howard 9999 / 1948 / "'Pepper bush'"

Brazil / J. C. Morais 957 / 1953 / "'murta velame'"

925   *Petiveria alliacea* Honduras / P. C. Standley 52967 / 1927–28 / "Has strong skunk odor. Used medicinally" / "'Ipacina'"

Mexico / G. Martínez-Calderón 21 / 1940–41 / "Se usa para curar a las personas que se encuentren ravioso"

Brazil / B. A. Krukoff 7638 / 1936 / "Used in curare of Tecuna Indians"

926   *Microtea* indet. W.I.: Dominica / W. H. & B. T. Hodge 3376 / 1940 / "...weed used for tea" / "'demoiselle'"

927   *Stegnosperma watsonii* Mexico / C. Lumholtz 9 / no year / "(Said... to be a splendid remedy for hydrophobia)" / "Chapa color"

928   *Phytolacca icosandra* Mexico / W. H. Camp 2335 / 1936 / "Used instead of soap by natives"

929   *P. rivinoides* Brit. Honduras / W. A. Schipp 125 / 1929 / "Caribs boil leaves and use as tea"

Peru / F. Woytkowski 5761 / 1960 / "...medicinal leaves for washing produce foam; leaves eaten cooked like spinach; tonic" / "'Jaboncillo' or 'Airampo'"

Colombia / R. E. Schultes 3458 / 1942 / "Used with Phyllanthus sp. as fish poison" / "Nombre kofán: 'un-shum-bey' 'altusa'"

930   *P. acinosa* China / Cheo & Yen 214 / 1936 / "Root for medical use"

931   *P. esculenta* Japan / E. Elliott 33 / '46 / "Food use: Leaves boiled in water and eaten mixed with rice. Medicinal use: Root and seed diuretic; also said to be good for kidney trouble, pleuritis and pneumonia.... Some kind of root poisonous"

AIZOACEAE

932   *Trianthema portulacastrum* Siam / Wichian 309 / 1946 / "...used as vegetables"

PORTULACACEAE

933   *Talinum paniculatum* Peru / J. West 8003 / 1936 / "Used medicinally"

934   *Calandrinia caulescens* Mexico / G. B. Hinton 4923 / 44 / "eaten boiled & fried" / "Chivitas"

935   *C. acaulis* Peru / C. Vargas 9844 / 1939 / "The indigenes eat the roots" / "'Ckapacio'"

936   *Portulaca perennis* Peru / Mr. & Mrs. F. E. Hinckley 1 / 1920 / "...used for urinal troubles"

937   *P. quadrifolia* Caroline Is. / C. C. Y. Wong 386 / 1948 / "This plant is eaten by the people of the atolls around the Yap Is. The leaves are used as feed for young turtle and pigs" / "N.v. gogus (gagil); kamot pacibak"

BASELLACEAE

938   *Basella alba* Indochina / Poilane 532 / 1919 / "Feuilles tendres et comestible" / "Ate.: Cây dây mang to'i"

939   *Anredera diffusa* Peru / Mr. & Mrs. F. E. Hinkley 78 / 1920 / "Use, medicinal" / "'Lloto del cirro'"

CARYOPHYLLACEAE

940   *Sclerocephalus arabicus* Kuwait / H. Dickson 53 / 61 / "Eaten by camels" / "Ah Tharaisa"

941   *Drymaria cordata* El Salvador / S. Calderón 266 / 1921 / "'Comida de Canario'"

942   *D. gracilis* Mexico / H. S. Gentry 1338 / 1935 / "Medic. cook in water and wash for 'yaza'" / "Cadenilla, Mex. Kueypali, W."

943   *D. villosa* El Salvador / P. C. Standley 22194 / 1922 / "Remedy for coughs" / "'Poleo'"

944   *D. pauciflora* Ecuador / W. H. Camp E2624 / 1945 / "Said to be medicinally more potent than other 'Drimarias'"

945   *Silene* indet. Afghanistan / E. Bacon 83 / 1939 / "Used for soup"

946   *Polycarpaea gaudichaudii* China / S. K. Lau 462 / 1932 / "Used as a cure for wound in the head by crushing root, stem, leaves with wine" / "Mun tin sing ts'o"

947   *P. repens* Saudi Arabia / H. Dickson 80 / 61 / "Eaten by animals"

948   *Stellaria media* China / W. T. Tsang 21998 / 1933 / "... food for pigs"
China / W. T. Tsang 21860 / 1933 / "Eatable when cooked"

949   *Lepyrodiclis holosteoides* India: Punjab / W. Koelz 8454 / 1936 / "Used: vegetable"

950   *Gypsophila acutifolia* China / C. Y. Chiao 2618 / 1930 / "Used as vegetable by natives"

951   *G. capillaris* Kuwait / H. Dickson 22 / 61 / "Eaten by camels" / "Eshet al Dhabbi"

952   *Acanthophyllum* indet. Afghanistan / E. Bacon 101 / 1939 / "One of the few types of plants left untouched by grazing sheep"

NYMPHAEACEAE

953   *Nelumbo speciosum* China / C. Schneider 3775 / 1914 / "... in temple ..."

China / C. Y. Chiao 3039 / 1930 / "Sepals, seeds & rhizomes *edible*" / "'Lotus'"

China / Steward & Cheo 789 / 1933 / "Seeds and roots edible" / "'Ho Hwa'"

954   *Brasenia* indet. China / C. W. Wang 81210 / 1936 / "Young shoots lvs. clad with a layer of transparent gelatinous sheath very sticky flowers... lvs. perhaps edible"

955   *Euryale ferox* China / F. A. McClure 20548 / 1937 / "The seeds are sold in food shops and in drug shops under the name Ch'i Sat... as a Pu-pin... that is, one of the many foods especially esteemed by the Chinese as restoring or preserving health. It is listed in Chinese Materia Medica"

956   *Nymphaea odorata* var. *gigantea* U.S.: Okla. / G. W. Stevens 2386 / 1913 / "... tubers eaten by Ottawa Indians and called Pok-shicken-i-uck"

957   *Nymphaea* indet. Mexico / H. S. Gentry 7104 / 1944 / "Bulbs are dug and eaten during the dry season"

958   *N. ampla* Brit. Honduras / W. A. Schipp 640 / 1930 / "... sweetly perfumed leaves ..."

RANUNCULACEAE

959   *Coptis teeta* China / W. T. Tsang 24881 / 1935 / "... used for medicine" / "Fu Wong Lin"

China / W. T. Tsang 25065 / 1935 / "... fr. white, edible" / "Wong Lin"

960   *Aquilegia skinneri* Mexico / W. P. Hewitt 151 / 1946 / "... roots cooked are used as remedy for bruises" / "Pericos"

961   *Aconitum carmichaeli* China / Steward, Chiao & Cheo 738 / 1931 / "Medicinal"

962   *A. delavayi* China / H. Wang 41683 / 1939 / "... root aromatic, officinal"

963   *A. japonicum* Japan / E. Elliott 132 / 1940 / "Poisonous plant.... Rhizomes famous materia medica to cure tumor and cancer" / "Tori-kabuto"

964   *A. sinomonianum* China / W. P. Fang 1026 / 1928 / "One kind of medicine"

965   *A. transsectum* China / J. F. Rock 5377 / 1922 / "Poisonous"

966   *A. volubile* China / H. Wang 41685 / 1939 / "... root very poisonous"

967   *A. heterophyllum* Kashmir / W. Koelz 9339 / 1936 / "Root used medicinal"

968   *Clematis dioica* Guatemala / J. Steyermark 42182 / 1942 / "Used for drawing heat out of skin" / "'Barba venado'"

Mexico / R. Cardenas 339 / 1911 / "Cura el catarro a los animales principal el caballo, y caustico poderosa usandolo algunos Medicos en sus enfermos" / "Barba de Chivo"

969 *C. drummondii* Mexico / A. Lopez 59 / 1948 / "'Barbas de Chivos' Bueno para los granos"

970 *C. alpina* Japan / E. H. Wilson s.n. / 1914 / "Whole plant said to be poisonous"

971 *C. brevicaudata* China / S. W. Williams s.n. / 1878 / "Very strong rank smell when flowering"

972 *C. rhederiana* China / W. T. Tsang 20847 / 1932 / "Used for medicine" / "Muk Tung T'ang"

973 *C. meyeniana* var. *granulata* Indochina / Poilane 1043 / 1920 / "...les feuilles servent à faire un beverage que les femmes prennent aprés l'accouchement"

974 *C. papuasica* Solomon Is. / S. F. Kajewski 2379 / 1930 / "The leaves are crushed and applied by the natives to their heads when suffering from fever" / "Quala-hu-dum"

975 *C. gouriana* Philippine Is. / G. E. Edaño 3770 / 1953 / "Leaves when pounded used in the treatment of wounds"

976 *C. javana* Philippine Is. / P. Añonuevo 244 / 1950 / "The juice of the triturated leaves is applied to open wound" / "Kanding kanding Bis."

977 *Ranunculus scleratus* China / H. T. Feng 125 / 1925 / "For chinese medicine"

China / W. T. Tsang 23429 / 1925 / "For feeding pigs" / "Shui Yueng Mooi Choi"

978 *Adonis chrysocyathus* Kashmir / R. R. Stewart 14813A / 1935 / "Said to be poisonous. Avoided by animals"

LARDIZABALACEAE

979 *Holboellia fargesii* China / Fan & Li 612 / 1935 / "...fr....edible"

BERBERIDACEAE

980 *Dyosma hispidum* China / W. T. Tsang 22885 / 1933 / "...used as medicine" / "Pat Kak Lin"

981 *Berberis longipes* Mexico / W. P. Hewitt 30 / 1945 / "Fruit crushed mixed with water to prepare cooling drink" / "Coyaira"

982 *B. trifoliata* Mexico / W. P. Hewitt 103 / 1946 / "...berries prepared like cranberries, make delicious jelly" / "Aigrito or Palo amarillo"

983 *B. heterophylla* Argentina / Eyerdam, Beetle & Grondona 24517 / 1939 / "Berries eaten by children and Yaghana" / "Calefate"

984 *Berberis* indet. Burma / F. G. Dickason 8530 / 1939 / "Bark use medicinally for stomach trouble"

MENISPERMACEAE

985 *Cocculus thunbergii* China / T. C. Huang 2332 / 1961 / "... medical plant"

986 *C. trilobus* Nippon or vicinity / K. Uno s.n. / 1944 / "Medicine"

987 *Stephania brachyandra* China / C. W. Wang 76571 / 1936 / "Root stock thick edible"

988 *S. salomonium* Solomon Is. / S. F. Kajewski 2258 / 1930 / "The leaves are rubbed on the skins of the natives to allay pain" / "'Karkar-mukui'"

989 *Cissampelos ovalifolia* Brit. Guiana / A. C. Smith 3229 / 1938 / "Roots one of the chief ingredients of Macusi Urari (Information from Father Mather, S.J., from a recently deceased old Macusi)"

990 *Abuta grandifolia* Columbia / G. Klug 1692 / 1931 / "'Calentura caspi'; remedy for malaria"

991 *A. selloana* Brazil / Y. Mexia 4456 / 1930 / "Fruit green; said to be ... edible" / "'Baco Pary'"

992 *Fibraurea chloroleuca* Malaya / G. anak Umbai for A. H. Millard (K.L. 1453) / 59 / "Phytochemical Survey of Malaya" / "Medicinal"

993 *Fibraurea* indet. Celebes / H. Curran 293 / 40 / Fruits "contain milky juice and mucilaginous pulp, not edible ... "

994 *Tinospora crispa* Philippine Is. / E. Fénix 284 / 1939 / "... very bitter sap ... "

995 *T. rumphii* Philippine Is. / C. O. Frake 846 / 1958 / "Vine scraped and soaked—drunk for malaria" / "'Melibutigan'" Sub.

996 *Tinospora* indet. Philippine Is. / Edaño & Gutierrez 237 / 1957 / "Medicinal——(leaves?)—for baby bath" / "'bang agang'"

997 *T. reticulata* Philippine Is. / C. O. Frake 796 / 1958 / "Leaves burnt and applied for pinworms" / "'glingumelibutigan'"

Philippine Is. / M. D. Sulit 1889 / 1947 / "Roots sliced, mixed with Kamote—used as bait for wild pigs. When pig eats the Kamote it becomes drowsy and can be caught easily" / "'Kisai'" Bis.

998 *Anamirta cocculus* Philippine Is. / M. D. Sulit 1477 / 1947 / "Fruits when mature are roasted, powdered, then mixed with baits and thrown into the sea—as fish poisoned when baits are eaten" / "'Lagtang'" Tag.

Philippine Is. / Ramos & Edaño 641 / 1932 / "Fruits are used to poison fish" / "'Langtang'" Belaan

India / R. D. Anstead 156 / 1927 / "Source of a drug known as Cocculus indicus"

Ceylon / collected in Royal Botanic Garden (Econ. Plants of the World 121) / 1907 / "The seeds are very bitter and poisonous. Called Titta-wel in Ceylon" / (This collection has been identified also as *A. paniculata*.)

999 *Arcangelisia loureiri* China / C. W. Wang 80948 / 1935–36 / "Edible"

1000 *Arcangelisia* indet. Philippine Is. / A. L. Zwickey 113 / 1938 / "fish poison" / "'Tigau' (Lan.); 'Lagtan' (Bis.)"

1001 *Hypserpa monilifera* Solomon Is. / S. F. Kajewski 2125 / 1930 / "The bark of this vine is taken off and pounded, being applied to the head to relieve headaches" / "'Luckali'"

1002 *Pycnarrhena manillensis* Philippine Is. / M. D. Sulit 1029 / 1946 / "Medicinal—Decoction of roots—tonic, cures stomach trouble. Very bitter" / "'Ambal'" Tag.

1003 *Pycnarrhena* indet. Philippine Is. / G. E. Edaño 1900 / 1949 / "Medicine for stomachache. The bark is ground into fine particles, boiled and the infusion is given for stomachache" / "'Unatunat'" Ma

1004 *Hyperbaena tonduzii* El Salvador / P. C. Standley 20888 / 1922 / "'Cuero del diablo'"

El Salvador / P. C. Standley 23216 / 1922 / "'Huevo del diablo'"

1005 *H. brevipes* Haiti / E. C. & G. M. Leonard 15251 / 1929 / " . . . ripe fruit red, astringent . . . "

1006 *H. cubensis* Cuba / E. L. Ekman 9452 / 1918 / "'Picha jutía'" or "'Mavinga'"

1007 *H. longiuscula* Cuba / E. L. Ekman 2594 / 1914 / "'Picha jutia'"

1008 *Chondrodendron toxicoferum* Brazil / B. A. Krukoff 7535 / 1935 / "Component of Curare of Tecuna Indians"

MAGNOLIACEAE

1009 *Magnolia obovata* China / C. Y. Chiao 2597 / 1930 / "Medicinal"

ILLICIACEAE

1010 *Illicium brevistylum* China / W. T. Tsang 26260 / 1936 / " . . . fr. . . . edible"

1011 *I. dunnianum* China / W. T. Tsang 25476 / 1935 / " . . . fr. . . . edible" / "Shan Pat Kok Shue"

1012 *I. majus* China / Steward, Chiao & Cheo 516 / 1931 / "Fruits used for seasoning food"

1013 *I. pachyphyllum* China / W. T. Tsang 24278 / 1934 / " . . . fr. edible" / "Ching Pat Kok Shue"

SCHISANDRACEAE

1014 *Schisandra chinensis* China / J. Hers 2154 / 1922 / "medicinal" / "wu wei tze"

1015 *S. henryi* China / Steward, Chiao & Cheo 132 / 1931 / "Fruit edible"

1016 *S. sphenanthera* China / J. Hers 860 / 1919 / "Chinese name: wu wei tze 'the five-tastes plant' fruit said to be similar to grapes, but red; edible"

1017 *Kadsura heteroclita* China / W. T. Tsang 17016 / 1928 / "...fr. green; medicine" / "Kwo Shan Lung T'ang"

1018 *K. oblongifolia* China / C. I. Lei 205 / 1932 / "...fr. red, edible" / "Fan Luen Tang"

1019 *K. longepedunculata* China / H. H. Hu 565 / 1920 / "...berry... edible"

1020 *K. scandens* Philippine Is. / C. O. Frake 805 / 1958 / "Roots boiled and drunk for general malaise" / "malamala Sub."

Philippine Is. / C. O. Frake 696 / 1958 / "Agricultural ritual. Leaves applied to ulcers" / "Mezgarid"

Philippine Is. / C. O. Frake 578 / 1958 / "Medicine for puerperium" / "gantungantung Sub."

WINTERACEAE

1021 *Drimys granadensis* Ecuador / W. H. Camp E4370 / 1945 / "'Cascarilla picante'"

1022 *D. granadensis* var. *grandiflora* Colombia / F. W. Pennell 6945 / 1922 / "(Canela del parâmo)"

1023 *Belliolum gracile* Solomon Is. / S. F. Kajewski 2630 / 1931 / "They plant this tree when growing Taro as a superstition the Taro will grow well" / "Ses-a-vere"

1024 *B. haplopus* Solomon Is. / S. F. Kajewski 1994 / 1930 / "If the pigs have a skin disease, the leaves are pounded and rubbed on the place to cure it" / "Oigu"

ANNONACEAE

1025 *Neouvaria merrillii* N. Borneo / B.N.B. For. Dept. 3110 / 1933 / "... fruit...used for treating fever in children" / "Kayu bissing"

1026 *Uvaria microcarpa* China / Y. W. Taam 1826 / 1940 / "...fruit black, edible"

China / C. I. Lei 298 / 1932 / "Leaves used for making wine ferment cakes" / "Tzou Beng Kwo"

1027 *U. purpurea* China / H. Fung 20029 / 1932 / "...fr. edible" / "Shan Pe Tsiu"

1028 *U. cauliflora* N. Borneo / B.N.B. For. Dept. 3102 / 1933 / "Water from this vine good to drink" / "Akar pitudong (Kedayan)"

1029 *U. confertiflora* N. Borneo / G. Pascual 1048 / 1929 / "Fruit like bananas"

1030 *U. grandiflora* Malaya / Ishmail & Millard KL232 / 1958 / "Sudorific and also used for coughs. Freshly cut roots have an odour reminiscent of Eau de Cologne" / "Akar larak"

1031 *U. ovalifolia* N. Borneo / B.N.B. For. Dept. 2727 / 1932 / "...native medicine" / "Sagombong Dusun"

1032 *U. rosenbergiana* Solomon Is. / S. F. Kajewski 2257 / 1930 / "The leaves are dried and hung in the native armlets, often being mixed with cocoanut oil on account of their scent" / "Poie-ma"

1033 *U. rubra* Philippine Is. / F. M. Salvoza 1043 / 1946 / " . . . fruit edible" / "Saging-bulag Tag."

1034 *U. sorsogonensis* Philippine Is. / M. D. Sulit 365 / 1945 / "Fruit edible when ripe" / "Ulagak Dialect Bic"

1035 *Anomianthus heterocarpus* Annam / Poilane 8153 / 23 / " . . . fruits . . . comestible" / "Annamite: Cây du de trâu"

1036 *Cananga odorata* Solomon Is. / S. F. Kajewski 2318 / 1930 / " . . . petals very strongly scented and used by the natives as a perfume. The leaves are also crushed and applied to boils"

1037 *Desmos cochinchinensis* China / W. T. Tsang 24606 / 1934 / "Chau Pan Tang"

1038 *D. hancei* Indochina / Poilane 1184 / 1920 / " . . . les Indigènes. Ils emploient les rameaux et feuilles pour faire un breuvage qu'ils font prendre au femme aprés accouchement pour augmenter la secrétion lactée; ils donnerait également des fruits comestibles" / "Annte: Cây Công tây"

1039 *D. mindorensis* Philippine Is. / Sulit & Conklin 5178 / 1953 / "Branch obtained & put in crossing of path of wild pig. Wild pig will take path where 'bolatik' is set, hence it will be caught" / "Dalokdok-Buladlad (Mang.)"

1040 *Polyalthia consanguinea* China / S. K. Lau 224 / 1932 / " . . . fr. . . . edible"

1041 *P. beccarii* Sarawak / P. W. Richards 1354 / 1932 / "Bark . . . smelling of guava jelly" / "Serebeh"

1042 *P. elongata* Philippine Is. / Malana & Alviar 5 / 1917 / "Seed glutinous. Itchy when chewed"

1043 *Polyalthia* indet. Philippine Is. / M. D. Sulit 4095 / 1951 / "Bark bitter" / "Kalimatas Tag."

1044 *Henicosanthum* indet. Philippine Is. / C. O. Frake 820 / 1958 / "Agric. ritual" / "Puyu nutuna Sub."

1045 *Popowia pisicarpa* Philippine Is. / P. Añonuevo 43 / 1950 / "Leaves are powdered and rubbed on the skin to prevent being stung by honey bees" / "Marapohayon Dialect Manobo"

1046 *Phaeanthus ebracteolatus* Philippine Is. / C. O. Frake 802 / 1958 / " . . . stem applied to pain in waist" / "pav'ian"

1047 *Orophea polycarpa* Malaya / Ishmail & Millard 191 / 1958 / "Roots . . . with ginger-like odour. Sudorific and employed for coughs" / "Malay name: Subang Intan"

1048 *Goniothalamus suluensis* N. Borneo / A. Cuadra A2253 / 1949 / "Natives used the wood as a charm against wild animals in the jungle" / "limpanas (kayu) (Brunei) tak (kayu) (Brunei)"

1049 *G. tapis* Sarawak / J. A. R. Anderson S12596 / 61 / "Bark used as repellant against mosquitoes"

1050 *G. grandiflorus* Solomon Is. / S. F. Kajewski 2432 / 1931 / "The natives have a superstition if a little piece of this tree is placed in a taro set when planting, it will grow well"

1051 *Goniothalamus* indet. Philippine Is. / M. D. Sulit 3445 / 1949 / "Decoction of all parts of the plant taken in, cures fever" / "Sangkulai Dialect Bukid."

1052 *Xylopia nitida* Venezuela / Wurdack & Monachino 39687 / 1955 / "'Fruta de burro montanera'"

1053 *Enantia kummeriae* Tanganyika Terr. / P. J. Greenway 921 / 1928 / "...bright yellow wood used as medicine for cuts by the Wasambaa"

1054 *Drepananthus carinatus* var. *deltoideus* Sarawak / P. M. Synge 1651 / 1932 / "Has gum. Sap said to be good to drink when tree is not in flower"

1055 *Fissistigma glaucescens* China / S. K. Lau 804 / 1932 / "Roots for poisoning fishes" / "Ue T'ang"

1056 *F. polyanthum* China / W. T. Tsang 26503 / 1936 / "...fr....edible" China / W. T. Tsang 22151 / 1933 / "Used to make malt" / "Chau Ping T'ang"

1057 *F. uonicum* China / W. T. Tsang 23072 / 1933 / "...fruit...edible" / "Kok Yeuk T'ang"

1058 *F. pallens* Annam / Poilane 1122 / 1920 / "...les fruits seraient comestibles" / "Ate.: Dây bŭ Tru chúõi"

1059 *F. schefferi* Annam / Poilane 1423 / 1920 / "...les fruits sont comestibles" / "Ate.: Cây bō hō"

1060 *Annona ambotay* Brazil / B. A. Krukoff 7547 / 1935 / "Component of Curare of Tecuma Indians"

1061 *A. tessmannii* Peru / Y. Mexia 6070 / 1931 / "...ripe fruit...edible" / "'Anonilla'"

1062 *Eupomatia laurina* Australia: New S. Wales / C. T. White 12843 / 1945 / "fruits green with a spicy rather burning flavor"

MYRISTICACEAE

1063 *Compsoneura sprucei* Brit. Honduras / P. H. Gentle 3949 / 1942 / "'Wild pecan'" or "'Wild almond'"

1064 *Iryanthera juruensis* Brazil / B. A. Krukoff 1298 / 1931 / "...Seeds used for extraction of oil" / "'Ucuuba rana'"

1065 **Virola calophylloides** Colombia / Schultes & Cabrera 12872 / 1951 / "The source of a narcotic snuff in the Vaupés Commissary of Colombia" / "'Ya-kee'" Puinave

1066 **V. divergens** Brazil / B. A. Krukoff 1120 / 1931 / "Seeds are used for extraction of oil" / "'Ucuuba'"

1067 **V. macrocarpa** Colombia / A. E. Lawrance 675 / 1933 / "The ripe albumen of this fruit is crimson red and the birds named Pekong with huge beaks were eating this fruit. These birds are called locally 'Ya-tarro' and not Pekong"

1068 **V. sebifera** Venezuela / J. A. Steyermark 60758a / 1944 / "...inner bark is dried and smoked by witch doctors for smoking at dances when curing fevers; it is very strong" / "'wircawei-yek'"

Venezuela / J. A. Steyermark 58565 / 1944 / "Indians boil bark and use to drive away evil spirit, Piassám" / "'erika-bai-yek'"

Brazil / B. A. Krukoff 1008 / 1931 / "...seeds used for extraction of oil in commercial quantities" / "'Ucuuba'"

1069 **V. surinamensis** Brazil / B. A. Krukoff 1000 / 1931 / "Seeds used for oil extraction (in commercial quantity)" / "'Ucuuba branca'"

1070 **Gymnacranthera** indet. Philippine Is. / Sulit & Conklin 5047 / 1953 / "Bark boiled & decoction is medicine for spitting of blood" / "'Dugan-tagabas'" (Mang.—Hanunoo)

1072 **Myristica** indet. Philippine Is. / C. O. Frake 876 / 1958 / "Young leaves applied for internal pain" / "'Gupaw ulangan'" Sub.

1073 **M. cimicifera** var. **insipida** Australia: N. Terr. / C. E. F. Allan s.n. / 1933 / "...nut edible, relished by natives and pigeons; wood yellow, used in spear and throwing sticks"

1074 **Knema curtisii** Malaya / G. anak Umbai for A. H. Millard (K.L. 1582) / 1959 / "Phytochemical Survey of Malaya" / "Poisonous fruits" / "'Sengkuning'" Temuan

1075 **K. glaucescens** var. **patentinervia** Malaya / G. anak Umbai for A. H. Millard (K.L. 1496) / 1959 / "Phytochemical Survey of Malaya" / "Medicinal" / "'Penuju' or 'Penujoh'" Temuan

1076 **K. laurina** Malay Peninsula / Gadoh (K.L. 1471) / 1959 / "Phytochemical Survey of Malaya" / "'Mendarah'" Temuan

1077 **K. glomerata** Philippine Is. / C. O. Frake 569 / 1958 / "Roots boiled and drunk during puerperium" / "'Beneb'" Sub.

1078 **Knema** indet. Philippine Is. / C. O. Frake 700 / 1958 / "Bark applied to internal pain" / "'Gelangit'" Sub.

1079 **Knema** indet. Philippine Is. / Sinclair & Edaño 9586 / 1958 / "Pericarp both sweet and bitter, astringent"

MONIMIACEAE

1080   *Peumus boldus* Chile / F. W. Pennell 12903 / 1925 / "Leaves fragrant; infusion used for troubles of stomach and liver"

1081   *Hedycarya solomonensis* Solomon Is. / S. F. Kajewski 2561 / 1931 / "The leaves are macerated and applied to sores" / "Maroi"

Solomon Is. / S. F. Kajewski 2384 / 1931 / "The leaves are rubbed on sore places by the natives who think a lot of this treatment" / "Undie"

1082   *Mollinedia guatemalensis* Guatemala / J. A. Steyermark 44664 / 1942 / "Drink an infusion of the boiled leaves, a remedy for stomach ache" / "'sac-e-yen,' 'anyac' (kekchi)"

1083   *Matthaea ellipsoidea* Philippine Is. / C. O. Frake 803 / 1958 / "Stems scraped and applied for headache" / "selimbwang Sub."

1084   *Siparuna nicaraguensis* Brit. Honduras / P. C. Gentle 2107 / 1937 / "'Wild coffee'"

Mexico / G. B. Hinton 10814 / 37 / "Smells of lemons.... Uses medicinal" / "Limoncillo"

1085   *S. guianensis* Brazil / Y. Mexia 5079 / 1930 / "Tea of leaves used as stimulant in stomach disorders" / "'Congonha'"

Venezuela / L. Williams 11728 / 1939 / " ... se usan las hojas para el reumatismo" / "Hoja de Danta"

Brazil / Y. Mexia 4145 / 1929 / "Infusion of leaves used in mange or other skin diseases in man" / "'Coerana'"

Peru / F. Woytkowski 6145 / 1961 / " ... medicinal for 'aire'" / "Curu-huinci-sacha (the true one)"

1086   *Siparuna* indet. Ecuador / A. Rimbach 39 / no year / "Infusion of leaves used as a tea. Said to be good against pain of stomach and sterility of women (?)" / "Guayusa"

1087   *Monimiacea* indet. New Guinea: Papua / Gray & McDonald 7144 / 53 / "Single seed edible when mature"

LAURACEAE

1088   *Cinnamomum burmanni* Hainan / H. Fung 20173 / 1932 / " ... twigs used as ingredient in incense powder"

Hainan / W. T. Tsang 82 / 1928 / " ... fr. black; used for incense powder"

1089   *C. cassia* China / W. T. Tsang 26799 / 1936 / "fr. black, edible"

China / R. C. Ching 8289 / 1928 / "A commonly cultivated tree for its bark and leaves for medical purposes known as 'Noo (sp.?) Kwei'"

1090   *C. glandulifera* China / W. T. Tsang 22037 / 1933 / "For making oil"

1091   *C. pseudopedunculatum* Bonin Is. / E. H. Wilson s.n. / 1917 / "'Sassafras'" / "'Tea wood tree'"

1092   *C. sancaurium* China / A. Henry 11598 / 1938 / "Said to be the tree yielding the precious Laos cinnamon"

1093   *C. sericans* China / A. Henry 11785A / no year / "leaves used to smoke incense"

1094   *C. sieboldii* Formosa / A. Henry 451 / no year / "(Root-bark is very fragrant and used locally as Cassie bark)"

1095   *Cinnamomum* indet. China / W. T. Tsang 22270 / 1933 / "Used to make camphor"

1096   *C. tavoyanum* Burma / R. N. Parker 2234 / 1924 / "Bark aromatic"

1097   *C. tetragonum* Cambodia / Poilane 378 / 1919 / "Forte de lauries les feuilles écrassées ainsi que le bois dégage une odeur agréable, les feuilles servent à préparer un breuvage stimulant et recomfortant bon bois dur qui peut également servir à préparer un breuvage . . . "

1098   *Cinnamomum* indet. N. Burma / F. Kingdon-Ward 20788 / '53 / " . . . aromatic foliage"

1099   *Cinnamomum* indet. India or Burma / F. Kingdon-Ward 18762 / 1949 / "Wood very aromatic. Leaves used by the Khasis in curry"

1100   *Cinnamomum* indet. India or Burma / F. Kingdon-Ward 18466 / 1949 / "Leaves . . . very aromatic when crushed, as is the young stem when cut"

1101   *Cinnamomum* indet. Siam / T. Smitinand 249 / 51 / "Yng. frs. . . . aromatic" / "'Maha-prarb'"

1102   *C. iners* Malaya / Gadoh (K.L. 1239) / 59 / "Phytochemical Survey of Malaya" / "'Medang Tenyo'"

        Malaya / Gadoh for A. H. Millard (K.L. 1434) / 59 / "Phytochemical Survey of Malaya" / "Re-collected in bulk. See KL. 393" / "'Medang Tenyo'"

        Philippine Is. / C. O. Frake 614 / 1957 / "Bark applied for headaches" / "'Glingag'" Sub.

1103   *Cinnamomum* indet. W. MALAYSIA / Kostermans & Anta 584 / 1949 / "Wood smelling of aniseed . . . "

1104   *Cinnamomum* indet. W. MALAYSIA / A. Kostermans 9622 / 1954 / "Wood and bark with strong nutmeg smell"

1105   *C. solomense* Solomon Is. / Waterhouse 126 / 1932 / "Small native spice tree with scented bark . . . " / "'Enu'"

        Solomon Is. / S. F. Kajewski 2137 / 1930 / "The leaves are bruised and applied to the head for headache" / "'Uri-arku'"

1106   *Cinnamomum* indet. New Guinea: Papua / C. E. Carr 12127 / 35 / "Bark strongly fragrant" / "'API API'"

1107   *Cinnamomum* indet. New Guinea: Papua / C. E. Carr 11457 / 35 / "The bark contains an essential oil" / "'API API'"

1108   *Cinnamomum* indet. New Guinea: Papua / M. F. C. Jackson 4600 / 54 / "Bark . . . strong spicy smell" / "Wood . . . strong spicy odour"

1109   *C. mercadoi* Philippine Is. / C. O. Frake 781 / 1958 / "Bark and leaves applied to pain in the joints" / "'Gliyugelingag'" Sub.

Philippine Is. / M. D. Sulit 1475 / 1947 / "Bark medicinal—sliced and boiled as substitute for tea" / "'Kalingãg,'" Tag

Philippine Is. / Ahern's collector (Decades Phil. For. Flora 9) / 1904 / " . . . the bark much used by the natives in the practice of medicine" / "'Calíñgag,' 'Caníñgag'"

1110   *Cinnamomum* indet. Philippine Is. / M. D. Sulit 3412 / 1949 / "Bark with very strong odor"

1111   *C. leptopus* Fiji Is. / A. C. Smith 6867 / 1947 / " . . . leaves and wood aromatic"

1112   *C. pallidum* Fiji Is. / B. E. Parham 1082 / 38 / "Bark used for scenting coconut oil" / "'MACOU'"

1113   *Neocinnamomum caudatum* Sumatra / W. N. & C. M. Bangham 1189 / 1932 / "Lvs. have spicy fragrance"

Sumatra / W. N. & C. M. Bangham 815 / 1932 / "Leaves have pungent, spicy odor. Natives dry bark to make spice" / "'Kaiae manis'"

1114   *Persea palustris* U.S.: Va. / Fernald & Long 3944 / 1934 / " . . . fragrance of *Benzoin*"

U.S.: Ga. / M. G. Henry 3214 / 1941 / " . . . fragrant foliage"

U.S.: Tex. / A. Traverse 80 / 1956 / "Sweet Bay odor"

1115   *P. americana* var. *drymifolia* Mexico / R. L. Dressler 2024 / 1957 / " . . . leaves anise-scented . . . " / "'Aguacatillo'"

1116   *P. hintonii* Mexico / Hinton et al. 14195 / 39 / "Leaves fragrant"

Mexico / G. B. Hinton 2980? / 33 / "'laurel cimarron'"

1117   *P. podadenia* Mexico / C. H. Muller 2736 / 1939 / "'Salsafras'"

1118   *P. standleyi* Guatemala / J. A. Steyermark 31491 / 1939 / "Twigs and leaves slightly aromatic"

1119   *P. glaberrima* Brit. W.I. / J. S. Beard 417 / 1944 / "'Sweetwood'"

1120   *P. americana* Venezuela / H. Pittier 12854 / 1928 / "cult." / "'Curo'"

1121   *P. mutisii* Venezuela / F. Tamayo 2392 / 1942 / "'Curo de páramo'"

1122   *Alseodaphne perakensis* Malaya / G. anak Umbai for A. H. Millard (K.L. 1578) / 1959 / "Phytochemical Survey of Malaya" / "Poisonous" / "Fruits: very strong odor on cutting" / "'Medang' (Temuan)"

Malaya / G. anak Umbai for A. H. Millard (K.L. 1482) / 1959 / "Phytochemical Survey of Malaya" / "Poisonous fruits" / " 'Medang Kuning' (Temuan)"

1123   *Alseodaphne* indet. N.E. Borneo / A. Kostermans 9214 / 1954 / "The bark contains fine needle-like, itching particles"

1124   *Alseodaphne* indet. Philippine Is. / C. O. Frake 749 / 1958 / "Leaves applied to skin eruptions" / "'Panslan'" Sub.

1125   *Machilus ichangensis* China / W. Y. Chun 5230 / 1922 / "Leaves ... aromatic"

1126   *M. areophila* China / H. H. Chung 911 / 1922 / " ... bark used as incense"

China / H. H. Chung 1129 / 1923 / " ... bark containing an aromatic resin used as incense" An additional note, signed R. K., dated 1927, states: "This species may be the same to *Machilus longipaniculata* Hay.... The bark of this kind used for making incense sticks. The bark does not contain very aromatic resin at all, but it has mucilage stuff which makes good ingredient for mixing various kinds of incense materials"

China / H. H. Chung 3569 / 1925 / "Bark aromatic, wood full of mucilage, leaf infected with Lichens and leaf spots"

1127   *M. thunbergii* Japan / I. Hurusawa 82 / 1946 / "Wood for carvings, furniture" / "'Tabu'"

China / R. C. Ching 1679 / 1924 / "lvs. dried and powdered used as incense"

1128   *M. thunbergii* var. *japonica* Japan / T. Tanaka 44 / 1924 / "Use: wax"

1129   *M. odoratissima* Cambodia / Poilane 14625 / 1928 / " ... l'écorce pourtant pas odorante serait utilisée par les bonzes pour brûler dans les cérémonies réligieuses"

1130   *Machilus* indet. Philippine Is. / C. O. Frake 792 / 1958 / "Economic uses: Agric. ritual" / "'Gunteling'" Sub.

1131   *Phoebe bourgeauviana* Mexico / Popenoe & Williams 14392 / 1948 / "'Aguacatillo'"

1132   *P. effusa* El Salvador / S. Calderón 1386 / 1922 / "'Pimiento,' 'Canelito'"

1133   *P. ehrenbergii* Mexico / G. B. Hinton 309 / 32 / "'Aguacatillo'"

Mexico / W. P. Hewitt 274 / 1948 / "edible fruit" / "At Tenariba (Cajon—see 270) a leaf similar to this used to poison fish" / "'Ara'"

1134   *P. helicterifolia* Mexico / Schultes & Reko 726 / 1939 / "Fruit edible"

1135   *P. mexicana* Honduras / P. C. Standley 54735 / 1927–28 / "'Aguacatillo'"

Brit. Honduras / P. H. Gentle 3549 / 1941 / "'Wild Pear'"

1136   *P. pittieri* Panama / C. & W. von Hagen 2031 / 1940 / "'Agua Catillo'"

1137   *P. salvini* Guatemala / A. F. Skutch 1679 / 1934 / "'Wild Avocado'"

1138   *P. glaziovii* Bolivia / M. Cárdenas 1988 / 1921 / "Bark very aromatic"

1139   *P. henryi* Hainan / W. T. Tsang 81 / 1928 / "... used for incense powder"

Hainan / S. K. Lau 173 / 1932 / "Cook leaf to heal stomachache" / "'Jia thun doe' (Lois)"

1140   *P. nanmu* China / W. Y. Chun 3554 / 1922 / "... aromatic"

1141   *P. neurantha* China / W. Y. Chun 3876 / 1922 / "... aromatic"

1142   *Phoebe* indet. New Guinea: Papua / L. J. Brass 28446 / 1956 / "leaves ... aromatic"

1143   *Caryodaphnopsis tonkinensis* E. Borneo / A. Kostermans 5905 / 1921 / "Heartwood ... with cedarwood smell. ... Fr. ... with soft bitter-aromatic, outer layer. ... Buttresses smell ..."

1144   *Pleurothyrium densiflorum* Peru / C. Klug 2908 / 1933 / "'Canela muena'"

1145   *Ocotea atirrensis* Costa Rica / W. A. Dayton 3124 / 1943 / "'Tiquissaró'"

1146   *O. bernoulliana* Brit. Honduras / P. H. Gentle 3351 / 1940 / "'Timber Sweet'"

Honduras / P. C. Standley 53129 / 1927–28 / "'Aguacatillo'"

1147   *O. cernua* Brit. Honduras / P. H. Gentle 3310 / 1940 / "'Timber Sweet'"

Nicaragua / A. Molina R. 2208 / 1949 / "'Canelo macho'"

1148   *O. effusa* Brit. Honduras / P. H. Gentle 2926 / 1939 / "'Timber Sweet'"

1149   *O. ira* Panama / Cooper & Slater 309 / ca. 1927 / "'Aguacatón'"

1150   *O. skutchii* Costa Rica / A. F. Skutch 3755 / 1938 / "'Ira rosa'"

1151   *O. stenoneura* Costa Rica / W. R. Barbour 1033 / 46 / "'Ira'"

1152   *O. veraguensis* Guatemala / J. A. Steyermark 42164 / 1942 / "'Pimiento'"

El Salvador / S. Calderón 1390 / 1922 / "'Pimiento,' 'Canelito'"

El Salvador / S. Calderón 2258 / 1925 / "Wood very fragrant" / "'Pimientillo'"

1153   *O. jamaicensis* Brit. W.I. / W. R. Maxon 8973 / 1926 / "'Sweetwood'"

1154   *Ocotea* indet. Brit. W.I. / J. S. Beard 535 / 1945 / "... leaves and wood aromatic" / "'Laurier canelle'"

1155   *O. aciphylla* Brazil / Reitz & Klein 3780 / 1956 / "'Canela amarela'"

1156   *O. acutifolia* Brazil / Y. Mexia 5787 / 1931 / "'Canélla sassafráz'"

1157   *O. catharinensis* Brazil / Reitz & Klein 3152 / 1956 / "'Canela preta'"

1158   *O. calophylla* Colombia / Phillip & Smith 18153 / 1927 / "'Loto'"

1159   *O. costulata* Brazil / Ducke 2000 / 1946 / "'Louro cánfora'"

1160   *O. cujumary* Colombia / G. Klug 1958 / 1931 / "'Arbol de aguarras' (turpentine tree)"

1161   *O. guianensis* var. *subsericea* Brit. Guiana / D. B. Fanshawe F2478 / 45 / "bark faintly aromatic" / "'White Silverballi'"

1162   *O. indecora* Brazil / Y. Mexia 5124 / 1930 / "'Canélla sassafráz'"

1163   *O. kuhlmanni* Brazil / Reitz & Klein 5953 / 1957 / "'Canela burra'"

1164   *O. obovata* Ecuador / J. N. Rose 23430 / 1918 / "'Canelo deloga'"

1165   *O. pauciflora* Brazil / B. A. Krukoff 1252 / 1931 / "'Louro Chumbo'"

1166   *O. pretiosa* Brazil / Y. Mexia 5300 / 1930 / "'Canélla preta'"

      Brazil / Reitz & Klein 6217 / 1958 / "'Canela sassafrás'"

1167   *O. puberula* Brit. Guiana / D. B. Fanshawe F2428 / 45 / "... bark and wood faintly aromatic"

      Brazil / Y. Mexia 5030 / 1930 / "'Canella cheirósa'"

      Brazil / Y. Mexia 4486 / 1930 / "... wood scented" / "'Canella Sassafraz'"

1168   *O. teleiandra* Brazil / Reitz & Klein 5719 / 1957 / "'Canela pimenta'"

1169   *O. tenuiflora* Brazil / Y. Mexia 4707 / 1930 / "'Canella Babona'"

1170   *Ocotea* indet. Colombia / G. Klug 1958 / 1931 / "'Arbol de aguarras' (turpentine tree)"

1171   *O. usambarensis* Kenya / E. H. Wilson 94 / 1921 / "'African Camphor'"

1172   *Umbellularia californica* U.S.: Calif. / G. Thurber 509 (or 309) / 12 / "... fragrance of Cinnamon"

      U.S.: Calif. / J. F. Collins 8? / 1918 / "... very spicy fragrant when alive"

      U.S.: Calif. / W. R. Dudley s.n. / ca. 1894? / From a letter: the root of this tree is "... like the bark of the stem, aromatic..." "The fragrance of the root-bark is that of the bruised leaf, a 'bay' fragrance. The flavor is in part that of *Sassafras,* in part that of *Asarum*" "There is an almost ineradicable belief among a certain set of Californians that it (the tree) furnishes the extract used in 'Bay rum.' The W. India tree cuts no sort of figure with them. This class of persons, mostly recent arrivals, call this tree the 'Bay tree.' 'California nutmeg' is the common name with some of the older settlers"

1173   *Nectandra coriacea* U.S.: Fla. / L. J. Brass 18088 / 1947 / "... leaves aromatic"

      Jamaica / W. Harris 5210 / 94 / "'Small-leaved Sweet-wood'"

1174 *N. belizensis* Brit. Honduras / P. H. Gentle 3281 / 1940 / "'Timber sweet'"

1175 *N. ambigens* Honduras / P. C. Standley 55406 / 1927–28 / "'Aguacatillo'"

1176 *N. gentlei* Guatemala / J. A. Steyermark 44827 / 1942 / "...fruit... with spicy odor"

Brit. Honduras / W. A. Schipp 164 / 1929 / "...wood has a faint odor of cinnamon"

1177 *N. globosa* Honduras / P. C. Standley 55448 / 1927–28 / "'Aguacatillo'"

El Salvador / S. Calderón 1419 / 1922 / "'Canelón'"

El Salvador / P. C. Standley 22225 / 1922 / "'Aguacate del monte'"

Nicaragua / A. Molina R. 2263 / 1949 / "'Aguacatillo'"

1178 *N. latifolia* Panama / L. H. & E. Z. Bailey 92 / 1931 / "aromatic"

1179 *N. lundellii* Honduras / P. C. Standley 54797 / 1927–28 / "'Aguacatillo'"

Brit. Honduras / P. H. Gentle 3343 / 1940 / "'Timber sweet,' 'Wild pear'"

1180 *N. panamensis* Honduras / P. C. Standley 54734 / 1927–28 / "'Aguacatillo'"

1181 *N. perdubia* Mexico / G. B. Hinton 13668 / no year / "'Aguacatillo'"

Brit. Honduras / P. H. Gentle 3543 / 1941 / "'Bastard timber sweet'"

1182 *N. reticulata* Nicaragua / A. Molina R. 2441 / 1949 / "'Aguacatillo'"

Nicaragua / A. Molina R. 2189 / 1949 / "'Canelo'"

1183 *N. savannarum* Brit. Honduras / W. A. Schipp 79 / 1929 / "...fruits like cinnamon"

Guatemala / J. A. Steyermark 44428 / 1942 / "leaves...with medicinal odor when crushed"

1184 *N. splendens* Guatemala / A. F. Skutch 1980 / 1934 / "The hollow branches are occupied by small stinging ants"

1185 *N. salicifolia* Brit. Honduras / P. H. Gentle 2232 / 1938 / "'Timber sweet'"

1186 *N. sinuata* El Salvador / P. C. Standley 19959 / 1922 / "'Trompito,' 'Aguacate de mico'"

Guatemala / A. F. Skutch 1978 / 1934 / "'tepeaguacate'"

El Salvador / S. Calderón 1356 / 1922 / "'Cachimbo'"

El Salvador / S. Calderón 2012 / 1924 / "'Aguacate amarillo'"

1187 *N. tabascensis* Mexico / G. B. Hinton 10589 / 37 / "'Aguacatillo blanco'"

1188 *N. woodsoniana* El Salvador / S. Calderón 1117 / 1922 / "'Tepeagua-cate'"

1189 *N. antillana* Jamaica / Fawcett & Harris 7027 / 1898 / "'Long-leaved Sweetwood'"

1190 *N. membranacea* W.I.: Antigua / J. S. Beard 284 / 1944 / "...leaves strongly aromatic" / "'Sweetwood'"
Jamaica / Fawcett & Harris 7027 / 1898 / "'Long-leaved Sweetwood'"

1191 *N. patens* Jamaica / Maxon & Killip 1503 / 1920 / "'Sweetwood'"
Jamaica / W. Harris 7081 / 1898 / "'Loblolly Sweetwood'"

1192 *Nectandra* indet. Brit. W.I. / J. S. Beard 405 / 1944 / "Leaves aromatic" / "'Pitchpine Sweetwood'"

1193 *N. elaiophora* Brazil / Ducke 38 / 1935 / "'Sassafras,' 'Inamuhy'"

1194 *N. pichurim* Brazil / Y. Mexia 5279 / 1930 / "'Canélla'"

1195 *N. rigida* Brazil / Reitz & Klein 6591 / 1958 / "'Canela garuva'"

1196 *N. oppositifolia* Brazil / Y. Mexia 4461 / 1930 / "'Canella cheirosa'"

1197 *Nectandra* indet. Brazil / Y. Mexia 4486 / 1930 / "'Canella Sassa-fraz'"

1198 *Actinodaphne pilosa* Hainan / F. A. McClure (Canton Christian Coll. Herb. 7674) / 1921 / "the sap is used by women on their hair" / "'Pau fa,' 'Cha kau shu'"
Hainan / C. I. Lei 433 / 1933 / "Wood soaked in water produces sticky substance which women use for hair fixer"

1199 *A. sesquipedalis* Malaya / Umbai for A. H. Millard (K.L. 1524) / 1959 / "Phytochemical Survey of Malaya" / "Strong somewhat unpleasant odor" / "Fruits said to be stupefying but eaten by birds... etc."

1200 *Actinodaphne* indet. Solomon Is. / S. F. Kajewski 2198 / 1930 / "The natives have two superstitions in regard to this plant. When a man is very sick, he is taken close to water, the branches of this tree is dipped in water and sprayed on the sick man" / "If there is too much rain a branch is broken and hung up in another tree to make the rain stop" / "'Nagia'"

1201 *Actinodaphne* indet. New Guinea / P. van Royen 3507 / 1954 / "The wood after carbonizing is used as an astringent" / "'Mansaoh'" Biak

1202 *Neolitsea confertifolia* China / W. Y. Chun 3784 / 1922 / "leaves... aromatic"

1203 *N. levinei* China / W. T. Tsang 21498 / 1932 / "...bark used to make medicine"

1204 *N. obtusifolia* Hainan / Chun & Tso 44337 / 1932–33 / "tree, aromatic..."

1205   *N. dealbata* Australia: Queensland / S. T. Blake 14949 / 1943 /
"...yellow when cut, with spicy scent"

1206   *Litsea glaucescens* var. *subsolitaria* Mexico / R. L. Dressler 1854 /
1957 / "'Laurel,'...aromatic, used for 'té de laurel'"

Mexico / G. B. Hinton 15418 / 40 / "Leaves are boiled to make tea"
/ "'Laurel'"

Mexico / G. B. Hinton 12365 / 38 / "Tree...fragrant" / "'Laurel'"

Guatemala / J. A. Steyermark 35862 / 1940 / "Leaves used for flavoring in cooking" / "'Laurel'"

Mexico / E. Palmer 2770 / 1891 / "Purchased in the market of Culican" / "Laurell leaves used in cooking as a flavoring"

Mexico / H. S. Gentry 2539 / 1936 / "Herbage used as a tea and
leaves for seasoning meats and other foods" / "'Laurel,' Mex."

1207   *L. chunii* China / T. T. Yü 19181 / 1938 / "aromatic..."

China / Steward, Chiao & Cheo 35 / 1931 / "Benzoin"

1208   *L. cubeba* China / Fan & Li 187 / 1935 / "Shrub...fragrant"

China / Y. L. Keng 241 / 1926 / "Fruit...aromatic"

China / Y. L. Keng 2634 / 1929 / "'Benzoin'"

China / H. H. Hu 903 / 1921 / "Leaves fragrant like oranges..."

China / H. H. Chung 3633 / 1925 / "Leaf pungently aromatic containing a Mucilage used in paper making locally. Seed aromatic; used
as stomach tonic"

China / H. Fung s.n. / 1937 / "Fruit edible"

Hainan / F. A. McClure (Canton Christian Coll. Herb. 7815) / 1921 /
"...roots used to make a medicine to be taken after child birth" /
"'Shan ch'ong shu'"

China / F. A. McClure 255 / 1921 / "drug plant" / "'To Sz Kun,
Shan Chong'"

China / W. T. Tsang 21768 / 1932 / "Use as medicine"

China / W. T. Tsang 22350 / 1933 / "Use roots for medicine"

China / W. T. Tsang 22903 / 1933 / "flower light yellow, fragrant,
edible; used as medicine"

China / W. T. Tsang 21535 / 1932 / "Flower yellow; used as medicine; roots cooked with pork for food"

Formosa / A. Henry 114 / no year / "Called 'mountain pepper' by
natives"

Indochina / Poilane 959 / 1920 / "...écorce...très odorante..."

Indochina / Poilane 988 / 1920 / "Port ordinaire, l'écorce est très
odorante elle est employée pour faire des parfums-offert au Boudda"

Sumatra / W. N. & C. M. Bangham 700 / 1932 / "Lvs. and stems have strong sassafras odor"

1209 *L. euosma* China / C. W. Wang 73391 / 1935–36 / "seeds edible"

N. Burma / F. Kingdon-Ward 9103 / 31 / "Leaves very aromatic"

1210 *L. forrestii* China / J. F. Rock 24250 / 1932 / "flowers . . . fragrant, as is the wood"

1211 *L. glutinosa* var. *brideliifolia* Hainan / F. A. McClure (Canton Christian Coll. Herb. 8984) / 1922 / " . . . bark used in making incense" / "'Ye kau Shu'"

1212 *L. mollis* China / Steward, Chiao & Cheo 349 / 1931 / "Fruits used in making incense"

1213 *L. panamonja* China / C. W. Wang 76369 / 1936 / "Leaf aromatic"

1214 *L. variabilis* Hainan / C. I. Lei 365 / 1933 / " . . . fruit green, poisonous"

1215 *Litsea* indet. China / J. F. Rock 14729 / 1926 / " . . . bark, leaves, etc. fragrant . . . "

1216 *Litsea* indet. China / J. F. Rock 7996 / 1923 / " . . . fragrant"

1217 *Litsea* indet. China / J. F. Rock 7661 / 1922 / "branches etc. very fragrant of balsam"

1217A *L. cambodiana* var. *acutifolia* INDIA / Poilane 10232 / 24 / "Donnerait des fruits comestibles . . . "

1218 *L. glutinosa* INDIA / Poilane 10494 / 24 / " . . . les femmes chique les fleurs avec le bétel . . . "

1219 *L. polyantha* Siam / Bloembergen & Kostermans 290 / 1946 / "flower . . . stunning scented"

1220 *L. rubescens* N. Burma / F. Kingdon-Ward 9586 / 31 / " . . . aromatic leafed tree . . . "

1221 *Litsea* indet. Siam / native collector DI 152 Royal For. Dept. / 46 / "Wood used by native for malarial remedy" / "'Mued Kon Dong'"

1222 *Litsea* indet. Siam / T. Smitinand 872 / 51 / "Lvs. aromatic, bitterly hot. Fr. . . . aromatic, bitterly hot. Febrifugous bark" / "'Dee-ngoo Ton'"

1223 *Litsea* indet. N. Burma / F. Kingdon-Ward 20375 / '53 / "Leaves aromatic"

1224 *Litsea* indet. N. Burma / F. Kingdon-Ward 21421 / 1953 / "Smells powerfully of lemon grass"

1225 *Litsea* indet. N. Burma / F. Kingdon-Ward 20490 / '53 / " . . . tree aromatic . . . " / "Smells strongly of lemon grass"

1226 *L. spathacea* Malaya / Gadoh (K.L. 1442) / 59 / "Phytochemical Survey of Malaya"

Malaya / G. anak Umbai for A. H. Millard (K.L. 1499) / 59 / "Phyto-chemical Survey of Malaya"

1227 **L. tuberculata** Sumatra / W. N. & C. M. Bangham 1003 / 1932 / "Fruits ... pungent odor"

1228 *Litsea* indet. Brit. N. Borneo / Md. Tahir 796 / 1928 / "Medicinal"

1229 *Litsea* indet. S. Borneo / A. Kostermans 8057 / 1953 / "Wood ... with faint Cinnamomum porrectum smell"

1230 *Litsea* indet. E. Borneo / A. Kostermans 6069 / 1951 / "Bark and wood slightly aromatic"

1231 *Litsea* indet. N.E. Borneo / A. Kostermans 9176 / 1953 / "Wood white with faint pungent smell"

1232 *Litsea* indet. S. Borneo / A. Kostermans 8025 / 1953 / " ... bark ... smelling of manggo"

1233 *Litsea* indet. Borneo / A. Kostermans 7429 / 1952 / "Bark a little aromatic with smell like Cinnamomum porrectum"

1234 **L. solomonensis** Solomon Is. / S. F. Kajewski 2470 / 1931 / "The bark of this tree is macerated with water and applied to sore legs" / "'Arli Arli'"

1235 *Litsea* indet. New Guinea / H. A. Brown 299 / 1953 / "The sticky sap is said to be very irritating to the skin" / "'Kave Merava' (Elema, Moviave)"

1236 *Litsea* indet. New Guinea: Papua / R. D. Hoogland 4425 / 1954 / "Wood ... aromatic" / "'Waureh' (Onjob language, Koreaf), 'Gaina' (Bausa language, Oi-ai)"

1237 *Litsea* indet. New Guinea / A. Havel (NGF. 7547) / 55 / "Turpentine smell in cut fruit"

1238 *Litsea* indet. New Guinea: Papua / R. D. Hoogland 4267 / 1954 / "inner bark ... with strong scent"

1239 **L. hutchinsonii** Philippine Is. / C. O. Frake 864 / 1958 / "Bark scraped and applied to wounds" / "'Punawan'" Sub.

1240 **L. luzonica** Philippine Is. / C. O. Frake 492 / 1958 / "Agricultural ritual" / "'Mentuken'" Sub.

Philippine Is. / C. O. Frake 695 / 1958 / "Boils applied to wounds" / "'Geliyan'" Sub.

Philippine Is. / C. O. Frake 639 / 1958 / "Leaves applied to waist pain" / "'betibud ulangan'" Sub.

Philippine Is. / P. Añonuevo 155 / 1950 / "Fruit can be eaten when ripe" / "'Balasinong'"

1241 **L. perrottetii** Philippine Is. / A. L. Zwickey 775 / 1938 / " ... chewed lvs. used in treating boils" / "'Wakan' (Lan.)"

1242   *L. quercoides* Philippine Is. / C. O. Frake 711 / 1958 / "Leaves applied for headache" / "'Gambusati'" Sub.

Philippine Is. / C. O. Frake 735 / 1958 / "Leaves applied to ulcers" / "'Gelungus'" Sub.

1243   *L. tomentosa* Philippine Is. / C. O. Frake 523 / 1958 / "Leaves applied to headache" / "'Mangupung-ulangan'" Sub.

1244   *Litsea* indet. Philippine Is. / C. O. Frake 540 / 1958 / "Roots applied for rigid abdomen" / "'Stilala'" Sub.

1245   *Litsea* indet. Philippine Is. / C. O. Frake 659 / 1958 / "Bark applied for rigid abdomen" / "'dulianutung'"

1246   *Litsea* indet. Philippine Is. / C. O. Frake 829 / 1958 / "Leaves applied to skin eruptions" / "'belebanen'" Sub.

1247   *Litsea* indet. Philippine Is. / C. O. Frake 822 / 1958 / "Leaves applied for pin worms" / "'gebindang'" Sub.

1248   *Litsea* indet. Philippine Is. / C. O. Frake 679-A / 1958 / "Leaves applied to headache" / "'Manuling'" Sub.

1249   *Litsea* indet. Philippine Is. / C. O. Frake 681 / 1958 / "Agric. ritual—roots boiled and drunk for sickness and sudden onset" / "'genigus'" Sub.

1250   *Litsea* indet. Philippine Is. / C. O. Frake 378 / 1957 / "Roots boiled and drunk for dark urine" / "'Kumbilangan'" Sub.

1251   *L. mellifera* Tonga Is. / T. G. Yuncker 15431 / 1953 / "...fruit eaten by birds"

1252   *L. vitiana* Fiji Is. / O. Degener 15457 / 1941 / "Stomach medicine" / "'Seti'" or "'Siti'" Ra

1253   *Beilschmiedia pendula* Brit. W.I. / J. S. Beard 438 / 1944 / "Leaves faintly aromatic" / "'Red Sweetwood'"

1254   *B. sulcata* Ecuador / E. L. Little, Jr. 6657 / 1943 / "'Aguacatillo de montaña'"

1255   *B. roxburghiana* Hainan / H. Fung 20213 / 1932 / "...fr....bittersweet in flavor"

India / R. N. Parker 3238 / 1932 / "Leaves minutely gland-dotted smelling of mango when crushed"

1256   *Beilschmiedia* indet. Indo-China / W. T. Tsang 26926 / 1936 / "bark—sharp odor—more pungent than most Laurac....Oil dots in leaves"

1257   *B. madang* N.E. Borneo / A. Kostermans 8628 / 1953 / "...bark...astringent"

1258   *B. maingayi* E. Borneo / A. Kostermans 9576 / 1954 / "Fr. a little aromatic in taste"

1259   *B. obtusifolia* Australia: Queensland / C. T. White 8815 / 1933 / "flowers cream with rather unpleasant (excrement-like) odour"

1260 *Lauromerrillia appendiculata* Hainan / S. K. Lau 6 / 1932 / "Water in which the bark has been boiled can be used to heal boils"

1261 *Dehaasia cuneata* E. Borneo / A. Kostermans 9540 (9534) / 1954 / "Wood . . . with very strong Cedarwood smell"

1262 *D. firma* S. Borneo / A. Kostermans 9052 / 1953 / "Heartwood . . . with strong Cedarwood smell"

1263 *D. triandra* Philippine Is. / C. O. Frake 838 / 1958 / "Leaves applied to ulcers" / "'Kumbel'" Sub.

1264 *Aiouea costaricensis* Costa Rica / E. L. Little, Jr. 6023 / 1943 / "'Ira Rosa'"?

1265 *A. demerarensis* Brit. Guiana / D. B. Fanshawe L480? / 45 / ". . . tree . . . with faint aromatic smell"

1266 *Aniba excelsa* Brit. Guiana / D. B. Fanshawe F1362 / 43 / "bark and wood sweetly aromatic"

1267 *Endlicheria poeppigii* Peru / Y. Mexia 6329 / 1931 / "pale yellow flower; strong disagreeable odor" / "'Moéna'"

1268 *Cryptocarya griffithiana* Malaya / G. anak Umbai for A. H. Millard (K.L. 1634) / 59 / "Phytochemical Survey of Malaya" / "Poisonous"

1269 *C. cordata* Solomon Is. / S. F. Kajewski 2213 / 30 / "The leaves are heated and applied to sore eyes by natives" / "'Tembu'"

1270 *Cryptocarya* indet. New Guinea: Papua / M. Allan (N.G.F. 3278) / 49 / "Wood has mango like odour" / "'Wau'"

1271 *Cryptocarya* indet. Solomon Is. / S. F. Kajewski 2031 / 30 / "The fruits are very immature, but I am putting the specimens in on account of the natives using them as an antidote for their poisons, which are mostly sorcery, but sometimes they have accidentally stumbled on a virile poison" / "'Kabaku'"

1272 *Cryptocarya* indet. New Guinea: Papua / L. J. Brass 23304 / 1953 / ". . . fruits . . . aromatic"

1273 *Cryptocarya* indet. New Guinea: Papua / R. D. Hoogland 4604 / 1954 / ". . . bark . . . aromatic" / "'Waureh'" (Onjob)

1274 *Cryptocarya* indet. Brit. New Guinea / L. J. Brass 6490 / 1936 / "bark . . . slightly aromatic"

1275 *Cryptocarya* indet. Solomon Is. / S. F. Kajewski 2413 / 1931 / "The bark of this tree has a superstitious value. It is taken and stored in the houses until dry. Its use is to scatter it about to ward off sickness" / "'Cooreu'"

1276 *Cryptocarya* indet. Solomon Is. / S. F. Kajewski 2098 / 1930 / "The natives say that possums eat fruit" / "'Magei'"

1277 *Cryptocarya* indet. New Guinea: Papua / L. J. Brass 1072 / 1926 / ". . . bark smelling slightly of cinnamon" / "Fruit . . . cinnamon scented"

1278   *C. medicinalis* Solomon Is. / F. S. Walker B.S.I.P. 243 / 1946 / " . . . bark . . . odour when cut sweet and fragrant. Bark used as a native cough medicine"

1279   *C. turbinata* Fiji Is. / A. C. Smith 1528 / 1934 / "Bark grated and used to scent coconut oil" / " 'Mbatho' "

1280   *C. hypsopodia* Australia / C. T. White 12752 / '45 / " . . . bark, with a faint 'sassafras' odour . . . "

Australia / C. T. White 10904 / 1937 / " . . . cultivated as Laurus australis"

1281   *Endiandra celebica* Celebes / cultivated in Hort. Bog. Sub. XI.B.SVII.-133 / 1950 / "Fleshy part of fruit eaten by bats"

1282   *E. rubescens* Sumatra / W. N. & C. M. Bangham 1062 / 1932 / "Fruits . . . with resinous odor"

1283   *Endiandra* indet. New Guinea / McVeigh & Ridgwell (N.G.F. 7325) / 1955 / "Leaves . . . with a laurel smell"

1284   *E. aneityensis* New Hebrides Is. / S. F. Kajewski 955 / 1929 / "Fruit eaten by natives" / " 'Incihraif' "?

1285   *E. muelleri* Australia: Queensland / C. T. White 12829 / 1945 / " . . . bark . . . with a markedly rose odour when cut"

1286   *Licaria campechiana* Guatemala / J. A. Steyermark 51270 / 1942 / " 'granadilla' "

1287   *L. cervantesii* Mexico / G. B. Hinton 10273 / 37 / " 'Aguacatillo' "

1288   *Licaria* indet. Costa Rica / W. R. Barbour 1032 / 46 / "Wood has spicy sassafras odor . . . " / " 'Quina' "

1289   *L. triandra* W.I.: San Domingo / J. Schiffino 140 / 1944 / " 'Cigua prieta' "

1290   *L. cavennensis* Brit. Guiana / D. B. Fanshawe F1290 / 43 / " . . . bark strongly aromatic, very similar to *L. canella* but sweeter" / " 'Wabaima' "

1291   *L. limbosa* Ecuador / J. N. Rose 23442 / 1918 / " 'cenela' "

1292   *Licaria* indet. Venezuela / H. Pittier 10986 / 1922 / "Wood aromatic; gives in medular canal an aromatic oil"

1293   *Lindera benzoin* U.S.: Mich. / C. K. Dodge s.n. / 92 / " 'Spice-bush,' 'Benjamin's-bush,' 'Wild allspice,' 'Fever-bush' "

U.S.: Ohio / J. A. Sanford 2447 / 79 / "Medicinal" / " 'Wild Allspice' "

1294   *L. angustifolia* China / Y. L. Keng 1519 / 1928 / "Aromatic plant, rare"

1295   *L. cercidifolia* China / J. F. Rock 9483 / 1923 / " . . . branches fragrant"

1296 *L. chunii* China / W. T. Tsang 24061 / 1934 / " . . . fr. black; used for the joss stick powder"

1297 *L. communis* China / J. F. Rock 7977 / 1923 / "A sort of oil is made of the fruit flesh"

China / H. H. Hu 213 / 1920 / "Medicine for rheumatism" / "'Nai Dung Shu'"

China / H. T. Tsai 51670 / 1933 / "reported oil-producing"

1298 *L. fruticosa* China / W. Y. Chun 4027 / 1922 / "Lvs. aromatic"

1299 *L. glauca* Japan / S. Suzuki (Herb. 88) / 1951 / "Leaves odorous" / "'Yama koobasi'"

China / R. C. Ching 4847 / 24 / "Fruit . . . with odor"

China / Y. L. Keng 1520 / '28 / "Aromatic, rare"

1300 *L. kariensis* China / F. T. Wang 20663 / 1930 / "fragrant"

1301 *L. membranacea* Japan / S. Suzuki (Herb. 85?) / 1951 / "Leaves odorous" / "'Yama-koobashi'"

1302 *L. nacusua* Tibet / J. F. Rock 11384 / 1923 / " . . . fruits red aromatic"

1303 *L. obtusifolia* Japan / S. Suzuki 85 or 88 / 1951 / "Leaves odorous"

1304 *L. pulcherrima* var. *attenuata* China / Steward, Chiao & Cheo 363 / 1931 / "Lves. used in making incense" / "'Shui Mei Yah'"

1305 *L. sericea* var. *tenuis* Japan / S. Suzuki SS-484004 / 1952 / "Leaves odorous" / " 'Usuge-kuromoji'"

1306 *L. umbellata* Japan / E. H. Wilson 6437 / 1914 / "Wood used for making toothpicks"

Japan / S. Suzuki 282 / 1937 / "Leaves and timber odorous specially" / "'Kuromoji'"

1307 *L. caesia* L'Indochine / Poilane 975 / 1920 / " . . . les indigènes chique les feuilles avec le betel, le bois est . . . très odorant . . . "

1308 *L. eberhardtii* Indochina / Poilane 927 / 1920 / " . . . les racines odorantes sont employées pour combattre diverses maladies. Ils font bouillir les racines et boivent cette tisane"

1309 *L. pipericarpa* Malaya / K. M. Kochummen (K.L. 66467) / 58 / "Phytochemical Survey of Malaya"

1310 *Parabenzoin praecox* Japan / S. Suzuki AA1285 / 1951 / "Leaves odorous" / "'Abura char'"

1311 *Cassytha filiformis* W.I.: Cayman Is. / G. R. Proctor 15226 / 1956 / "'Old man berry'"

Solomon Is. / S. F. Kajewski 2282 / 30 / "When a native has a cold, he eats the fruit with betel nut to cure it" / "'Bikor'"

1312 *Lauracea* indet. Guatemala / J. A. Steyermark 42846 / 1942 / "'aguacate'"

HERNANDIACEAE

1313   *Hernandia moerenhoutiana* Cook Is. / Yuncker & Fosberg 9638 / 1940 / "Uses: for medicine and for scenting coconut oil" / "'pipi' or 'huni'"

1314   *H. ovigera* Caroline Is. / C. C. Y. Wong 133 / 1947 / "The flowers are used as medicine for stiffness of joints, the bark is used for *säfein sät* (sea medicine), see Wong no. 120. (Elbert says the grated fruit with grated coconut and turmeric is used as woman's scalp treatment" / "N.v. ökyrang (Elbert akurang))"

PAPAVERACEAE

1315   *Macleya cordata* China / H. Fung 21074 / 1937 / ". . . root poisonous" / "P'ao T'ung Kuan"
      Japan / E. Elliott 117 / '46 / "Very poisonous" / "Chanpogiku"

1316   *Bocconia arborea* Mexico / Y. Mexia 524 / 1926 / "Wood and inner bark orange; tanning" / "'Grano de Oro' 'Palo Judas' 'Chicalote'"

1317   *B. frutescens* Argentina / S. Venturi 9131 / 1929 / ". . . venenoso"

1318   *B. pearcei* Bolivia / J. Steinbach 8686bis / 1928 / "Beeren . . . giftig, werden . . . flussabwärts beim Fischfang (Wasservergiftung) gebraucht" / "Japará"

1319   *Argemone gracilenta* Mexico / C. Lumholtz 22 / 1910 / "boiled and used for eye diseases"

1320   *A. ochroleuca* Mexico / H. S. Gentry 1337 / 1935 / "Medic. sap used for sore eyes" / "Cardo, Mex. Tachina, W."

1321   *A. sanguinea* Mexico / G. B. Hinton 16600 / 44 / "In witchcraft, a brew of this plant is used to treat insanity" / "'Toloache'"

1322   *A. subfusiformis* ssp. *subfusiformis* Bolivia / H. Cutler 7387 / 1942 / "The petals are dried and used in a tea for lung diseases"

1323   *Corydalis aurea* ssp. *occidentalis* Mexico / W. P. Hewitt 13 / 1945 / ". . . a tea considered good for women directly after childbirth" / "Cylandrillo"

1324   *C. edulis* China / H. T. Feng 138 / 1925 / "For medicine"

1325   *Fumaria media* Chile / E. E. Gigoux s.n. / 1886 / "'vivara; Flor de la culebre'"

1326   *F. muralis* Brazil / R. Reitz C1248 / 45 / "Medicinal"

CRUCIFERAE

1327   *Aschersoniodoxa mandoniana* Peru / E. Cerrate 1514 / 1952 / "Toman en infusion para las afecciones del corazon" / "'Ckorhuackack'"

1328   *Sisymbrium magellanicum* Argentina / Y. Mexia 7949 / 1936 / "The Yaghan Indians made a sort of bread of the seeds" / "'Ti-yu'"

1329 *Descurainia halictorum* Mexico / P. C. Standley 1303 / 1935 / "Seeds put in cold water, sugared and eaten as remedy for the liver. Leaves eaten as greens. Carlotta's mother used to gather and sell to Botica (drug store)" / "Pamita, Mex."

1330 *Lepidium lasiocarpum* var. *orbiculare* Mexico / R. McVaugh 10118 / 1949 / "'hierba del pajarito'"

1331 *L. virginicum* U.S.: Mass. / E. F. Fletcher s.n. / 1899 / "'Wild Pepper-grass'"

Mexico / G. B. Hinton 3459 / 33 / "Concoction for enemas" / "Me-cheche"

Venezuela / H. Pittier 7293 / 1917 / "Smells like watercress"

1332 *L. bipinnatifidum* Ecuador / W. H. Camp E-2462 / 1945 / "When there is a sickness causing cold hands or feet the plant is toasted over the fire and the cold parts are rubbed very hard with it; infusion also used to expell air from the body" / "'Chichira'"

1333 *L. bidentatum* Society Is. / E. H. Quayle 230 / 1922 / "...lvs. pungent tasting like pepper-cress. Eaten as salad and rather good" / "'nau u'"

S. Pacific: Tuamotu Archipelago / H. St.John 14373 / 1934 / "Used for salad and as garnish, cooked with fish"

1334 *Erucaria hispanica* Kuwait / H. Dickson 127 / 60 / "Eaten by all animals" / "'Seliyh'"

1335 *Diplotaxis harra* Kuwait / H. Dickson 46 / 61 / "Grazed by camels" / "Al Kashain"

1336 *Raphanus macropoda* var. *miyashige* Japan / T. Tanaka 125 / 1924 / "Use important food (root)"

1337 *R. raphanistrum* Australia: Capital Terr. / R. Pullen 3023 / 1962 / "This plant eaten by sheep"

1338 *Dryopetalon runcinatum* Mexico / H. S. Gentry 1250 / 1935 / "Warihos report medicinal; seeds mixed with grease and rubbed on back to relieve back pains. Leaves cooked as greens" / "Mastasa, Mex. Wachelai, W."

1339 *Cardamine fulcrata* Mexico / R. McVaugh 10091 / 1949 / "...plant with some acrid taste"

1340 *C. macrophylla* China / W. P. Fang 1016 / 1928 / "One kind of medicine"

1341 *Physaria newberryi* U.S.: Ariz. / J. G. Owens 5 / 1891 / "Root boiled, liquid used as an emetic, especially in the Snake Dance" / "Ho-ho-young-uh"

1342 *Lesquerella fendleri* U.S.: Tex. / O. T. Solbrig 3203 / 1960 / "...eaten by animals"

1343   *Eremobium aegyptiacum* Saudi Arabia / H. Dickson 83 / 61 / "Eaten by all animals"

1344   *Capsella bursa-pastoris* China / H. T. Feng 50 / 1925 / "Eat as vegetable when it is young"

India: Punjab / W. Koelz 8457 / 1936 / "Used; earliest potherb"

1345   *Farsetia aegyptia* Kuwait / H. Dickson 66 / 61 / "Eaten by camels" / "Albana"

1346   *Matthiola oxyceras* Kuwait / H. Dickson 103 / 62 / "Grazed by all animals" / "Shigara"

CAPPARIDACEAE

1347   *Cleome gynandra* Mexico / E. Langlassé 948 / 1899 / "Yerba del cuche"

1348   *C. pilosa* Mexico / G. B. Hinton 10972 / 37 / "Smells like garlic"

1349   *C. viscosa* El Salvador / P. C. Standley 21930 / 1922 / "'Tabaquillo'"

1350   *C. rutidosperma* Burma / P. Khant 511 / 48 / "Leaves are edible"

1351   *Gynandropsis pentaphylla* Indochina / Poilane 1437 / 1920 / "Employée pour fair des médicaments (contre le mal des reins)" / "Ate.: Cay mang mang"

1352   *Polanisia icosandra* Indochina / Poilane 72 / 1919 / "Donne des gousses comestibles"

1353   *Crateva tapia* Jamaica / Maxon & Killip 1692 / 1920 / "N.v. 'wild pear'"

1354   *C. nurvala* Indochina / Poilane 1147 / 1920 / "...feuilles...les fleurs seraient comestibles" / "Ate.: Cay bung"

1355   *Crateva* indet. Neth. New Guinea / P. van Royen 4612 / 1954 / "The smell of the fruits fills the whole forest with a soury scent not unlike durian" / "baloen (Bian dialect)"

1356   *C. religiosa* Solomon Is. / S. F. Kajewski 1969 / 1930 / "The bark is used to rub the skin for ... inflammation" / "Olou"

Solomon Is. / S. F. Kajewski 4201 / 1931 / "This is a very important article of medicine among the natives for constipation. The bark is macerated in water and the liquid drunk and is said to be very effective. This is the first medicine I have come across that the natives use as a purgative" / "Bambai-movu"

Solomon Is. / S. F. Kajewski 2339 / 1930 / "The leaves of this tree are held in much regard by the natives for earache. The leaves are heated and applied to the ear"

1357   *Capparis acutifolia* ssp. *sabiaefolia* Burma / F. Kingdon-Ward 17323 / 1948 / "The leaves are cooked and eaten"

1358   *C. micracantha* Indochina / Poilane 170 / 1919 / "...les infusions de racines grillées servire a combattre la toux"

Celebes / H. Curran 3409 / 1940 / "...fruits...strong odor...of citrus rind"

1359   *C. lasiantha* Australia: Queensland / C. T. White 12330 / 1943 / "...fruit said to be of a very sweet 'custard apple' (Annona) flavour"

1360   *Buchholzia coriacea* S. Nigeria / J. D. Kennedy 1661 / no year / "...edible fruits"

1361   *Stixis scandens* China / H. Fung 20151 / 1932 / "...fr., edible"

1362   *Forchhammeria matudei* Guatemala / J. A. Steyermark 33860 / 1940 / "'comida de pasha' (pasha is a bird which feeds on the fruit...)"

MORINGACEAE

1363   *Moringa oleifera* W.I.: St. Kitts / G. R. Cooley 8799 / 1962 / "Some parts of plant are boiled to produce a relief for colds"

Sumatra / W. N. & C. M. Bangham 613 / 1931 / "Fls. & Lvs. sweet, natives cook, eat and call it 'Kalor.' Pods green eaten when young"

DROSERACEAE

1364   *Drosera* subgenus *Rorella* Mexico / Schultes & Reko 836a / 1939 / "Chinantec name = o-ku-tu-ru; used in decoction for toothache and for intestinal troubles"

CRASSULACEAE

1365   *Kalanchoë pinnata* Virgin Is. / W. G. D'Arcy 26 / 1965 / "Local name 'Love Bush'"

Bahama Is. / A. E. Wright 21 / 1905 / "'Life Plant'"

Philippine Is. / C. Frake 352 / 1957 / "Ritual—leaves applied for ulcers and headache" / "'tubu' tubu Sub.'"

1366   *Kalanchoë* indet. Burma / F. Kingdon-Ward 19028 / 1949 / "The juice of the leaves is used locally as a cure for snake bites—it is rubbed into the wound"

SAXIFRAGACEAE

1367   *Saxifraga flagellaris* INDIA: Chamba State / W. Koelz 9497 / 1936 / "Fls. yellow; used medicinal"

1368   *Hydrangea serratifolia* Chile / Padre A. Pirion 195 / 1929 / "nom. vulg. 'Canelilla'"

1369   *Dichroa febrifuga* N. Sumatra / J. A. Lörzing 13537 / 1928 / "Fl.: petals and several other parts cyanose"

1370 *Valdivia gayana* Chile / J. West 4851 / 1935 / "Used medicinally" / "'Planta del leon'"

1371 *Escallonia herrerae* Peru / J. West 3788 / 1935 / "...aromatic scented foliage" / "'Pauca'"

1372 *Polyosma* indet. Philippine Is. / M. D. Sulit 3443 / 1949 / "Decoction of roots given to women who gave birth—after delivery women can immediately walk—according to native" / "At-at Bukid."

1373 *Ribes magellanicum* Argentina / Y. Mexia 7966 / 1936 / "...fruit eaten by Indians"

1374 *R. grossularia* India: Punjab / W. Koelz 8389 / 1936 / "Fr....very acid. Used as pickles"

1375 *Saxifragacea* indet. China / S. Y. Hu 1271 / 1939 / "...used as drug for cow disease" / "T'u-pa-kou"

PITTOSPORACEAE

1376 *Pittosporum baileyanum* China / W. T. Tsang 24400 / 1934 / "Used for medicine"

1377 *P. confertum* China / H. Fung 20566 / 1932 / "...fruit yellow, edible" / "Him Mai Shue"

1378 *P. moluccanum* Philippine Is. / M. D. Sulit 3353 / 1949 / "Juice of fruits medicine for wounds; or decoction of fruits used for cleaning wounds" / "Libaganon Bukid."

1379 *P. resiniferum* Philippine Is. / M. D. Sulit 1782 / 1947 / "The petrolium gas extracted from the fruit is medicinal for stomachache and cicitrizant" / "Amimis Bic."

1380 *P. arborescens* Fiji Is. / O. Degener 14586 / 1941 / "Fijians boil fruits and use liquid as a fish poison; said to be very effective" / "'Saranga' (Sabatu)"
Fiji Is. / Commander Burrowes 1010 / 38 / "...suspected poison" / "Duva Kalou"
Fiji Is. / O. Degener 14939 / 1941 / "Leaves crushed in water as remedy for stomach troubles"

1381 *P. brackenridgei* Fiji Is. / A. C. Smith 4617 / 1947 / "...fruit boiled and the resultant liquid used as a fish-poison" / "'Konakona'"
Fiji Is. / O. Degener 15332 / 1941 / "Used by Fijians as fish-poison. They boil the fruits and use the liquid. Fijians crush leaf in water and drink liquid as cough medicine. Weak for fish-poison in comparison to red seeded one" / "'Duva Kora' in Tailevu"

1382 *P. pickeringii* Fiji Is. / O. Degener 14430 / 1941 / "...fruits boiled and liquid used as fish-poison" / "'Mundu' 'Saranga' (Sabatu)"

1383   *P. rhytidocarpum* Fiji Is. / L. Reay 20 / 1941 / "Natives boil the fruit then pound it in the river; it stuns and kills the fish. It is a deadly poison" / "Tuva"

      Fiji Is. / O. Degener 15355 / 1941 / "Leaf and scraped bark mixed in water, strained and the resulting liquid drunk by women after child-birth" / "Nduva (Ra)"

      Fiji Is. / O. Degener 15097 / 1941 / "Fruit boiled and liquid used as a fish-poison" / "Nduva Serua"

CUNONIACEAE

1384   *Weinmannia* indet. Bolivia / J. Steinbach 9226 / 1929 / "Ich sah wie ein Indianermedizinmann einem Weissen einen verrenkten Fuss mit zerstossenem Blättern umwickelte. Der Weisse (Gutsvorsteher) be-hauptet, dass die Pflanze wirkt ... gute Dienste leiste"

HAMAMELIDACEAE

1385   *Liquidambar styraciflua* Guatemala / J. A. Steyermark 51726 / 1942 / "Used here for treating wounds and sore gums and toothache" / "'tzo-té'"

1386   *L. formosana* China / J. Hers 608 / 1921 / "medicinal" / "feng siang"

1387   *Altingia chingii* China / W. T. Tsang 21257 / 1932 / " ... root used to heal wounds" / "Poon Fung Hoh"

1388   *Corylopsis griffithii* India: Assam / L. F. Ruse 3 / 1923 / "Flowers are cooked and eaten by natives" / "Pu-uir symrang"

ROSACEAE

1389   *Physocarpus alternans* U.S.: Nev. / R. McVaugh 6090 / 1941 / "'Buck brush'"

1390   *Chamaebatiaria millefolium* U.S.: Ariz. / S. Braem s.n. / 1927 / "'desert sweet'"

1391   *Cotoneaster affinis* var. *bacillaris* India: Punjab / W. Koelz 3139 / 1931 / " ... fruit blue-black, insipid then bitter"

1392   *Pyracantha crenato-serrata* China / Tsiang & Wang 16116 / 1939 / " ... frts. red, edible"

      China / Steward, Chiao & Cheo 8 / 1931 / "Fruit eaten in time of famine" / "'Hung tze'"

1393   *Osteomeles anthyllidifolia* Bonin Is. / E. H. Wilson 8214 / 1917 / " ... fruit edible" / "Wild Plum"

1394   *Hesperomeles nitida* Colombia / Killip & Smith 18063 / 1927 / " ... fruit ... edible" / "Mortina"

1395   *Chaenomeles cathayensis* China / W. T. Tsang 27610 / 1937 / "Used for medicine"

1396   *C. speciosa* China / Cheo & Yen 137 / 1936 / "Fr. . . . for medical use"

1397   *Docynia delavayi* China / A. Henry 10036 / no year / "Fruit when ripe is edible & like an apple"

1398   *Pyrus kansuensis* China / W. T. Tsang 20943 / 1932 / ". . . fruit . . . edible"

1399   *P. calleryana* China / Cheo & Yen 176 / 1936 / "Fruit edible after frost"

1400   *P. serotina* China / W. T. Tsang 27690 / 1937 / ". . . fr. . . . edible"

1401   *P. pashia* India: Punjab / W. Koelz 1852 / 1931 / "Fr. . . . flesh black and mealy when edible, sweet" / "Kulu name 'Shegel'"

1402   *Malus hupehensis* China / E. H. Wilson 451 / 07 / "Leaves used as a substitute for tea"

       China / Steward, Chiao & Cheo 239 / 1931 / "Fruit edible, sour to taste" / "Shan Ch'a tze"

1403   *M. sieboldii* China / Fan & Li 697 / 1935 / ". . . fr. . . . edible"

1404   *M. yunnanensis* China / W. P. Fang 3728 / 1928 / "Fruit . . . edible"

1405   *Sorbus lanata* INDIA: Gozang / W. Koelz 10204 / 1936 / ". . . fruit . . . sweet" / "'Ja'"

1406   *S. thianshanica* Kashmir / Webster & Nasir 6253 / 1955 / ". . . bark with strong odor of bitter almonds"

1407   *Raphiolepis indica* China / W. T. Tsang 28529 / 1938 / ". . . fr. edible" / "Chun Fa Shue"

1408   *R. lanceolata* China / W. T. Tsang 22007 / 1933 / "Fr. edible" / "Ch'un Fa Shue"

1409   *Eriobotrya cavalieriei* China / W. T. Tsang 22631 / 1933 / "Fr. . . . edible" / "Kom Kwo Shue"

1410   *Photinia davidsoniae* China / W. T. Tsang 23639 / 1934 / ". . . fr. edible" / "Chock Tsz Shue"

1411   *P. parvifolia* China / W. T. Tsang 27906 / 1937 / ". . . fr. edible" / "Po Lei Tsz Shue"

1412   *P. arguta* var. *wallichii* India: Assam / L. F. Ruse 426 / 1924 / ". . . fruit sold in bazaars and eaten as dessert" / "Dun-soh-rynkha-mum"

1413   *Rubus glaucus* Guatemala / J. A. Steyermark 48556 / 1942 / ". . . fruit turning black, somewhat sweeter than most of the other species in Guatemala"

1414   *R. trivialis* Mexico / J. G. Schaffner 105 / 1896 / "'Planta preciosissima'"

1415   *R. vulcanicola* Costa Rica / A. Smith P2028 / 1939 / "Fruit agreeably sub-acid"

1416   *R. braziliensis* Brazil / L. O. Williams et al. 5244 / 1945 / "Used as a gargle for sore throat. All of plant. Boil and use soup"

1417   *R. boliviensis* Argentina / T. Meyer 3596 / 1941 / "...fruto comestible"

1418   *R. urticifolius* Peru / Goodspeed & Stork 11495 / 1939 / "Ripe fruit has agreeable, somewhat sweet taste"
        Peru / Stork, Horton & Vargas 10649 / 1939 / "...fr. black, of average sweetness....Fruit prized by the Indians"

1419   *R. ampelinus* China / Steward, Chiao & Cheo 243 / 1931 / "Fruit said to be edible" / "Mo Kou Tze"

1420   *R. alceaefolius* S. China / Levine & Groff 143 / 1916 / "...very juicy and good flavor"

1421   *R. corchorifolius* China / E. H. Wilson 151 / 07 / "fruit red, good flavor"

1422   *R. platysepalus* China / W. T. Tsang 24404 / 1934 / "...fr. edible" / "Tai Pau Tsz Lak"

1423   *R. tsangorum* China / H. H. Hu 445 / 1920 / "...berry edible"

1424   *R. amabilis* China / R. C. Ching 762 / 1923 / "...fruit...very sweet"

1425   *R. alpestris* India: Almora / R. N. Parker 2056 / 1923 / "Fruit...acid"

1426   *R. assamensis* India: Assam / L. F. Ruse 162 / 1923 / "...fruit eaten by natives" / "shiat khunai"

1427   *R. moluccanus* India: Assam / L. F. Ruse 111 / 1923 / "...the fruit is eaten by natives who also boil the roots and use the extract for stomach complaints"

1428   *R. niveus* India: Assam / L. F. Ruse 71 / 1923 / "...fruits sold in Bazaar"

1429   *R. acuminatus* Sumatra / W. N. & C. M. Bangham 879 / 1932 / "Fruits...Edible, sour"

1430   *R. glomeratus* Sumatra / W. N. & C. M. Bangham 685 / 1932 / "Fruits...edible"

1431   *R. brassii* Solomon Is. / S. F. Kajewski 2580 / 1931 / "The fruits are eaten by the natives" / "Chi Chi"

1432   *R. dendrocharis* Solomon Is. / S. F. Kajewski 2377 / 1930 / "The natives eat the fruit of this vine" / "Fara-cow"

1433   *R. ledermannii* New Guinea / L. J. Brass 4932 / 1933 / "...sweet red fruit"

1434   *R. fraxinifolius* Philippine Is. / G. L. Alcasid 3 / 1946 / "...fruit edible"

1435   *Duchesnea indica* Argentina / Eyerdam & Beetle 22242 / 1938 / "...good to eat but natives prefer it with sauce or oil" / "'Frutilla sylvestre'"

1436 *Potentilla rubra* Mexico / G. B. Hinton 4391 / 33 / " . . . root masticated to clean and strengthen teeth" / "Sangre de Grado"

1437 *P. thurberi* Mexico / H. S. Gentry 1929 / 1955 / "Medic. Decoction made from the roots for stomach ailments" / "Yerba Colorado, Mex."

1438 *Fallugia paradoxa* Mexico / R. M. Stewart 626 / 1941 / " 'Yerba del Pasmo' "

1439 *Cercocarpus breviflorus* U.S.: Tex. / R. McVaugh 8010 / 1947 / " . . . much grazed by deer"

1440 *Alchemilla orbiculata* Ecuador / W. H. Camp E2718 / 1945 / " 'Arifuela'—said to be good pasture for animals"

1441 *A. pinnata* Argentina / A. Burkart 7122 / 1935 / "Planta forrajera muy util llamada" / " 'pasto de oveja' "

1442 *Sanguisorba canadensis* Newfoundland / Fernald, Long & Dunbar 26803 / 1924 / "Tobacco Leaf".

   Japan / E. Elliott 4 / 1946 / "Food use: leaves are boiled, washed with water to remove bitterness, and eaten with salt and Miso. Medicinal use: Root is decocted and taken for diarrhoea and haemorisis, also good to stop vomitting and abdominal pain" / "Waremoko"

1443 *S. minor* Peru / F. L. Herrera 1483 / 1927 / "Medicinal"

1444 *Sanguisorba* indet. China / H. T. Feng 81 / 1925 / "For medicine"

1445 *Margyricarpus pinnatus* Chile / P. Aravena 18045 / 1939 / "Medicinal"

1446 *Acaena integerrima* Chile / P. Aravena 33330 / 1943 / " . . . poisonous"

1447 *A. macrostemon* Argentina / A. L. Cabrera 8805 / 1945 / "Planta medicinal, diuretica n.v. 'Cadillo' "

   Chile / I. M. Johnston 4904 / 1925 / " . . . used as a blood tonic in tea" / " 'Cadillo' "

1448 *Rosa californica* U.S.: Calif. / W. A. Dayton 584½ / 1913 / "Forage value good browse"

1449 *R. cymosa* China / W. T. Tsang 21155 / 1932 / " . . . fruit black, edible"

1450 *R. henryi* China / S. K. Lau 4362 / 1934 / " . . . fruit edible. Can be made as sugar" / "Tong Im Tsz"

1451 *R. laevigata* China / W. T. Tsang 28558 / 1938 / " . . . fr. edible" / "Tong Ang Tze Tang"

1452 *R. longicuspis* India: Assam / L. F. Ruse 112 / 1923 / " . . . in case of mad dog bites the leaves and flowers are used by the natives for binding over the wound. Extract from the leaves and flrs. are drunk for this purpose" / "Mishih-khlim"

1453 *Neurada procumbens* Kuwait / H. Dickson 51 / 61 / "Best food for fattening sheep" / "Saadan"

1454 *Grielum humifusa* S. Africa: Transvaal / L. Hunter 19 / '20 / "Said to increase milk supply in cows"

1455 *Osmaronia cerasiformis* U.S.: Ore. / R. McVaugh 6224 / 1941 / ". . . fr. . . . juicy, bitter-sweet"

1456 *Pygeum* indet. Dutch E.I.: Lesser Sunda Is. / Kostermans 18649 / 1961 / ". . . bark . . . with strong HCN smell"

1457 *Pygeum* indet. New Guinea: Papua / Womersley & Simmonds 5090 / 55 / "Twigs with distinct but rather faint benzaldehyde smell"

1458 *P. vulgare* Philippine Is. / C. O. Frake 847 / 1958 / "Leaves pounded and applied to burns" / "Tangalektin Sub."

1459 *P. turnerianum* Australia: Queensland / S. F. Kajewski 1221 / 1929 / "Bark . . . strong almond flavour when cut"

1460 *Prunus emarginatus* var. *mollis* U.S.: Idaho to Calif. / F. Binns s.n. / no year / ". . . the fruit . . . extremely bitter"

1461 *P. virginiana* var. *melanocarpa* U.S.: Colo. / D. H. Andrews 15 / 1911 / "Fruit only slightly astringent"

1462 *P. barbata* Mexico / E. Matuda 4672 / 1941 / ". . . fruits edible" / "'cerezo'"

1463 *P. brachybotrya* Mexico / G. B. Hinton 12676 / 38 / "Poisonous to cattle"

1464 *P. capuli* Mexico / Kenoyer & Crum 2687 / 48 / "Medicine tree. Fruit puckering"

1465 *P. cortapico* Mexico / H. S. Gentry 5572 / 1940 / "Fruit eaten by natives" / "Cortapico, Bebelama de la Sierra"

1466 *P. salassii* Guatemala / P. C. Standley 59958 / 1938–39 / "Fruit dark red, very bitter" / "'Carreto'"

1467 *P. serotina* Mexico / G. B. Hinton 3684 / 33 / "Fruit poisonous" / "Capulín"

1468 *Prunus* indet. Mexico / G. B. Hinton 13737 / 39 / "Leaves poisonous to cattle?" / "Ucaz"

1469 *P. myrtifolia* Jamaica / J. R. Perkins 13309 / 1917 / "'Wild Cassava'"

1470 *P. boliviana* Bolivia / O. Buchtien 242 / 1921 / ". . . Frucht ist wohlschmeckend"

1471 *P. buergeriana* Japan / U. Mizushima 1868 / 1952 / "Wood strongly scented"

1472 *P. dielsiana* China / A. N. Steward 5270 / 24 / ". . . fruit . . . eaten" / "'Yeh-ying'"

1473   *P. glandulosa* China / W. T. Tsang 2694 / 1936 / "...fr. edible" / "Chuk Lei Tze"

1474   *P. japonica* China / Tso 47 / 1926 / "Fruit...edible"

1475   *P. mume* China / W. T. Tsang 26195 / 1936 / "...fr....edible" / "Suen Mui Shue"

1476   *P. perulata* China / R. C. Ching 2810 / 1925 / "Fruit—bitter taste"

1477   *P. afghana* India: Puri Valley / E. Bacon 86 / 1939 / "Berries eaten fresh" / "čaka"

1478   *P. puddum* India: Assam / L. F. Ruse 31 / 1923 / "The fruit is eaten" / "Dieng soh-iong-kum"

1479   *P. cornuta* India: Punjab / W. Koelz 8504 / 1936 / "Fr. edible"

1480   *P. wallichii* INDIA: Ukhrul / F. K. Ward 18276 / 1948 / "The fruit is edible, but very dull-flavoured"

1481   *P. javanica* N. Borneo / Kandilis B.N.B. For. Dept. 10323 / 1938 / "Bark used as medicine against worms of kerbau" / "memot (Malay)"

1482   *Prinsepia uniflora* var. *serrata* China / J. F. Rock 13504 / 1925 / "...fruits red, juicy edible"

1483   *P. utilis* China / J. F. Rock 3281 / 1922 / "...seed collected—common—used for cooking oil"
India: Punjab / W. Koelz 10292 / 1936 / "...seeds gathered for oil" / "'Bekar'"

1484   *Chrysobalanus icaco* Mexico / Loesener 1782 / 1896 / "Früchte roh und aufgekocht gern gegessen. Rind und Wurzeln Gerbstoffhaltig, zum beizen der Fischnetze verwandt" / "'xicaco'"

1485   *Licania platypus* Honduras / W. A. Schipp 308 / 1929 / "...fruits much sought after 'by Tapir & Peccary'"

1486   *L. ternatensis* W.I.: Dominica / W. H. & B. T. Hodge 2067 / 1940 / "...seed sweet edible, by people as well as by wild pigeons and parrots"

1487   *L. parinarioides* Peru / J. M. Schunke 151 / 1935 / "...fruit red edible" / "'Parinari'"

1488   *Hirtella americana* Costa Rica / A. F. Skutch 3948 / 1939 / "Called gerrapato from appearance of black, sweetish, edible fruit"

1489   *H. triandra* W.I.: Dominica / J. S. Beard 657 / 1946 / "...fr....edible" / "'Bouis poil'"

1490   *H. collina* Brazil / B. A. Krukoff 1244 / 1931 / "'Canella de Velho'"

1491   *Couepia dodecandra* Honduras / W. H. Schipp 569 / 1930 / "Known locally as 'Baboon cup' fruits...edible"

1492   *Parinari excelsa* Brazil / B. A. Krukoff 1501 / 1931 / "...oil from pulp is used for oiling hair (perfume)" / "'Uchirana'"

1493 *P. glaberrimum* N. Borneo / Madin B.N.B. For. Dept. 1644 / 1932 / "Poisonous" / "Torog (Orang Sungei, K'tangan)"

CONNARACEAE

1494 *Rourea minor* Philippine Is. / M. Adduru 45 / 1917 / " . . . wood poisonous. Pounded and boiled to mix with dog's food" / "Paroganwe Ib."

1495 *Cnestis palala* Burma / C. E. Parkinson 13948 / 32 / "Dogs are said to go mad if they eat the fruit"

LEGUMINOSAE

1496 *Affonsea bullata* Brazil / F. C. Hoehne 3408 / 1919 / "'Inga'"

1497 *A. edwallii* Brazil / F. C. Hoehne 738 / 1917 / "Nome Vulgar: 'Ingá liso'"

1498 *Inga calderoni* Mexico / E. Matuda 4236 / 1941 / " . . . fruit edible"

El Salvador / P. C. Standley 20186 / 1922 / "'Zapato de mico'"

1499 *I. hintoni* Mexico / G. B. Hinton et al. 8978 / 36 / "Uses: fruit edible" / "Vernac. Name: 'Jacanicuil'"

Mexico / G. B. Hinton 4182 / 33 / "Uses: Jacanicuil edible" / "Vernac. Name: 'Jacanicuil'"

1500 *I. laurina* Costa Rica / A. F. Skutch 3705 / 1938 / "'Cuajiniquil'"

Mexico / Martínez-Calderón 171 / 1940 / "Nombre Castellano: 'Bainilla.' Nombre Chinanteco: 'Chág réhg'"

W.I.: St. Vincent / G. R. Cooley 8475 / 1962 / "Seed edible; 'Like almonds' a native said"

Puerto Rico / Schubert & Winters 289 / 1954 / "Sample flowers and leaves for chemical analysis"

W.I.: Grenada / J. S. Beard 232 / 1944 / "'Cocolay'"

W.I.: St. Vincent / J. S. Beard 231 / 1944 / "'Spanish ash'"

W.I.: Dominica / W. H. & B. T. Hodge 3524 / 1940 / "'Pois doux'"

W.I.: Dominica / W. H. & B. T. Hodge 3367 / 1940 / "'Pois doux'"

Dominican Republic / E. J. Valeur 716 / 1931 / "'Jina'"

1501 *I. leptoloba* Honduras / J. B. Edwards P-436 / 193. / "Vernac. name: 'Petiña'"

1502 *I. pinetorum* Brit. Honduras / P. H. Gentle 4149 / 1942 / "'Tamatama'"

1503 *I. paterno* El Salvador / P. C. Standley 23563 / 1922 / "'Paterno'"

El Salvador / P. C. Standley 22326 / 1922 / "'Guama,' 'Paterno'"

Honduras / Williams & Molina R. 9032 / 1946 / "A large tree cultivated for the legume which is used in the diet"

1504   *I. punctata* Honduras / P. C. Standley 54532 / 1927–28 / "'Cuajini-quil'"

Panama / G. P. Cooper 492 / 1928 / "'Guava'"

Nicaragua / Shank & Molina R. 4743 / 1951 / "'Guavo Colorado'"

El Salvador / S. Calderón 1574 / 1923 / "'Pepeto'"

Mexico / Y. Mexia 9258 / 1938 / "... fruit said to be large legume, seeds cooked and eaten" / "'Puih' (Miji)"

1505   *I. rodrigueziana* Guatemala / P. C. Standley 60269 / 1938 / "Used as shade tree" / "'Paterna'"

Guatemala / P. C. Standley 62255 / 1939 / "Planted abundantly as coffee shade" / "'Cushín'"

1506   *I. sapindoides* El Salvador / P. C. Standley 21803 / 1922 / "... fruit eaten ..." / "'Cujín,' 'Cujinicuil'"

El Salvador / S. Calderón 1454 / 1923 / "'Quijinicuil'"

El Salvador / S. Calderón 117 / 1922 / "'Nacaspilo,' 'Quijiniquil'"

El Salvador / S. Calderón 117 / 1921 / "'Quijinicuilé'"

El Salvador / P. C. Standley 19109 / 1921–22 / "'Cuijinicuil'"

El Salvador / P. C. Standley 19198 / 1921–22 / "Pulp of fruit much eaten ..." / "'Cujinicuil'"

1507   *I. vera* ssp. *eriocarpa* Mexico / G. B. Hinton et al. 9089 / 36 / "Fruit edible" / "'Jacaniquil'"

Mexico / G. B. Hinton et al. 9248 / 30 / "Fruit edible" / "'Jecaniqui'"

Mexico / no collector s.n. / 1898 / "Nom indigène: 'Bainillo'"

1508   *I. vera* ssp. *spuria* Guatemala / P. C. Standley 79177 / 1940 / "'Cushe'"

Mexico / H. Harms 2044 / 1896 / "'cuajinicuil'"

El Salvador / P. C. Standley 21228 / 1922 / "'Pepete'"

El Salvador / P. C. Standley 22047 / 1922 / "'Cujin'"

El Salvador / P. C. Standley 21931 / 1922 / "... fruit edible ..." / "'Nacaspilo,' 'Pepeto,' 'Pepetillo'"

El Salvador / P. C. Standley 23236 / 1922 / "'Pepito'"

El Salvador / P. C. Standley 22151 / 1922 / "... fruit eaten ..." / "'Pepeto,' 'Cujin'"

El Salvador / P. C. Standley 20935 / 1922 / "'Cuajinicuil'"

Mexico / H. E. Moore, Jr. 2905 / 1947 / "'chalahuitle del monte'"

Mexico / no collector s.n. / 1888 / "'Chalahuitl'"

Brit. Honduras / H. O'Neill 8601 / 1936 / "Pulp of legume white, edible"

Nicaragua / A. Molina R. 2123 / 1949 / "'Guava'"

1509   *I. ingoides* W.I.: Dominica / W. H. Hodge 2352 / 1940 / "'Pois doux marron'"

W.I.: Martinique / H. & M. Stehlé 6023 / 1942 / "Nom vulgaire: 'Pois doux caracoli'"

W.I.: Dominica / J. S. Beard 654 / 1946 / "'Pois doux marron'"

W.I.: St. Vincent / J. S. Beard 604 / 1945 / "'Spanish Ash'"

Venezuela / L. Williams 12935 / 1940 / "... the fruit is twisted, containing an edible pulp ..." / "'Guamo'"

1510   *I. vera* ssp. *vera* Puerto Rico / P. Sintenis 47 / 1884 / "'Guara'"

Santo Domingo / J. Schiffino 166 / 1944 / "'Guamá'"

1511   *I. affinis* Brazil / no collector 623 / 17 / "'Ingá'"

Brazil / G. Edwall 13063 / no year / "'Ingazeiro'"

1512   *I. barbata* Brazil / F. C. Hoehne 739 / 1917 / "'Ingá pilludo'"

1513   *I. cylindrica* Bolivia / I. Steinbach 6509 / 1924 / "Fruta comible tiene alguna semejanza con las perlas del rosario" / "'Pacay de rosario'"

Brazil / F. C. Hoehne 28332 / 1931 / "'Ingá Mirim'"

1514   *I. lopadadenia* Peru / G. Klug 2128 / 1931 / "Huitoto Indian Name: 'Mitiño'"

1515   *I. mathewsiana* Peru / J. M. Schunke 39 / 1935 / "... fruit ... edible" / "'Shimbillo'"

1516   *I. marginata* Brazil / Y. Mexia 4133 / 1929 / "Fruit ... pulp surrounding seeds edible" / "'Ingá'"

Bolivia / J. Steinbach 6362 / 1924 / "Fruta madura ... con semilla negra, envuelta en pulpa blanca, que es comible i de agradable gusto ..." / "'Pacai de los rios'"

1517   *I. nobilis* Bolivia / J. Steinbach 6383 / 1924 / "... fruta comible" / "'Pacay'"

1518   *Inga* indet. Peru / R. Kanehira 361 / 1927 / "Fruit edible" / "'Pacai'"

1519   *Inga* indet. Bolivia / J. Steinbach 5368 / 1921 / "La semilla envuelta en una pulpa blanca muy agradable para comer"/ "'Pacaÿ de rosaria'"

1520   *Samanea saman* W.I.: Grenadines / R. A. Howard 10939 / 1950 / "... animals like pods" / "'coco tamarind'"

W.I.: Tobago / W. E. Broadway 3565 / 1910 / "Pods ... eaten by cattle" / "'Saman,' 'Cow-bean'"

1521   *Pithecellobium saman* Philippine Is. / C. O. Frake 857 / 1958 / "Bark applied to abdomen for diarrhea—leaves applied to head for headache" / "'Kalia'" Sub.

1522   *P. albicans* Mexico / C. L. & A. A. Lundell 7887 / 1938 / "... bark used for tanning" / "'Chucum'"

1523   *P. johanseni* Honduras / P. C. Standley 53741 / 1927–28 / " . . . the pulp . . . juicy, sweet"

1524   *P. jupunba* W.I.: Grenada / P. Beard 1259 / 1945 / "Useful timber" / " 'Dalmaré' "

W.I.: Dominica / W. G. A. Ramage s.n. / 1889 / " 'Bois pipirit' "

W.I.: Grenada / J. S. Beard 220 / 1944 / " . . . timber and shingles from wood" / " 'Savonette' "

W.I.: Tobago / W. E. Broadway 4795 / 1914 / " 'Soapwood' "

W.I.: Dominica / W. H. & B. T. Hodge 1013 / 1940 / " . . . fruit much sought after by parrots" / " 'Pipiri' "

W.I.: Dominica / W. H. / B. T. Hodge 2074 / 1940 / " 'Bois pipiri' or 'Bois ciceron' "

1525   *P. unguis-cati* W.I.: Grenada / J. S. Beard 214 / 1944 / "Economic uses: aril" / " 'Bois crabbe' "

W.I.: Grenada / P. Beard 1326 / 1945 / " 'Bread & cheese' "

1526   *Pithecellobium* indet. Haiti / W. J. Eyerdam 82 / 1927 / " 'Tamarind moru' "

1527   *Pithecellobium* indet. Brazil / Y. Mexia 4858 / 1930 / "Used for firewood" / " 'Farinha seca,' 'Barba timao' "

1528   *Pithecellobium* indet. Colombia / Herb. Lehmanianum 8887 / no year / " 'Chiparo' "

1529   *Pithecellobium* indet. Venezuela / Curran & Haman 1013 / 1917 / " 'Quebracho' "

1530   *Pithecellobium* indet. Venezuela / Curran & Haman 562 / 1917 / " 'Cogicillo' "

1531   *Pithecellobium* indet. Brazil / R. Froes, under direction of B. A. Krukoff 1887 / 1932 / "Bark used for tanning" / " 'Angico branco' "

1532   *P. auriculatum* Brazil / B. A. Krukoff 7938 / 1936 / " 'Faveira' "

1533   *P. cauliflorum* Brazil / Ducke 503 / 1937 / " 'Ingá-rana' "

1534   *P. daulense* Ecuador / J. N. & G. Rose 23588 / 1918 / " . . . funny Pitheco" / " 'Nance' "

1535   *P. dulce* Venezuela / Killip & Tamayo 37041 / 1943 / "Used for fuel" / " 'Taguapire' "

Colombia / E. L. Little, Jr. 7701 / 1944 / "payandé"

1536   *P. forfex* Colombia / Killip & Smith 14338 / 1926 / " 'Tiraco' "

1537   *P. gonggrijpii* Surinam / G. Stahel 88 / '42 / " 'Water-Tamarinde' " / " 'manaliballi' (A.); 'kala eipjo,' 'kleipjo' (K.)"

1538   *P. huberi* Brazil / R. Froes, under direction of B. A. Krukoff 1760 / 1932 / " 'Pau de Formiga' "

1539   *P. incuriale* Brazil / A. Gehrt 3473 / 1919 / "'Páo de cortiça'"

1540   *P. langsdorffii* Brazil / Reitz & Klein 3.912 / 1956 / "Lenho na xilo-teca" / "'Pau gambá'"

      Brazil / F. C. Hoehne 29842 / 1932 / "'Raposeira'"

1541   *P. latifolium* Peru / Y. Mexia 6515 / 1932 / "'Shimbillo'"

1542   *P. polycephalum* Venezuela / H. pittier 7860 / 1918 / "'Caro hueso de pescado'"

      Venezuela / Curran & Haman 850 / 1917 / "'Camburi chiquito'"

1543   *P. pubescens* Venezuela / H. Pittier 10537 / 1922 / "'Maiz cecido'"

1544   *P. racemosum* Surinam / G. Stahel 72 / '42 / "'Manaliballi tataro' (A.), 'apakaniran' (K.)"

1545   *P. clypearia* China / W. T. Tsang 24391 / 1934 / "'Sai Yeung Lu Kung Chok Muk'"

      China / W. T. Tsang 28703 / 1938 / "'Lu Kung Chok Shue'"

      Hainan / S. K. Lau 1892 / 1932 / "'Niu Tung Kung'"

      China / W. T. Tsang 22063 / 1933 / "'Loi Chok Shue'"

      Borneo / Moh. Enoh 298 / 1948 / "'poko baje'"

      Sumatra / R. S. Toroes 1867 / 1932 / "'kaju si boesoek'"

      Sumatra / R. S. Toroes 2390 / 1932 / "'Kaju pandia'"

      Philippine Is. / G. E. Edaño 2007 / 1950 / "The bark and leaf good medicine for snake-bite. Scrape the bark and apply to the wound" / "'Aguyangyang'" Bis.

1546   *P. ellipticum* Indonesia: Bangka / Kostermans & Anta 664 / 1949 / "'Djengkol hutan'"

      Sumatra / R. S. Toroes 4889 / 1933 / "'Kaju longgajan'"

      Sumatra / R. S. Toroes 5204 / 1933 / "'Kaju djoring mandi'"

      Sumatra / R. S. Boeea 7659 / 1934 / "'Kaju si mardjoring-djoring'"

      Sumatra / C. G. G. J. v. Steenis 9339 / 1937 / "root used as fishpoi-son" / "'toeba koboh'"

      Sumatra / R. S. Boeea 7027 / 1934–35 / "'Kaju pandia djoring'"

      Philippine Is. / L. E. Ebalo 923 / 1941 / "'Magasaluka'" Yakan

      Philippine Is. / L. E. Ebalo 385 / 1940 / "'Lamacao'" Tagbanua

1547   *P. kunstleri* S. Borneo / P. Buwalda 7941 / 1940 / "'Talinan'"

1548   *P. microcarpum* Central E. Borneo / F. H. Endert 1841 / 1925 / "'Meloré'"

      Brit. N. Borneo / B.N.B. For. Dept. 3396 / 1933 / "'Langgir hantu' (Kedayan)"

      E. Borneo / Sabana 6 Herb. Bog. / 1954 / "'djaring burung'"

1549   *P. rosulatum* S.E. Borneo / E. G. Sauveur 82 / 1951 / "'Djaring han-tu'"

S.E. Borneo / E. G. Sauveur 47 / 1951 / "'Girik' (Bandjar)"

1550   *P. mindanaense* Philippine Is. / C. O. Frake 687 / 1958 / "Roots boiled and drunk for cough" / "'Tiatak'" Sub.

Philippine Is. / L. E. Ebalo 919 / 1941 / "'Balangkuya'" Yakan

Philippine Is. / L. E. Ebalo 1116 / 1941 / "'Indang'" Maranao

1551   *P. scutiferum* Philippine Is. / C. O. Frake 765 / 1958 / "Leaves applied to skin eruptions" / "'Pandalaga'" Sub.

1552   *Zygia jiringa* Brit. N. Borneo / B.N.B. For. Dept. 1979 / 1932 / "Edible fruit" / "'Jaring'" Brunei

Borneo / J. & M. S. Clemens 22286 / 1929 / "...frs. eaten"

Sumatra / B. A. Krukoff 4266 / 1932 / "...fruits edible" / "'Djedoe'"

1553   *Cedrelinga catenaeformis* Brazil / B. A. Krukoff 8789 / 1936 / "'Cedrorana'"

1554   *Ortholobium yunnanense* China / W. T. Tsang 24577 / 1934 / "'Ma Ling Kwo Muk'"

1555   *Wallaceodendron celebicum* Philippine Is. / H. C. Conklin 18727 / 1953 / "'Parukpúk'"

Philippine Is. / M. D. Sulit 4941 / 1952 / "'Kayu sampaga'"

1556   *Albizzia caribaea* Honduras / Allen & Trafton 6591 / 1952 / "'Tejo'"

El Salvador / P. H. Allen 7212 / 1958 / "'Conacaste blanco'"

W.I.: Grenada / J. S. Beard 205 / 1944 / "...timber useful" / "'Tantacayo'"

1557   *A. adinocephala* Honduras / A. Molina R. 2945 / 1950 / "'Madre cacao'"

1558   *A. idiopoda* Brit. Honduras / P. H. Gentle 254 / 1931–32 / "Used for tanning" / "'salem'"

1559   *A. lebbeck* El Salvador / P. C. Standley 22391 / 1922 / "'Canjuro'"

India / J. Fernandes 405 / 1949 / "Fruits sometimes serve as rattles for children"

India / J. Fernandes 1363 / 1950 / "'Siris' (Marathi)"

Fiji Is. / A. C. Smith 6874 / 1947 / "'Vaivai'"

1560   *A. chinensis* Burma / F. G. Dickason 7663 / 1938 / "Bark used as fish poison; pound up in water. Wood used as rice pounders"

Philippine Is. / Sulit & Conklin 5086 / 1953 / "Bark scraped off and put in water; sliced nami (korot) mixed in solution for 2 days; taken and washed thoroughly in river for another 2 days. The nami is ready for the table" / "'Taganhuk'" Hanunoo

1561 *A. occidentalis* Mexico / H. S. Gentry 4982 / 1939 / "'Palo Fierro'"
Baja Calif. / H. S. Gentry 4440 / 1939 / "'Palo Escopete'"

1562 *A. berteriana* W.I.: St. Kitts / G. R. Cooley 8796 / 1962 / "Feed to stock" / "Called 'Bread and Cheese'"

1563 *A. corniculata* Hainan / C. I. Lei 102 / 1932 / "'Tang shan si'"
China / W. T. Tsang 25733 / 1935 / "'Fong Pei Tang'"

1564 *A. julibrissin* Japan / K. Uno s.n. / 1951 / "'Nemunoki'"

1565 *A. saponaria* Philippine Is. / P. Añonuevo 199 / 1950 / "Medicinal. Scrape the bark, squeeze and apply the juice on the hair to remove dandruff"
Philippine Is. / L. E. Ebalo 757 / 1940 / "'elongigue'" Sub.
Philippine Is. / K. I. Pelzer 79 / 1950 / "Bark is squeezed, sap is drunk by man who is impotent"

1566 *A. basaltica* Australia: Queensland / S. L. Everist 1568 / 1938 / "Excellent sheep fodder" / "'Dead Finish'"

1567 *A. procera* Australia: Queensland / C. T. White 12198 / 1935 /
" . . . regarded as a good cattle feed and as a sign of good country for farming (sugar cane)" / "'Acacia'"

1568 *Calliandra calothyrsis* Guatemala / P. C. Standley 23846 / 1922 / "'Yaje'"

1569 *C. confusa* Honduras / P. C. Standley 55072 / 1928 / "'Cabello de ángel'"

1570 *C. gentryi* Mexico / H. S. Gentry 6524 / 1941 / "Not used as fuel because smoke will cause women to loose hair" / "'Huaje'"

1571 *C. emarginata* Mexico / J. T. Howell 8492 / 1932 / "'Clabellina'"

1572 *C. penduliflora* Mexico / R. L. Dressler 2006 / 1957 / "'Barba del chivo'"
El Salvador / P. C. Standley 21818 / 1922 / "'Barbón montañés'"

1573 *C. tetragona* Mexico / Y. Mexia 1079 / 1926 / "'Cola de Iguana'"

1574 *Calliandra* indet. Mexico / Y. Mexia 8942 / 1937 / "'Chapuli'"

1575 *Calliandra* indet. Mexico / R. M. Stewart 384 / 1941 / "'Tabardillo'"

1576 *C. tergemina* W.I.: Martinique / H. & M. Stehlé 5720 / 1945 / "'Bois mirette,' 'bois patate'"
W.I.: St. Lucia / J. S. Beard 195 / 1943 / "'Bois patate'"

1577 *C. amazonica* Peru / Y. Mexia 6336 / 1931 / "'Machete vaina,' 'Shingata' (Aguaruna Indian dialect)"

1578 *C. angustifolia* Peru / Y. Mexia 6313 / 1931 / "'Bobinsana'"
Ecuador / Y. Mexia 8464 / 1936 / "'Chipero'"

1579 *C. marginata* Colombia / Killip & Smith 14368 / 1926 / "'Veranero'"

1580 *C. medellinensis* Colombia / J. M. Duque 1815 / no year / "'Carbonero morado'"

1581 *C. tweedii* Brazil / N. Andrade 861 / 17 / "'Mandavaré'"

1582 *Lysiloma aurita* Mexico / G. B. Hinton et al. 10132 / 37 / "'Quitaz' "
El Salvador / S. Calderón 340 / 1922 / "'Cicaguite,' 'Sicahuite'"

1583 *L. bahamensis* Mexico / E. C. Stewart 231 / 1935 / "'Tzalám'"
Cuba / R. A. Howard 6594 / 1941 / "'Tengue'"

1584 *L. acapulcensis* Mexico / G. B. Hinton 2784 / 32 / "'Tepehuaje'"
El Salvador / P. C. Standley 21650 / 1922 / "'Quebracho colorado'"
Mexico / Y. Mexia 8884 / 1937 / "'Parotillo'"

1585 *L. candida* Mexico / H. S. Gentry 3705 / 1938 / ". . . bark used for tanning" / "'Palo Blanco'"

1586 *L. divaricata* Baja Calif. / I. L. Wiggins 15334 / 1959 / "'Manto'"
Mexico / Y. Mexia 8733 / 1937 / "Bark used for tanning . . ." / "'Cuitás'"
Mexico / Y. Mexia 944 / 1926 / "'Tepemesquite'"
Mexico / H. S. Gentry 4846 / 1939 / "Used for tanning" / "'Mauuta'"
El Salvador / P. C. Standley 20654 / 1922 / "'Carbon'"
El Salvador / S. Calderón 7011 / 1922 / "'Quebracho'"

1587 *L. tergemina* Mexico / G. B. Hinton 929 / 32 / "'Pata de Venado'"

1588 *L. watsoni* Mexico / H. S. Gentry 4787 / 1939 / "Bark used for tanning & chewed for teeth ailments" / "'Tepeguaje'"

1589 *Acacia pennata* India / J. Fernandes 986 / 1950 / ". . . Hanuman monkeys were splitting the pods and eating the seeds"

1590 *Acacia* indet. Siam / native collector S 417 Royal For. Dept. / 49 / "Leaves sour tasted. Flowers . . . sour-tasted. Edible young shoots" / "'Som Poi'"

1591 *Acacia* indet. India / J. Fernandes 726 / 1949 / "Very prickly climber bearing pods; used as soap, the lather acting as an insecticide"

1592 *Acacia* indet. India / J. Fernandes 626 / 1949 / "'Sembo'"

1593 *Acacia* indet. New Guinea / E. Gray NGF 7165 / 55 / "Bark collected for tannin"

1594 *Acacia* indet. Mexico / G. B. Hinton 4744 / 33 / "Bark used for tanning" / "'Timbre'"

1595 *Acacia* indet. Mexico / G. B. Hinton 7162 / 34 / "Uses: tanning"

1596 *Acacia* indet. Mexico / Y. Mexia 8930 / 1937 / "'Cuin de la Parotilla'"

1597 *Acacia* indet. Mexico / G. B. Hinton 6843 / 34 / "Fruit edible" / "'Guaje chiquito'"

1598 *Acacia* indet. Bolivia / J. Steinbach 6249 / 1924 / "El polvo de la cascara se ocupa para secar llagas purulentas"

1599 *A. pennata* Philippine Is. / M. D. Sulit 1480 / 1947 / "Heartwood source of commercial tannin known as 'cutch'" / "'Cutch'"

1600 *A. concinna* India / J. Fernandes 911 / 1950 / "...the pods are emulsified in water and the resulting soapy solution used as insecticide. Fruit is called 'shikakai' and is sold in the bazaars as such"

Burma / C. E. Parkinson 13979 / 32 / "Fr. used to make a hail wush. Leaves and fl. eaten"

1601 *A. farnesiana* U.S.: Fla. / A. A. Eaton 780½ / 1903 / "'Stinking bean'"

Mexico / Y. Mexia 757 / 1926 / "Eagerly eaten by goats"

Mexico / I. L. Wiggins 7454 / 1934 / "'Vignorama'"

Mexico / H. S. Gentry 1237 / 1935 / "...may be regarded as life zone indicator. Medic. flowers mixed with grease and rubbed on head for headaches. Report Warihio"

Mexico / R. M. Stewart 2195 / 1941 / "'Huisache'"

Cuba / Wood, Jr., & Atchison 7386 / 1947 / "'Aroma creole'"

1602 *A. berlandieri* U.S.: Tex. / E. J. Palmer 33449 / 1928 / "'Huajilla'"

Mexico / G. B. Hinton 16576 / 44 / "Fodder for goats and cattle" / "'Guajillo'"

1603 *A. cornigera* El Salvador / P. C. Standley 22674 / 1922 / "Spines full of ants" / "'Iscaal,' 'Cutupito'" "'Cartapito'" "'Pico de Gurrión'"

Guatemala / P. C. Standley 24054 / 1922 / "'Iscanal,' 'Huiscanal'"

1604 *A. pennatula* Mexico / H. S. Gentry 1202 / 1934 / "Seeds of plants eaten by Warihios when they do not have corn, grind pod and all. A dye (color) made from the bark to dye leather" / "'Palo Garobo,' Mex. 'Yepowecha,' W."

Honduras / A. Molina R. 823 / 1948 / "'Bolita blanca'"

Mexico / G. B. Hinton 9986 / 37 / "'Tepamo'" / "'Tepame'"

1605 *A. cymbispina* Mexico / Y. Mexia 8885 / 1937 / "Legumes drop and eagerly eaten by stock" / "'Espino'"

Mexico / Y. Mexia 924 / 1926 / "'Huinole'"

1606 *A. hindsii* El Salvador / P. C. Standley 21027 / 1922 / "...pulp of fr. sweet, eaten" / "'Guascanal'"

1607 *A. melanoceras* Costa Rica / P. H. Allen 5997 / 1951 / "Hollow thorn inhabited by two quite different species of ants. One, possibly a species of Azteca stings very severely"

1608 *A. constricta* Mexico / Stanford, Retherford & Northcraft 174 / 1941 / " ... heavily grazed by goats"

Mexico / G. B. Hinton et al. 16501 / 44 / "'Huisachillo'"

1609 *A. igualensis* Mexico / G. B. Hinton 2500 / 32 / "Uses is tanning" / "'Timbre'"

Mexico / G. B. Hinton 5412 / 33 / "Uses: root used for tanning"

1610 *A. tequilana* Mexico / Y. Mexia 8850 / 1937 / "Thick root used for tanning" / "'Guasillo'"

1611 *A. acatlensis* Mexico / G. B. Hinton 5955 / 34 / "Flower buds sold in markets for food" / "'Saisqua'"

El Salvador / S. Calderón 1774 / 1923 / "'Quebracho'"

1612 *A. choriophylla* Bahama Is. / O. Degener 18727 / 1946 / " ... pulp of ripe legume edible and like sweet, fresh bread"

1613 *A. macrantha* W.I.: Grenada / W. E. Broadway s.n. / 1904 / "Goats are very fond of the ripe pods"

1614 *A. nilotica* W.I.: Antigua / J. S. Beard 262 / 1944 / " ... browsed by cattle"

1615 *A. glomerosa* Brazil / B. A. Krukoff 1088 / 1931 / " ... extract from the bark used as substitute for ink" / "'Coronha'"

Peru / Y. Mexia 6258 / 1931 / "'Pashaco'"

1616 *A. multiflora* Brazil / R. Froes 1855 / 1932 / "Bark used for tanning ... "

1617 *A. paniculata* Brazil / Y. Mexia 4460 / 1930 / "Leaves said to be used in tanning"

1618 *Leucaena confusa* Mexico / Y. Mexia 8820 / 1937 / " ... fruit a long pod with edible seeds" / "'Guaje colorado'"

Mexico / G. B. Hinton 1965 / 32 / "Fruit edible common in markets" / "'Guaje'"

Mexico / Y. Mexia 736 / 1926 / "Fruit ... ill-smelling pod, occasionally eaten" / "'Huaje'"

1619 *L. guatemalensis* Guatemala / P. C. Standley 73562 / 1940 / "'Yaje'"

1620 *L. macrophylla* Mexico / G. B. Hinton 7493 / 35 / "Fruits sold in markets edible" / "'Guaje'"

1621 *L. shannonii* El Salvador / S. Calderón 2105 / 1924 / "'Cascahuite'"

1622 *Leucaena* indet. Mexico / G. B. Hinton 2256 / 32 / "Fruits sold in markets" / "'Guaje'"

1623 *L. leucocephala* Haiti / R. J. Seibert 1756 / 1942 / "Cattle and pigs eating this plant are said to lose their hair" / "'Oriman'"

Bahama Is. / R. A. & E. S. Howard 9984 / 1948 / "'Jimbo bean'"

1624 *L. trichodes* Peru / Y. Mexia 8024 / 1936 / "Foliage fed to stock" / "'Chamba'"

1625 **Mimosa palmeri** Mexico / H. S. Gentry 1583 / 1935 / "Bark chewed to harden gums" / "'Chopa'"

1626 *M. pudica* W.I.: St. Kitts / G. R. Cooley 8822 / 1962 / "Root boiled to relieve or aid passage of urine" / "'Jump up and kiss me'" / "'Piss-a-bed'"
Colombia / Killip & Smith 14237 / 1926 / "'Cierra-de-puta'"

1627 *M. coroncoro* Colombia / Dugand & Jaramillo 3461 / 1943 / "'Coroncoro'"

1628 *M. malacocentra* Brazil / Y. Mexia 5624 / 1931 / "Tea made of leaves for pain" / "'Angiquin'"

1629 *M. myriadena* Brazil / Ducke 125 / 1936 / "'rabo de camaleão'"

1630 *M. pigra* Brazil / Y. Mexia 4181 / 1929 / "'Vassourinha'" / "'Malicia'"
Brazil / B. A. Krukoff 1250 / 1931 / "'Jigury'"
Peru / Y. Mexia 6427 / 1932 / "'Pingahuisacha'"
Colombia / Killip & Smith 14706 / 1926 / "'Trupilla'"

1631 *M. velloziana* Brazil / Y. Mexia 4787 / 1930 / "'Malisia'" / "'Unha de gato'"

1632 *M. xinguensis* Peru / J. M. Schunke 46 / 1935 / "'Pashaquilla'"

1633 **Desmanthus bicornutus** Mexico / H. S. Gentry 2410 / 1936 / "Near leaf axils are sweet glands calling a legion of small ants to feast"

1634 **Piptadenia communis** Brazil / A. Gehrt 1235 / 1919 / "'Jacaré'"
Brazil / Y. Mexia 4488 / 1930 / "'Munjolo'"

1635 *P. contorta* Brazil / Y. Mexia 4438 / 1930 / "'Anjico Branco'"

1636 *P. flava* Peru / G. Klug 2034 / 1931 / "'Pashaguillo'"

1637 *P. micrantha* Brazil / Y. Mexia 4614 / 1930 / "'Unha de Gato'"

1638 *P. ovalifolia* Brazil / B. A. Krukoff 8082 / 1936 / "'Epinheiro'"

1639 *P. pterosperma* Brazil / Y. Mexia 4366 / 1930 / "'Farinha Seca'"

1640 *P. rigida* Argentina / Eyerdam & Beetle 22783 / 1938 / "'Tjarca'"

1641 *P. tocantina* Brazil / Froes & Krukoff 1875 / 1932 / "'Cuhuba'"

1642 *P. uaupensis* Brazil / Froes & Krukoff 11636 / 1939 / "'Sabia'"

1643 *P. africana* Ivory Coast / B. A. Krukoff 81 / 30 / "'Dabema'"
Nigeria / J. D. Kennedy 282 / no year / "'Ekhini'"
Liberia / G. P. Cooper 141 / 1928 / "'Gaw,' 'African Greenheart'"

1644 **Goldmannia constricta** El Salvador / P. C. Standley 21977 / 1922 / "'Lengua de vacca'"

1645 *Parkia ingens* Brazil / Krukoff & Froes 1909a / 1932 / "'Faveira grande'"

1646 *P. igneiflora* Brazil / Ducke 381 / 1937 / "'Arara-tucupy'"

1647 *P. multijuga* Brazil / B. A. Krukoff 1399 / 1931 / "... fruits used for preparation of soap" / "'Faveira'"

1648 *P. pendula* Surinam / G. Stahel 32 / 42 / "'Ipana'" / "'Koejali tapatje'"

1649 *P. speciosa* Brit. N. Borneo / H. G. Keith (B.N.B. For. Dept. 3133) / 1933 / "Fruits edible—when eaten cause urine to smell strongly" / "'Neup' (Idahan), 'Patag' (Murut), 'Petal' (Kedayan)"

1650 *P. versteeghii* New Guinea / R. D. Hoogland 4963 / 1955 / "'Mut' (Bembi), 'Mamangassi' (Kargorin), 'Bangsa' (Rawa), 'Kulubeling' (Jal)"

1651 *Dimorphandra* indet. Brit. Guiana / J. S. de la Cruz 1627 / 1922 / "'Mora'"

1652 *Cynometra zamorana* Colombia / R. E. Schultes 5429 / 1943 / "Fruit edible, sweet" / "'Coca'"

1653 *Copaifera pubiflora* Venezuela / Wurdack & Monachino 41122 / 1956 / "A commercial oil is said to be extracted from the wood" / "'Aceit'"

1654 *Sindora cochinchinensis* Cochinchine / D. Delcambre, Tan-Uyen-Binhoa s.n. / 1922 / "Bois de luxe: secrete un peu de résine non colorée. Arille chiguée avec betel"

1655 *Crudia choussyana* El Salvador / S. Calderón 1573 / 1923 / "'Chichipate'"

1656 *C. tomentosa* Brazil / Froes & Krukoff 1943 / 1932 / "'Castanha de burro'"

1657 *C. reticulata* Brit. N. Borneo / B.N.B. For. Dept. 1692 / 1932 / "Poisonous" / "'Cansay'"

1658 *Hymenaea courbaril* Brazil / Froes & Krukoff 1803 / 1932 / "Trees exude resin of considerable value, used for medicinal purposes" / "'Jatoba'"

1659 *H. stigonocarpa* Brazil / Williams & Assis 5855 / 1945 / "Gives good resin" / "'Jatoba do Campo'"

1660 *Peltogyne* indet. Peru / J. M. Schunke 85 / 1935 / "'Soliman'"

1661 *Peltogyne* indet. Peru / J. M. Schunke 329 / 1935 / "'Muchinmango'"

1662 *Brachystegia spiciformis* var. *latifoliolata* S. Africa: Transvaal / J. Borle 263 / '21 / "Leaves used as disinfectant for wounds"

1663 *Cryptosepalum curtisiorum* Angola / A. G. Curtis 207 / 1923 / "The sable antelope like to eat the leaves"

1664 *Intsia bijuga* Fiji Is. / O. Degener 15031 / 1941 / "As remedy against toothache. Fijians boil leaves with papaya root and hold liquid in mouth. Also good for yaugoua morfar" (". . . yangona morfor"?) "'Vesi'"

1665 *Macrolobium bifolium* Brazil / Y. Mexia 5935 / 1931 / "'Ipezeiro'"

1666 *Brownea* indet. Brit. Guiana / J. S. de la Cruz 1101 / 1921 / "Bark used for tea" / "'La Rosa montaña'"

1667 *Cercis occidentalis* U.S.: Calif. / W. A. Dayton 12 / 1912 / "Forage value good" / "'California red bud'"

1668 *C. canadensis* Mexico / R. M. Stewart 1510 / 1941 / "'Pata de vaca'"

1669 *Bauhinia kappleri* Nicaragua / D. Chaves 203 / 1926 / "'Astro de la china'"

1670 *B. columbiensis* Colombia / Killip & Smith 14692 / 1926 / "'Bejuco cadena'"

1671 *B. cuyabensis* Brazil / Y. Mexia 5671 / 1931 / "'Unha de boi'"

1672 *B. acuminata* Siam / G. W. Groff 308 / 1920 / ". . . called a lover's flower"

1673 *B. diptera* Brit. N. Borneo / B.N.B. For. Dept. 9254 / 38 / "Fruit beanlike—edible when roasted" / "'Lumapak' (Tenggara), 'Koripit' (Kedayan)"

1674 *B. finlaysoniana* Sumatra / P. Buwalda 6858 / '39 / "'Akar mangko-mangko'"

1675 *B. malabarica* var. *acidum* Dutch E.I.: Lesser Sunda Is. / Bloembergen 3338 / '39 / "Flowers . . . remedy for wounds"

1676 *B. megalantha* Brit. N. Borneo / Goklin B.N.B. For. Dept. 2812 / 1933 / "Native medicine" / "'Dadahop' (Dusun)"

1677 *B. merrilliana* Philippine Is. / C. Frake 379 / 1947 / ". . . medicine for spleenomegaly" / "'Kalambagay'" Sub.

1678 *B. semibifida* Philippine Is. / C. O. Frake 633 / 1958 / "Economic uses: animal ritual" / "'Kalenbangay menuti'" Sub.

1679 *Bauhinia* indet. Burma / F. G. Dickason 7354 / 38 / "Flowers edible" / "'Dallhla'" "'Sinzwe'"

1680 *Bauhinia* indet. Siam / native collector Royal For. Dept. 355 / 49 / "Roots used in local medicine" / "'Siew-dok-doeng'"

1681 *Bauhinia* indet. Philippine Is. / G. E. Edaño 2016 / 1950 / "Leaves good medicine for headache just apply on forehead"

1682 *Bauhinia* indet. Philippine Is. / R. B. Fox 81 / 1950 / "Young leaves used as 'paper' for cigarettes" / "'Takla-Takla'" Tagb.

1683 *Dialium schlechteri* S. Africa: Transvaal / J. Borle 63 / 19 / "Fruit edible"

1684   *Cassia alata* Philippine Is. / L. E. Ebalo 758 / 1940 / "...leaves used as medicine for skin disease" / "'Capis'" Sub.

1685   *C. bacillaris* Guatemala / J. Steyermark 45193 / 1942 / "'frijole cabro'"

1686   *C. cathartica* Brazil / L. O. Williams 5255 / 1945 / "'Senno' purgative, boil plant and take soup. About 1/3 of this specimen in 1 liter of water take at one time"

1687   *C. chrysocarpa* Peru / Y. Mexia 6501 / 1932 / "'Lluichu-vainilla'"

1688   *C. demissa* Mexico / Stanford, Retherford & Northcraft 229 / 1941 / "...heavily grazed by goats"

1689   *C. fastuosa* Brazil / Y. Mexia 5214 / 1930 / "...bark for tanning" / "'Canafistula'"

1690   *C. floribunda* Philippine Is. / A. L. Zwickey 374 / 1938 / "'Sinda' (Lan.); used as coffee, and like peas; for convulsions in babies, drink and rub crushed leaves on face"

1691   *C. grandis* Guatemala / J. Steyermark 45407 / 1942 / "...used as remedy for colds" / "'mukut'"

     Honduras / P. C. Standley 54489 / 1927–28 / "Fruit...with edible pulp" / "'Carao'"

1692   *C. hirsuta* El Salvador / P. C. Standley 21001 / 1922 / "Seeds used as coffee" / "'Frijolillo negro'"

1693   *C. laevigata* Guatemala / S. F. Blake 7668 / 1919 / "...plant steamed and applied to body for sickness especially for women" / "'Frijolillo'"

     Mexico / G. B. Hinton 662 / 32 / "They say that it is used cooked in a bath for hydrophobia patients" / "called 'Retama'"

1694   *C. multijuga* Brazil / B. A. Krukoff 1087 / 1931 / "...pounded beans are boiled and an ink is made" / "'Camunze'"

     Brazil / Y. Mexia 4218 / 1930 / "...leaves made into tea used in Asthma" / "'Fedegoso'"

     Brazil / Y. Mexia 4441 / 1930 / "'Farinha Senca'"

1695   *C. nicaraguensis* El Salvador / P. C. Standley 20676 / 1922 / "'Vainilla'"

1696   *C. nomame* Japan / E. Elliot 150 / '48 / "Food use: Boiled young leaves are eaten" / "Medicinal use: stem and leaves are used as a diuretic"

1697   *C. occidentalis* Mexico / T. C. & E. M. Frye 2575 / 1939 / "...browse plant"

     Brazil / B. A. Krukoff 1082 / 1931 / "...leaves boiled to make drink used as purgative; seeds used as a cure for liver" / "'Paramarioba'"

     Hainan / H. Fung 20260 / 1932 / "Mix leaves with rice to brew wine" / "'Lai Cha'"

Hainan / F. A. McClure 7616 / 1921 / " ... drug plant; boil the leaves with eggs good for headache" / "'Shan luk tau,' 'Chu kwat min'"

Brit. N. Borneo / B.N.B. For. Dept. A-82 / 1947 / "Leaves used for medicine by the natives for skin disease" / "'Manggarut Gelinggang' (Malay), 'Terong Suluk' (Brunei)"

Philippine Is. / C. Frake 618 / 1958 / "Ritual uses: Leaves applied to headache" / "'Sigbin' (Dialect Sub.)"

Philippine Is. / A. L. Zwickey 136 / 1938 / "'Kayoranti' or possibly 'Kayo a ranti'"

S. Africa: Transvaal / I. H. Pierce 4 J.N. / '20 / "Decoction from leaves used by natives for stomach pains. Seeds as fowl food"

1698 *C. racemosa* Peru / Y. Mexia 6074 / 1931 / "'Yacusaspi'"

1699 *C. reticulata* E. Salvador / P. C. Standley 20987 / 1922 / "Used as purgative ... " / "'Sambrán'"

Honduras / P. C. Standley 55141 / 1927–28 / "Roots used as purgative and for irregularities of menstruation. Said to be poisonous" / "'Baraja'"

Peru / J. M. Schunke 208 / 1935 / " ... medicinal" / "'Retama'"

1700 *C. tora* U.S.: Ga. / Wood, Jr., & Clement 7558 / 1947 / "called 'coffee bean' thereabouts. Natives say stock won't eat this plant"

1701 *Krameria cistoidea* Chile / R. Wagenknecht 18479 / 1939 / "The timber is exported to Germany to stain skins and makes an excellent aniline" / "'Pacul'"

1702 *Gleditsia horrida* N. China / J. C. Liu L.2249 / 1929 / "'Tsao Chia'"

1703 *G. sinensis* China / F. N. Meyer 35616 / 1913 / "A honey locust, of which the pods are used as a substitute of soap in washing the hairs and fine clothing" / "'Tsau Chiau Shu'"

1704 *Schizolobium amazonicum* Brazil / F. C. Hoehne 864 / 1917 / "Nome vulgar: 'Bacurubú,' 'Fayeiro'"

1705 *S. parahybum* Honduras / P. C. Standley 54629 / 1927–28 / "'Tambor'" / "'Zorra'"

1706 *Cercidium peninsulare* Baja Calif. / H. S. Gentry 4458 / 1939 / "Its branches are cut for cattle forage" / "'Palo Verde'"

1707 *C. torreyanum* Mexico / H. S. Gentry 1377 / 1935 / "Tree, sometimes exudes a sweet odorous secretion highly attractive to some insects"

1708 *Hoffmanseggia gracilis* Peru / Mr. & Mrs. F. E. Hinkley 8 / 1920 / " ... used for throat troubles" / "'Tarillita' or 'Tabaquillo'"

1709 *Caesalpinia cinclidocarpa* India: Assam / L. F. Ruse 421 / 1924 / " ... the extract from the leaves and the stem is said to be used by natives in cases of dog-bite" / "'Dieng-sia-khwai'"

1710 *C. coriaria* Venezuela / H. Pittier 10481 / 1922 / "The infusion used to strengthen teeth" / " 'Divi-divi' "

1711 *C. pulcherrima* Mexico / Y. Mexia 8704 / 1937 / "Flowers made into infusion for coughs" / " 'Ciriguanito' "

1712 *C. sappan* Brit. N. Borneo / B.N.B. For. Dept. 2296 / 1932 / "Wood for dying purposes"

1713 *C. spinosa* Peru / Y. Mexia 4048 / 1935 / "Used for tanning and to make a yellow dye"

1714 *C. tsoongii* China / F. A. McClure 13405 / 1925 / "vine said to be used to stupefy fish" / " 'To Ue T'ang' "

1715 *Caesalpinia* indet. Mexico / H. S. Gentry 7139 / 1945 / "Flowers eaten by deers"

1716 *Caesalpinia* indet. W.I.: Dominica / W. H. & B. T. Hodge 2937 / 1940 / " . . . seeds made into coffee for diabetes (caruque)"

1717 *Swartzia simplex* var. *dariensis* Panama / Stern, Dwyer, Chambers & Ebinger 203 / 1959 / " 'Naranjito' "

1718 *Ateleia arsenii* Mexico / G. B. Hinton 3421 / 33 / "food for deer"

1719 *Ormosia calavensis* Philippine Is. / C. Frake 640 / 1958 / "Agricultural ritual"

1720 *Alexa imperatricis* Venezuela / Wurdack & Monachino 39681 / 1955 / " 'Leche cochina' "

Brit. Guiana / A. Pinkus 171 / 1938–39 / "Bark used as fish poison" / " 'Aromata' " Arawak

1721 *Castanospermum australe* New Britain / A. Floyd 6507 / 54 / " . . . fruit not eaten in this area" / " 'La Kubele' " W. Nakanai

1722 *Sophora secundiflora* U.S.: Tex. / A. & R. A. Nelson 5105 / 1942 / "Seeds said to be poisonous"

U.S.: Tex. / E. N. Plank s.n. / no year / "Seeds a powerful narcotic poison"

U.S.: Tex. / S. D. McKelvey 1887 / 1931 / " 'Coral bean' "

1723 *S. flavescens* China / Cheo & Yen s.n. / 1936 / " . . . plant for medical use"

1724 *S. japonica* China / S. K. Lau 3967 / 1934 / " . . . starch in seed edible" / " 'Wai Shue' "

1725 *S. viciifolia* China / H. Wang 41425 / 1939 / " . . . fls. white, edible"

1726 *S. tomentosa* Fiji Is. / O. Degener 15112 / 1944 / "Pound up leaves, mix with coconut oil and tie around broken bone" / " 'Na(n)drala' " Serua

1727 *Gastrolobium crassifolium* Australia / E. H. Wilson 52 / 1920 / " 'Narrow leaf poison' "

1728   *G. oxylobioides* Australia / E. H. Wilson 214 / 1920 / "'Champion Bay Poison'"

1729   *G. parviflorum* Australia / E. H. Wilson 172 / 1920 / "'Burr Poison'"

1730   *G. spinosum* Australia / C. T. White 5291 / 1927 / "'Prickle Poison'"

1731   *G. villosum* Australia / E. H. Wilson 205 / 1920 / "'Crimp-Leaved Poison'"

1732   *Crotalaria stipularia* Mexico / J. Steinbach 5170 / 1920 / "Dicen las indigenas, que la raiz sirve contra picadas de serpientes. Las frutas suenan dentro de las vainas como la rola una cascabel al menor contacta" / "'Campos veneno' o 'Cascabelito'"

1733   *C. tuerckheimii* El Salvador / S. Calderón 4 / 1921 / "Used as diuretic" / "'Chinchin,' 'Cohetillo,' 'Espadilla'"

1734   *Crotalaria* indet. Mexico / W. P. Hewitt 125 / 1946 / "Cooked, used for rheumatism; dried, used as a shin wash" / "'Oreja del raton'"

1735   *C. nitens* Ecuador / Y. Mexia 8469 / 1936 / "'Barbaquilla'"

1736   *C. retusa* Brazil / R. Froes 1761 / 1932 / "...juice used for ink and a blue dye. Candido Mendez" / "'Anil'"

1737   *C. mitchellii* Australia: Queensland / C. T. White 11800 / 1941 / "...plant freely eaten by sheep"

1738   *Lupinus gentryanus* Mexico / H. S. Gentry 7183 / 1945 / "Plant decocted to delouse livestock"

1739   *L. grandis* Mexico / Y. Mexia 1719 / 1927 / "'Maiz negro'"

1740   *Spartium junceum* Uruguay / W. A. Archer 4969 / 1937 / "Used as medicinal remedy" / "'Cola de caballo'"

1741   *Indigofera suffruticosa* El Salvador / P. C. Standley 19278 / 1921–22 / "Decoction given to purify the blood"

1742   *I. lespedezioides* Bolivia / J. Steinbach 6305 / 1924 / "Se dice que es venenoso para el ganado" / "'Romerillo'"

1743   *Psoralea lutea* Chile / Y. Mexia 7855 / 1936 / "Used in infusion for stomach disorders of children" / "'Culen'"

     Chile / Worth & Morrison 16493 / 1938 / "In Valparaiso a delicious carbonated beverage is made from boiled leaves of this plant" / "'Aloja de culen,' 'Culen'"

1744   *P. pubescens* Peru / R. D. Metcalf 30259 / 1942 / "Indians used this for kidney troubles"

1745   *Eysenhardtia texana* Mexico / R. L. Dressler 2338 / 1957 / "...said to be medicinal for chickens" / "'Palo dulce'"

1746   *Dalea nutans* Mexico / Y. Mexia 1550 / 1927 / "'Escoba amarga'"

1747   *D. polygonoides* Mexico / H. S. Gentry 1951 / 1935 / "...eaten by cattle"

1748   *Tephrosia multifolia* El Salvador / P. C. Standley 19113 / 1921–22 / "Used to kill fish" / " 'Barbasco' "

1749   *Millettia lasiopetala* China / C. I. Lei 87 / 1932 / "The crushed root in water may kill the fish. Fisherman catches fish this way" / "Lo Tang"

1750   *Wisteria villosa* China / Cheo & Yen 219 / 1936 / " . . . fl. edible"

1751   *Gliricidia sepium* El Salvador / P. C. Standley 19263 / 1921–22 / "Lvs. used to kill rats & as poultice to bring boils to head. Fls. cooked & eaten" / " 'Madre de cacao' "

1752   *Willardia mexicana* Mexico / H. S. Gentry 7134 / 1945 / "The roots are used for poisoning fish. The nectar is reported to poison people when employed by bees in making honey" / " 'Nesco' "

1753   *Coursetia glandulosa* Mexico / W. P. Hewitt 27 / 1945 / "Economic uses: gum exuded by broken branches sold as 'Goma de Sonora'—eaten with chile and said to be good for stomach"

1754   *Cracca heterantha* Argentina / Meyer 31533 / 1940 / "Medicinal"

1755   *Sesbania grandiflora* Philippine Is. / A. L. Zwickey 35 / 1938 / "Flowers edible" / " 'Todi' (Lan.); 'Gaway-gaway' (Bis.)"

1756   *Astragalus* indet. Kashmir / R. R. Stewart 20372 / 1940 / "said to be poisonous"

1757   *Taverniera spartea* Saudi Arabia / H. Dickson 82 / '61 / "Grazed by camels"

1758   *Nissolia fruticosa* El Salvador / P. C. Standley 19123 / 1921–22 / "Remedy for Tamagaz bites" / " 'Hierba de Tamagaz' "

1759   *Chaetocalyx latisiliqua* Ecuador / Y. Mexia 8463 / 1936 / "Leaf used bruised for skin eruptions" / " 'Chupa-chupa' "

1760   *Ormocarpum cochinensis* New Britain / Waterhouse 414 / 1935 / "Leaf eaten as spinach" / "Kalawa"

1761   *O. glabrum* Caroline Is. / C. C. Y. Wong 453 / 1948 / "The plant is used in the medicine for sore throat (fely ligirä) or sores in the mouth, the leaves are chewed and the saliva is swallowed but in the case of sores in the mouth the saliva is spit out" / "N.v. Kenjic"

1762   *Aeschynomene amorphoides* Mexico / Y. Mexia 660 / 1926 / " 'Yerba del pajarito' "

1763   *A. sensitiva* W.I.: St. Lucia / G. R. Proctor 18150 / 1958 / "Shaved dry wood said to dissolve in water and lower the temperature" / " 'Manioc Chapelle' "

1764   *Zornia reticulata* El Salvador / P. C. Standley 20894 / 1922 / "Remedy for dysentery" / " 'Zornia, Barba de burro' "
        Mexico / Y. Mexia 758 / 1926 / " 'Yerba de la Vivora' "

1765 **Z. gibbosa** New Guinea / M. S. Clemens 10620 / 1939 / "Used with coconut grease for sorcery"

1766 **Desmodium barbatum** Brit. Honduras / P. H. Gentle 3785 / 1941 / "'wild senna'"

1767 **D. distortum** Guatemala / F. A. McClure 21668 / 1948 / "'Native name: Engorda caballo'"

1768 **D. nicaraguense** Honduras / P. C. Standley 13078 / 1948 / "Planted for forage" / "'Engorda-caballo'"

1769 **D. triflorum** Brit. Honduras / W. A. Schipp 702 / 1931 / "...all kinds of stock eat this plant greedily"

Haiti / R. J. Seibert 1767 / 1942 / "Said to be boiled as a tea and taken for rheumatism..." / "'Marlomin'"

Colombia / V. A. Plata 1 / 1939 / "Fodder plant of savanas" / "'Angelica'"

Burma / Herb. For. Res. Inst. 15055 / 32 / "Leaves eaten as a vegetable"

1770 **D. dunnii** China / S. K. Lau 222 / 1932 / "used as beans"

1771 **D. racemosum** Japan / E. Elliott 63 / '46 / "Food use: seeds pounded ...used to make ball, steamed and eaten" / "Nusubitonasi"

1772 **D. styracifolium** China / W. T. Tsang 499 / 1928 / "...medicine" / "Tung Tsiu Se T'so"

1773 **D. triquetrum** China / W. T. Tsang 25727 / 1935 / "...used as medicine" / "Pak Lo Sit"

Burma / F. G. Dickason 8134 / 1938 / "use to kill the warm" / "N.v. 'laught thay'"

N. Sumatra / J. A. Lörzing 13087 / 1928 / "Als heilmittel; siehe Lörzing 179"

1774 **Desmodium** indet. Siam / native collector Royal For. Dept. 3819 / 1948 / "Medicinal root used as neurotic" / "Ya-dab-kamlang-phra"

1775 **D. umbellatum** Solomon Is. / S. F. Kajewski 2452 / 1931 / "The young tender tips of this plant are chewed with betel nut and then placed in the mouth of a sick baby" / "Par-po"

Fiji Is. / O. Degener 15095 / 1941 / "If from drinking too much yanjona Fijian gets scaly skin, eats the leaves" / "'Sauthava' (Serua)"

1776 **D. laxiflorum** Philippine Is. / C. O. Frake 349 / 1957 / "Roots boiled and drunk for puerperium" / "direket Sub."

1777 **D. velutinum** Sierra Leone / G. F. S. Elliot 3985 / no year / "Shrub said to be poisonous"

1778 **Lespedeza cuneata** Japan / E. Elliott 45 / '46 / "Young plant boiled, removing bitterness by washing well and cooked to eat" / "Medohagi"

1779 *Dalbergia pinnata* N. Borneo / Tandom B.N.B. For. Dept. 4230 / 1933 / "...leaves can be used as medicine" / "Rumot (Dusun)"

1780 *D. mimosella* Philippine Is. / G. E. Edaño 2636 / 1950 / "Pound the bark and leaves and apply the juice on the wound" / "Nipotnipot Vis."

Philippine Is. / M. D. Sulit 2629 / 1948 / "Leaves crushed and applied to abdomen to cure dysentery"

1781 *Dalbergia* indet. Philippine Is. / C. O. Frake 836 / 1958 / "Bark applied to internal pain" / "Ganib Sub."

1782 *Dalbergia* indet. Philippine Is. / Sulit & Conklin 5147 / 1953 / "Decoction of root taken in, cures painful throat" / "Sampaga Mang."

1783 *Machaerium angustifolium* Bolivia / I. Steinbach 7525 / 1926 / "Se emplea en la fabricacion de jabon comun" / "Tusèque"

1784 *M. nigrum* Brazil / Y. Mexia 4446 / 1930 / "'Canella de Viado'"

1785 *Pterocarpus marsupium* India: Bombay Pres. / J. Fernandes 34 / 1949 / "Leaves are used as cattle fodder" / "Bibla"

1786 *P. indicus* New Guinea: Papua / L. J. Brass 6389 / 1936 / "...inner bark secretes red kino"

Philippine Is. / C. O. Frake 854 / 1958 / "Sap applied for thrush" / "Nala Sub."

Caroline Is. / C. C. Y. Wong 419 / 1948 / "The leaves are used in the medicine for ruptured vagina (fely ni mokur); the leaves are pounded to a very fine particle so as not to hurt the lining of the vagina..." / "larc"

1787 *P. esculentus* Guinée Francaise / M. Pobeguin 1263 / 1909 / "...fruits comestible. Racines et feuilles médicaments"

1788 *P. blancoi* Philippine Is. / E. Fénix 43 / 1938 / "Medicinal"

1789 *Platymiscium pinnatum* Venezuela / W. A. Archer 3213 / 1935 / "Stems smell strongly of rotenone" / "'Peraco'"

1790 *Platymiscium* indet. Brazil / Y. Mexia 5387 / 1930 / "'Canella cascudo'"

1791 *Lonchocarpus benthamianus* W.I.: Antigua / J. S. Beard 348 / 1944 / "...roots said to yield a fish poison" / "'Lady dogwood'"

1792 *L. caribaeus* W.I.: St. Lucia / G. R. Proctor 18047 / 1958 / "'Savonette 'Ti Feuille'"

1793 *L. chrysophyllus* Brit. Guiana / A. C. Smith 2823 / 1937 / "Stems and roots said to provide a very effective fish poison" / "'Ishel' (Wapisiana); 'Omaua' (Waiwai)"

1794 *L. martynii* Brit. Guiana / A. C. Smith 2834 / 1937 / "...roots sometimes used as fish poison but said to be less effective by Waiwais than no. 2823"

1795   *L. urucu* Colombia / Schultes & Cabrera 17243 / 1952 / "Cult. Fish poison" / "Barasana = ĕ-yoó; Nakuna = koo-na"

1796   *Pongamia pinnata* Philippine Is. / C. Frake 478 / 1958 / "Bark applied for splenomegaly" / "Bebelek ulangan Sub."

Philippine Is. / C. O. Frake 613 / 1958 / "Leaves applied to internal pains" / "bintulung Sub."

Fiji Is. / O. Degener 14970 / 1941 / "Fijians . . . crushed the leaves and made tea from them as a medicine" / "'naunau' (Serua)"

1797   *Muellera frutescens* Brit. Honduras / P. H. Gentle 3394 / 1940 / "'Madre cacao macho'"

1798   *Derris malaccensis* Malaya / G. anak Umbai for A. H. Millard K.L. 1613 / 59 / "Seeds, root and stem as fish poison" / "Tuba Tapa (Temuan)"

1799   *D. cebuensis* Philippine Is. / C. O. Frake 760 / 1958 / "Bark applied for internal pains" / "mantilaka Sub."

Philippine Is. / C. O. Frake 471 / 1958 / "Whole thing a fish poison"

Philippine Is. / H. C. Conklin 375 / 1953 / "stomach pawa? umum, bahi" / "Common name: Malagsing Dialect Káyu"

1800   *D. elliptica* Philippine Is. / C. O. Frake 577 / 1958 / "1) Stem applied for impetigo 2) Leaves applied for oxyuriasis" / "Tubalaw Sub."

1801   *Derris* indet. Caroline Is. / C. C. Y. Wong 457 / 1948 / "The roots . . . stupefy fish in three minutes. . . . The poison is used in killing human too by taking the bile of a large lizard (galuf), man's bile and another called fasuw, the effect of the poison causes the person to spit out blood after many moons" / "N.v. yup; up; yub"

1802   *Piscidia piscipula* U.S.: Fla. / A. A. Eaton s.n. / 1903 / "Dogwood; . . . bark used to poison fish"

1803   *Andira anthelmintica* Brazil / F. C. Hoehne 729 / '17 / "'Páo de Morcego'"

1804   *A. retusa* Brazil / R. Froes 1757 / 1932 / "Resin used as medicine" / "Angelim de Coco"

1805   *Dipteryx panamensis* Panama / P. H. Allen 4608 / 1947 / "'Almendro'"

1806   *D. alata* Bolivia / J. Steinbach 6691 / 1924 / "La pulpa de la fruta se come y de semilla se saca un aceite" / "Almendro"

1807   *D. tetraphylla* Brazil / R. Froes 1865 / 1932 / "Kernels used for oil and for heart illnesses" / "'Cumaru'"

1808   *Inocarpus edulis* Austral Is. / A. M. Stokes 72 / 1921 / "Uses: . . . seed . . . for food; boiled in coconut milk as medicine for uterine hemorrhage after childbirth" / "'mape'"

1809 *Cologania angustifolia* var. *stricta* Mexico / W. P. Hewitt 126 / 1946 / ". . . roots said to be purgative" / "Perrito"

1810 *Erythrina rubinervia* El Salvador / P. C. Standley 19396 / 1921–22 / "Fls. . . . eaten. . . . Cause sleep if much eaten" / " 'Pito' "

1811 *E. lanata* Mexico / Leavenworth & Hoogstral 1399 / 1941 / " . . . used as a source of poison by Indians" / " 'Tolorin' "
Mexico / Y. Mexia 1889 / 1927 / " 'Colorin' "

1812 *E. stricta* Burma / P. Khant 91 / 48 / "Leaves are eaten by men" / "Kathit"

1813 *E. variegata* var. *orientalis* Solomon Is. / S. F. Kajewski 2393 / 1930 / "The sap of the tree is used for making a cough medicine" / "Rara-rumbe"

1814 *E. subumbrans* Philippine Is. / C. O. Frake 611 / 1958 / "Bark boiled and drunk for splenomegaly" / "gerap Sub."

1815 *Strongylodon lucidus* Solomon Is. / S. F. Kajewski 2493 / 31 / "The leaves are heated then rubbed on boils" / "Low"

1816 *Butea frondosa* W. I.: Dominica / W. H. Hodge 2513 / 1940 / "source of 'Bengal Kino' "

1817 *Mucuna huberi* Brazil / B. A. Krukoff 1663 / 1931 / " . . . juice used for drinking, as a substitute for water"

1818 *M. cochinchinensis* China / F. A. McClure 242 / 1921 / " . . . fruit may be eaten" / "Kau Tsau Tau"

1819 *Mucuna* indet. China / H. T. Tsai 52797 / 1933 / "petal used as vegetable by natives" / "Yung-jen"

1820 *M. monosperma* India: Bombay / J. Fernandes 1106 bis / 1950 / "Used medicinally with other ingredients to relieve acute spasms" / " 'Son-Garpi' (Marathi)"

1821 *M. deeringiana* Philippine Is. / C. O. Frake 394 / 1957 / "edible pods" / "Alipantos Sub."

1822 *Spatholobus roxburghii* Burma / P. Khant 592 / 48 / "Leaves used for medicinal purposes" / "Dah-ma-nge"

1823 *Calopogonium* indet. Mexico / Y. Mexia 1208 / 1926 / "Good stock feed"

1824 *C. mucunoides* Caroline Is. / C. C. Y. Wong 340 / 1948 / "The plant is used in making medicine for general debility; (felynigof) the young leaves and shoots are pounded . . . . squeezed into a betel nut bowl, a young coconut water is added and the mixture is drunk" / "N.v. laga-thulip nuop"

1825 *Galactia glaucescens* Bolivia / J. Steinbach 6480 / 1924 / "Remedio casera contra . . . " / "n.v. Tamacu"

1826 *Galactia* indet. Bolivia / J. Steinbach 5630 / 1921 / "Remedio contra la toz" / "n.v. Orosu"

1827 *Pueraria gladiata* China / H. Fung 21135 / 1937 / "Pod is edible" / "Tao tou"

1828 *P. peduncularis* China / A. Henry 12483A / no year / "'root used to kill fish'"

1829 *Canavalia brasiliensis* Colombia / Killip & Smith 14205 / 1926 / "'Arrozcocoa'"

1830 *C. virosa* India: Bombay / J. Fernandes 10 / 1949 / "The seeds are eaten locally" / "Abai or Abi"

1831 *C. cathartica* Solomon Is. / S. F. Kajewski 2297 / 1930 / "The natives are highly immoral, if they want a woman they blow along this vine. Their offer is always universally accepted" / "Nambe"

1832 *C. ensiformis* Philippine Is. / C. Frake 399 / 1958 / "Bark boiled and drunk for sudden sickness" / "Kalicalis Sub."

1833 *Cajanus cajan* Brazil / Y. Mexia 5318 / 1930 / "Medicinal plant, an infusion of leaves used to bathe swellings and sores" / "'Faijão Andú'"

1834 *Rhynchosia precatoria* Mexico / G. B. Hinton et al. 14036 / '38 / "Cough medicine"

El Salvador / P. C. Standley 22246 / 1922 / "Lvs. used to rub dirt out of clothes" / "'Ojo de cangrejo'"

Mexico / P. C. Standley 1168 / 1934 / "Medic. seeds ground and mixed with grease making an ointment which is applied to various afflictions, rubbed on forehead for headache" / "Chanate-pusi (from Mayo: chanate, bird & pusi, eye)"

1835 *R. acuminatifolia* Japan / E. Elliott 134 / '46 / "The pod as well as the whole plant is pectorant" / "Tokirimamo"

1836 *Eriosema grandiflorum* Mexico / H. S. Gentry 5301 / 1939 / "... roots decocted ... and used as insecticide against fleas on dogs" / "Yerba de piojos"

1837 *Moghania strobilifera* Philippine Is. / C. Frake 368 / 1957 / "Magical —ward off cockroaches" / "Gipisipis Sub."

1838 *Phaseolus caracalla* Mexico / H. S. Gentry 2404 / 1936 / "Warihios employ the large root as a riser in making their fermented drink, 'batari'" / "Nawo, W."

1839 *P. ritensis* Mexico / H. S. Gentry 2523 / 1936 / "Root reported used as a fermenting agent in making the W. liquor 'batari'"

1840 *P. prostratus* Bolivia / J. Steinbach 6353 / 1924 / "La raiz dice el indigena se emplea ventajosamente contra picada de viboras" / "n.v. Contra-bivora"

1841 *P. aureus* China / F. A. McClure 20595 / 1937 / "Seeds an important food; Mungo bean . . . listed in earliest Chinese Materia Medica and is classed as one of the Pu-pin . . . that is, foods considered to be efficacious in promoting or restoring health" / "Lu-tou"

1842 *P. adenanthus* Caroline Is. / C. C. Y. Wong 302 / 1947 / "The women express the leaves . . . and drink the sap for labor pains, säfein uputiw (medicine for birth—to make birth easy)" / "onuuw"

1843 *Vigna marina* Solomon Is. / S. F. Kajewski 2229 / 1930 / "The natives have a superstition that if the leaves are rubbed on their fishing lines they will have a good catch" / "Pumborita"

1844 *Vigna* indet. Philippine Is. / A. L. Zwickey 676 / 1938 / "'Kodalis a bunbun' (Lan.); lvs. as cooked vegetable after childbirth"

1845 *Leguminosa* indet. Colombia / R. E. Schultes 3531 / 1942 / "Nombre kofán 'a-na-fa-sie' 'dormidero'"

1846 *Leguminosa* indet. Brazil / B. A. Krukoff 1774 / 1932 / "Fruit and wood yielding a poisonous substance" / "'Cutiuba'"

1847 *Leguminosa* indet. Brazil / R. Froes 12149 / 1941 / "Used by Indians in treating leprosy" / "'Jare'"

1848 *Leguminosa* indet. Brazil / R. Froes 12562/256 / 1942 / "Seeds eaten by Indians, rich in oil" / "Uucau"

1849 *Leguminosa* indet. Annam / B. Balansa 2158 / 1887 / ". . . les femmes annamites emploient la décoction des fruits comme insecticide" / "Annamite: Bau Kit"

GERANIACEAE

1850 *Geranium seemannii* Mexico / L. R. Stanford et al. 138 / 1941 / "Heavily grazed by goats"

1851 *G. pilosum* Australia: Queensland / L. S. Smith 552 / 1938 / "very good fodder"

1852 *Erodium cicutarium* U.S.: N.M. / A. L. Hershey 3123 / 1944 / "Prized as a range forage plant"

Mexico / Dr. Gregg s.n. / 1837 / "Alfilerillo. Decoction used for gargle etc. . . . "

Peru / R. S. Shepard 1 / 1919 / "Local name 'Aguja aguja' (Spanish) or 'Sanu sanu' (comb—Indian)"

1853 *E. deserti* Arabia / H. Dickson 62 / 61 / "Economic use: Eaten by all animals. N. vernac. Rugum"

1854 *Viviania revoluta* Chile / J. L. Morrison 16910 / 1938 / "'Oreganillo'"

1855 *V. marifolia* Chile / J. L. Morrison 16860 / 1938 / "'Oreganillo'; 'Té de burro'"

Chile / I. M. Johnston 5903 / 1926 / "Té de burro colorado"

OXALIDACEAE

1856   *Oxalis corniculata* Mexico / H. S. Gentry 2551 / 1936 / "Leaves eaten by native"

1857   *Biophytum dendroides* Colombia / Schultes & Villarreal 5125 / 1942 / "'Dormidero' nom. vulg."

1858   *Averrhoa bilimbi* Philippine Is. / C. O. Frake 550 / 1958 / "Leaves applied for swellings" / "Biba? Sub."

1859   *Sarcotheca glauca* Sarawak / J. A. R. Anderson 9798 / 60 / "Fruit . . . edible but very sour"

HUMIRIACEAE

1860   *Humiria balsamifera* Brit. Guiana / S. S. & C. L. Tillett 45600 / 1960 / "fruit . . . edible, slightly sweet, astringent to taste"

      Brit. Guiana / A. C. Smith 2423 / 1937 / "edible fruit. 'Umir' (Wapisiana)"

1861   *H. cuspidatum* Venezuela / Wurdack & Adderley 43353 / 1959 / "fruit said to be edible. 'Guaco'"

ERYTHROXYLACEAE

1862   *Erythroxylon carthagenense* W.I.: Trinidad / W. E. Broadway 9037 / 1932 / "'Wild cocaine'"

1863   *E. ovatum* W.I.: Trinidad / R. A. Howard 10400 / 1950 / "fruit black, tasty"

1864   *E. squamatum* W.I.: Dominica / G. A. Ramage s.n. / 1889 / "'ti feuille,' local name"

      W.I.: Martinique / M. Hahn 633 / 1869 / "Vulg. Bois Café"

1865   *E. gracilipes* Colombia / G. Klug 1961 / 1931 / "'Sacha maugua,' fruit edible"

1866   *Erythroxylon* indet. Bolivia / J. Steinbach 6455 / 1924 / "n.v. cascarilla, cascara astringente"

ZYGOPHYLLACEAE

1867   *Fagonia brugieri* Kuwait / H. Dickson 70 / 61 / "Lizards are said to eat it. . . . Vern. Name Al Jemba"

1868   *Guiacum officinale* Bahama Is. / O. Degener 18799 / 1946 / "'Lignum Vitae' gum is put in a little gin to dissolve, then a little of this tincture is mixed with water & drunk as a pain killer"

1869   *G. coulteri* Mexico / Y. Mexia 1926 / 1927 / "Remedy for colds etc. Flower for pneumonia, cough. 'Guayacan'"

1870   *Larrea tridentata* Mexico / Dr. Gregg s.n. / 46 / "leaves resinous—considered diuretic, and, in decoction said to be superior remedy in disuria etc." / "Gobemadora (called Guámas in the north)"

Mexico / A. Miranda 111 / 1948 / "util para echar humas y desinfectar las cazas" / "'Huamio'"

1871 *L. cuneifolia* Argentina / W. J. Eyerdam et al. 23562 / 1938 / "Heavily grazed by sheep & goats"

1872 *L. divaricata* Argentina / A. Krapovickas 822 / 1945 / "'jarilla'"

Argentina / J. West 6195 / 1936 / "creosote scented foliage. Used medicinally"

1873 *Tribulus terrestris* Australia: Queensland / C. T. White 11787 / 41 / "Said to be very good sheep feed, the animals being very fond of crunching up the spiny burrs"

1874 *Kallstroemia maxima* W.I.: Martinique / H. & M. Stehlé 7135 / 1946 / "'Pompier courant'"

1875 *K. brachystylis* Mexico / A. Lopez 32 / 1948 / "buena para el estomago. Golandrina"

1876 *K. rosei* Mexico / Leavenworth & Hoogstraal 1423 / 1941 / "'Yerba de la paloma'"

RUTACEAE

1877 *Zanthoxylum fagara* U.S.: Fla. / A. Traverse 628 / 58 / "Fruit spicy smelling when crushed"

El Salvador / P. C. Standley 20329 / 1922 / "'Salitrero'"

Mexico / G. B. Hinton 4559 / 33 / "'Pinzanillo'"

1878 *Z. anodynum* Honduras / A. Molina R. 3263 / 1950 / "'Duerme muelos'"

Honduras / A. Molina R. 4048 / 1951 / "Quiebra Muelos"

1879 *Z. kellermanii* Honduras / P. C. Standley 53115 / 1927–28 / "Crushed leaves have lemon odor" / "'Cedre espino'"

1880 *Z. simulans* var. *imperforatum* China / E. E. Maire 151 / before 1924 / "dit: poivre de Chine—fruit rouge, employés comme epise"

1881 *Z. acanthopodium* Burma / F. G. Dickason 7563 / 1938 / "'Mĕka' (Bur.)"

India: Assam / L. F. Ruse 73 / 1923 / "Economic use: Extract of fruit mixed with Dieng jaiur & Dieng jalew is drunk with water in cases of chlorea. The extract from the leaves of these three species is also mixed & used as a medicine" / "Vernac.: Dieng-shiah jaiur-Cum"

1882 *Z. alatum* India: Punjab / W. Koelz 8315 / 1936 / "Wood used by natives as tooth cleanser. Bark also used to clear throat in smallpox. Widely used in India & its properties are worth investigating"

India: Punjab / W. Koelz 10274 / 1936 / "bark very pungent & used to clean teeth" / "'Timbar' or 'Tirmira'"

1883   *Z. khasianum* India: Assam / L. F. Ruse 62 / 1923 / "The leaves are used on people bitten by mad dogs. Vern.: Mishiah-soh-sat"

1884   *Z. ailanthoides* Bonin Is. / E. H. Wilson s.n. / 1917 / "Poisonwood tree. . . . Vernacular name given by Rev. Gonzalez–Chichijima"

1885   *Fagara martinicensis* Haiti / W. Buch 1468 / 1917 / "Einheim. Name: Bois beni ou bini ou pini"

1886   *Euodia bodinieri* China / Tak & Chow 2984 / 1926 / "Common name (Cantonese) Chau To Kwo"

China / Herb. Lingnan Univ. 137.3993 / 1924 / "Name reported: Ch'a lat (Tea pungent)"

China / T. W. Tak (Herb. Canton Christian Coll. 9933) / 1922 / "Economic use: Medicine" / "(Cantonese) Cha Lat Shue"

1887   *E. lepta* China / W. T. Tsang 28824 / 1938 / "Sam Chai Fu Muk"

China / F. A. McClure 20096 / 1932 / "Leaves used by the Loi to brew a medicinal tea"

1888   *E. officinalis* China / W. T. Tsang 23111 / 1933 / "used as medicine" / "Cha Lat Muk"

1889   *E. lunura-ankende* Burma / F. G. Dickason 9107 / 1938 / "Leaves used medicinally"

1890   *E. elleryana* var. *tetragona* Solomon Is. / S. F. Kajewski 2323 / 30 / "Common name: Mar-mor. The sap of the tree is used by the natives to put on their sores"

1891   *E. hortensis* Solomon Is. / L. J. Brass 3198 / 32 / "Common name: N'glagobi. Very strongly perfumed"

Solomon Is. / S. F. Kajewski 2378 / 30 / "The leaves are crushed & applied by the natives to boils. The natives are also fond of the scent of the leaves. The essential oil may be valuable. Common name: For-orka"

1892   *E. peekelii* Solomon Is. / S. F. Kajewski 1682 / 30 / "Used by the natives to counteract poison, but more on account of the smell of the leaves than any other reason. It might stand investigation"

1893   *E. radlkoferiana* Solomon Is. / S. F. Kajewski 2361 / 30 / "If a native has a boil, the leaves are heated and applied to it. Common name: Bella-ha-ceen"

Solomon Is. / S. F. Kajewski 2516 / 31 / "The leaves and bark are boiled in hot water and applied to diseases of the skin. Common name: Mumborlo"

1894   *E. solomonensis* Solomon Is. / Walker & White 114 / 1945 / "leaves crushed and rubbed on body to relieve pain in headache and rheumatism. Native name 'Balu'"

Solomon Is. / L. J. Brass 3209 / 32 / "Com. name: Boboi"

Solomon Is. / J. H. L. Waterhouse 122 / 1932 / "Com. name: Hongoponipo"

1895 *Euodia* indet. New Guinea / Hoogland & Schodde 7535 / 1960 / "Local name: 'Kump' (Enga)"

1895A *Euodia* indet. Borneo / F. H. Endert 2350 / 1925 / "Nat. name: 'kajoe boekan'"

1895B *Euodia* indet. New Guinea / A. Kostermans 106 / 48 / "Nat. name: 'Breri' or 'Brorrie'"

1895C *Euodia* indet. New Guinea / L. J. Brass 29270 / 1959 / "attractive to lories"

1895D *Euodia* indet. New Guinea / J. C. Saunders 42 / 1954 / "Local name: Isilau (Onjob), Ariride (Minufia)"

1895E *Euodia* indet. Philippine Is. / L. E. Ebalo 381 / 1940 / "Name in Tag.: copang"

1895F *Euodia* indet. Philippine Is. / L. Navarro 26 / 1949 / "Common name: Matang-aran (Tag.)"

1895G *Euodia* indet. Siam / Royal For. Dept. 2854 / 49 / "Edible shoot"

1896 *E. confusa* Philippine Is. / C. O. Frake 764 / 1958 / "Bark applied for splenomegaly. glumawiy Dial. Sub."

1897 *E. kajewskii* New Hebrides Is. / S. F. Kajewski 300 / 28 / "Leaves heated and applied to abscesses or sores" / "neet-nung"

Solomon Is. / S. F. Kajewski 2406 / 31 / "When a man is sick with fever the leaves of this tree are rubbed all over the body. This tree is planted by the natives for this reason" / "Bombolu"

1898 *E. cuculata* var. *robustior* Fiji Is. / A. C. Smith 1041 / 1934 / "'Tokatolu'"

Fiji Is. / O. Degener 15047 / 1941 / "Fijians chewed fresh leaf for thrush or preferably squashed wilted leaves in water and drank liquid. They also drank liquid as remedy for colds and headache"

1899 *E. hortensis* f. *hortensis* Fiji Is. / O. Degener 14131 / 1941 / "Fijians crush leaves in water and bathe patients in it. Flowers boiled in coconut oil to scent it. Nat. n. Lauthi, Uthi"

1900 *E. hortensis* f. *simplicifolia* POLYNESIA: Tongapatu I. / T. G. Yuncker 15231 / 1953 / "leaves used in medicine. Uhi (Tongan)"

POLYNESIA: Lifuka / T. G. Yuncker 15713 / 1953 / "leaves and bark used in medicine. Uhi (Tongan)"

1901 *E. samoensis* Samoa / D. W. Garber 617 / 1921 / "Lupe food" / "'soopini'"

1902 *Melicope triphylla* New Guinea / L. J. Brass 1408 / 1926 / "Fruit peppermint scented. Com. name: Amaga"

1903   *M. monophylla* Philippine Is. / C. O. Frake 704 / 1958 / "Leaves applied to ulcers. Com. name: bayangan. Dial. Sub."

Philippine Is. / C. O. Frake 839 / 1958 / "Leaves burnt and applied to pimples. Com. name: Alangan (Sub.)"

Philippine Is. / C. O. Frake 746 / 1958 / "Young leaves applied for rigid abdomen. Com. name: Telianusaba"

Philippine Is. / C. O. Frake 674 / 1958 / "Roots pounded and drunk for puerpurium. Com. n.: Natu? (Sub.)"

1904   *Megastigma skinneri* Honduras / P. C. Standley 15398 / 1948 / "Crushed leaves have odor of lime"

1905   *Medicosma cunninghamii* Australia: Queensland / Goy & Smith 666 / 1939 / "leaves strongly mango scented"

Australia: Queensland / C. T. White 1939 / 1923 / "(local name 'Bone Wood')"

1906   *Ruta tuberculatum* Kuwait / H. Dickson 102 / 62 / "Used to cure scorpion stings. Nom. v. 'Al Messaicha'"

1907   *Thamnosma montanum* U.S.: Ariz. / E. Palmer 35 / 1867 / "Strongly scented of rue"

1908   *Boronia rivularis* Australia: Queensland / S. T. Blake 14389 / 1941 / "leaves . . . with sarsaparilla scent, flowers . . . with same scent"

1909   *Zieria minutiflora* Australia: Queensland / C. G. Briggs N.S.W. 56136 / 1961 / "Lemon scented foliage"

1910   *Phebalium woombye* Australia: Queensland / C. E. Hubbard 2577 / 1930 / "Leaves aromatic (rather like cinnamon)"

1911   *Rauia resinosa* Brazil / R. Froes 11966 / 1940 / "'Quina Branca'"

Brazil / R. Froes 11667 / 1937 / "Bark bitter used for curing fever. 'Quina da Mata'"

1912   *Erythrochiton brasiliensis* Bolivia / I. Steinbach 7279 / 1925 / "Nom. v. Chocolatillo"

1913   *Flindersia maculata* Australia: New S. Wales / S. Helms 687 / 1922 / "'Leopard tree'"

1914   *Balfourodendron riedelianum* Brazil / P. Dusén 16782 / 1915 / "'Guamixinga'"

Brazil / E. N. de Andrade 14548 / 1924 / "'Pau Marfim'"

1915   *Casimiroa edulis* El Salvador / P. C. Standley 19548 / 1921–22 / "Fr. unhealthy if much eaten. 'Matasano'"

1916   *C. sapota* Mexico / G. B. Hinton 2820 / 32 / "Fruit much eaten. Zopote blanco"

1917   *C. tetrameria* Honduras / A. Molina R. 5383 / 1955 / "'Matasano'"

1918   *Toddalia asiatica* China / W. T. Tsang 28641 / 1938 / "Fa Chiu Lak"

China / S. K. Lau 379 / 1932 / "edible. Gieng Diong (Lai)"

China / W. T. Tsang 23272 / 1933 / "fruit edible. Fa Tait Lat"

Siam / Kasin 370 / 1946 / "Young leaves edible. ta na re (Karieng)"

Sumatra / R. S. Boeea 10607 / 1936 / "'andor oette'"

Nyasaland / L. J. Brass 17769 / 1946 / "fruit orange, eaten by the natives. Nkandangfuku (Chinjanja)"

1919 *Acronychia pedunculata* China / R. C. Ching 7993 / 1928 / "Drupe aromatic, sweet in taste"

China / W. T. Tsang 359 / 1927 / "Tim Tong Shue"

China / W. T. Tsang 24099 / 1934 / "fr. edible. Ye Yau Kam Shue"

Malay Peninsula / F. Bidin 4193 / 1921 / "Nat. name: Je, Jaging"

Java / Monterie 50 / 1915 / "Nom. Ind.: ki ledja"

Borneo / Apostel B.N.B. For. Dept. 7679 / 1937 / "Nom. v.: paw (Dusan Rungus); marranggas (bajau Labuk)"

1920 *A. trifoliata* Dutch E.I.: Bali / Soegeng s 106 / 1958 / "Fruits somewhat acid like lemon. Tjĕrmen"

1921 *Hortia arborea* Brazil / R. Froes 12370/70 / 1943 / "Pau para tudo"

1922 *Hortia* indet. Brazil / R. Froes 12627 / 1942 / "Bark is used in treatment of malaria"

1924 *Amyris rekoi* Mexico / Y. Mexia 8847 / 1937 / "fruit eaten by birds. 'Comida de Chachalaca'"

1925 *Glycosmis pentaphylla* China / F. A. McClure 00519 / 1931 / "Tsau Pang Tsai (Cantonese)"

Philippine Is. / C. O. Frake 541 / 1958 / "Roots applied to chest pains. gulud (Sub.)"

Philippine Is. / C. O. Frake 650 / 1958 / "Roots applied to skin eruptions. bebegar (Sub.)"

Philippine Is. / H. C. Conklin 422 / 1953 / "Com. n. bakwit pūru (Dial. Arabwal)"

1926 *G. citrifolia* China / W. T. Tsang 21975 / 1933 / "fruit red, edible. Shan Kit Shue"

China / W. T. Tsang 156 / 1928 / "Tsau Peng Shue"

1927 *G. cochinchinensis* Hainan / C. I. Lei 429 / 1933 / "fruit edible. Shan Gut Shu"

1928 *G. craibii* var. *glabra* China / F. A. McClure 20139 / 1932 / "Fruits edible"

China / S. K. Lau 284 / 1932 / "Nang Ha kiu (Lai)"

1929 *Glycosmis* indet. China / S. K. Lau 308 / 1932 / "fruit pink, edible. Tim To"

1930 *Micromelum falcatum* China / S. K. Lau 7 / 1932 / "Kovahgoe (Lois)"
China / W. T. Tsang 24797 / 1934 / "Fruit edible. Ye Wong Pei Shue"
China / W. T. Tsang 66 / 1928 / "Medicine. Kai Lun Wong Shue"

1931 *M. inodorum* Philippine Is. / C. O. Frake 682 / 1958 / "Leaves applied for splenomegaly. Saipan. Dial. Sub."
Philippine Is. / C. O. Frake 665 / 1958 / "Bark applied to wounds. glubag. Dial. Sub."

1932 *M. minutum* Philippine Is. / M. D. Sulit 4938 / 1952 / "Com. n.: Buriñgut Malagli (Mang.)"
Fiji Is. / O. Degener 15395 / 1941 / "Leaf chewed for sore tongue. 'Ngingila' (Ra)"

1933 *Murraya paniculata* China / S. K. Lau 232 / 1932 / "Cook leaf to heal stomach-ache. Boe Sai Hau (Lai)"

1934 *M. crenulata* New Hebrides Is. / S. F. Kajewski 46 / 1928 / "Fruit eaten by natives"
New Hebrides Is. / S. F. Kajewski 789 / 29 / "Leaves have scent like Eucalyptus. Fr. . . . when ripe dotted with oil glands"

1935 *Clausena excavata* Hainan / H. Fung 20054 / 1932 / "Fruit edible"
Burma / F. G. Dickason 9192 / 1938 / "Good for stomach. Pyindan Thein"
N. Borneo / Puasa B.N.B. For. Dept. 3147 / 53 / "Leaves used for wounds, burned to frighten pudi birds" / "Imus (Idahan) Kialut (Kedayan) Pinisingan (Murat)"

1936 *C. dentata* India: Dindigul / R. D. Anstead 137 / 1926 / "Edible fruit like a white current"

1937 *C. platyphylla* Sumatra / R. S. Boeea 6935 / 39 / "Glands large caustic"

1938 *Clausena* indet. Solomon Is. / S. F. Kajewski 1915 / 30 / "The leaves & fruit are very aromatic and full of oil glands. This tree should be very valuable on account of the large & numerous oil bearing glands in the leaves & fruit"

1938A *Clausena* indet. New Guinea / L. J. Brass 28876 / 1956 / "Tree . . . grown for its aromatic leaves, which are worn in armbands, etc."

1938B *Clausena* indet. Philippine Is. / L. E. Ebalo 436 / 1940 / "'sibugay' (Tag.)"

1939 *Limonia crenulata* Siam / native Royal For. Dept. 2500 / 49 / "Bark used in perfumery. Loc. name Kra-chae"

1940 *Pleiospermium littorale* Annam / Poilane 4510 / 22 / "Les fruits sont comestible, les racines sont employées dans une sorte de mal (gale disentils) que les indigenes attrappent sous les pieds. Cay cam duong (Ant.)"

1941 *Triphasia trifolia* Philippine Is. / Sulit & Conklin 69S / 1953 / "Com. n.: Dayapdayáp (malagti)"

Philippine Is. / M. Adduru 36 / 1917 / "Fruit sweet, sour, edible. Limoncito"

1942 *Atalantia monophylla* Annam / Poilane 1454 / 1920 / "Mandarinier sauvage, il est Medicinal, employé contre les maladies des voies respirations, a cette effet prendre feuilles et rameaux les couper par morceau les greller comme du cafe, puis les faire bouillir et boire. Annte. Cây quit hôi"

1943 *Atalantia* indet. Indochina: Tonkin / W. T. Tsang 27319 / 1936 / "fr. green, edible"

1944 *Atalantia* indet. Fiji Is. / A. C. Smith 1293 / 1934 / "Molimoli"

1945 *A. simplicifolia* Annam / M. Poilane 1678 / 20 / " ... feuilles ... d'un bon effet elle sont employees pour combattre les maladies des voies respiratoire fruits comestible" / "Annte. Cay quit rung"

1946 *Fortunella hindsii* Hainan / F. C. How 72136 / 1935 / " ... fruit edible"

1947 *F. margarita* China / W. T. Tsang 24123 / 1934 / "Fruit edible. Shan Kam Kit Tsz"

1948 *Citrus decumana* Fiji Is. / E. H. Bryan 545 / 1924 / "Fr. rough-skinned lemon, eaten by natives. moli karokaro"

1949 *C. reticulata* Fiji Is. / O. Degener 15249 / 1941 / "fruit edible. Wichiwichi (Serua)"

SIMAROUBACEAE

1950 *Suriana maritima* Caroline Is. / C. C. Y. Wong 230 / 1947 / "The flowers are used as medicine for mwusocca (vomiting of blood) and feiseni (diarrhea) by pounding the flowers and mixing with coconut water & coconut milk, this mixture is given to adults and children" / "N.v.: won"

1951 *Samadera indica* New Britain / A. Floyd 6581 / 54 / "Seeds eaten by natives. La kilipa (W. Nak.) Aela (Pidgin)"

Solomon Is. / S. F. Kajewski 2459 / 31 / "This tree is largely used by the natives as a remedy for constipation. The bark is macerated & drunk with water" / "'Ungulu'"

Solomon Is. / L. J. Brass 2951 / 32 / "An infusion of the extremely bitter seeds is used by the natives as a febrifuge" / "'Saiesu'"

Solomon Is. / S. F. Kajewski 2247 / 30 / "The leaves are macerated and mixed with coconut oil and applied to the hair to kill lice" / "'Otsipu'"

1952 *Simarouba glauca* Honduras / P. C. Standley 52807 / 1927–28 / "Fruit edible" / "'Negrito'"

1953   *Simaba cedron* Brazil / Ducke 1435 / 1943 / "'Paratudo'"

Brazil / R. Froes 1867 / 1932 / "Bark used as a cure for malaria" / "'Pau de Gafauhoto'"

1954   *S. intermedia* Brazil / Froes 12651/17 / 1942 / "'Quina do mato'"

1955   *Eurycoma longifolia* N. Borneo / Nooridin B.N.B. For. Dept. 6263 / 1936 / "Roots boiled & decoction drunk to relieve pain in bones. Leaves boiled & decoction used for washing itches. Serirama (Brun.)"

1956   *Brucea javanica* China / H. V. Swa 5 / 1932 / "Leaves used to poison worms" / "'Yeung Kook Tsz Fa'"

China / S. K. Lau 534 / 1932 / "Used to heal cows from the worm disease. Shat Ngau Chung"

China / G. W. Groff 39 / 1919 / "Fruit green and poisonous. Fruit for poisoning fleas" / "'Á tám Shii'"

China / F. A. McClure 2373 / 1922 / "Said by the natives to be deadly poison. Used as a drug plant in treatment of sores on animals" / "'Lo Ah Tam'"

1957   *B. amarissima* Philippine Is. / C. O. Frake 663 / 1958 / "Leaves applied for splenomegaly and internal pains" / "'bakelan' (Sub.)"

1958   *Picrasma antillana* Leeward Is. / J. S. Beard 422 / 1944 / "Bark & wood containing bitter principle" / "'Bitter ash'"

1959   *P. javanica* New Guinea / P. van Royen 3520 / 1954 / "The bark is used as an antidotum after eating poisonous turtles" / "'annamoer' (Biak dialect)"

1960   *Ailanthus altissima* China / G. W. Groff 4051 / 19 / "Young leaves edible" / "'Chun Nga Muk'"

1961   *Soulamea amara* Solomon Is. / S. F. Kajewski 1556 / 30 / "Leaves are pounded up by the natives, afterwards being used as a hot fomentation"

Caroline Is. / C. Y. Wong 225 / 1947 / "The fruit is used as medicine for säfein sät (medicine against the bite of sea spirits, cönykken); the fruit is pounded and rubbed on the affected spot with the hands on legs or head" / "'märas'"

1962   *Amaroria soulameoides* Fiji Is. / O. Degener 15183 / 1951 / "One Fijian states fruit eaten; other, that it is poisonous" / "'Bau' (Serua)"

Fiji Is. / O. Degener 15056 / 1941 / "Fruit eaten by flying foxes"

1963   *Irvingia oliveri* Annam S. / Poilane P36 / 1919 / "fruit gros comme un oeuf de pigeon mangé par les cerfs" / "Annte: 'Cây Cây'"

1964   *Picramnia antidesma* Honduras / P. H. Allen 6378 / 1952 / "Bark bitter"

Honduras / J. B. Edwards 472 / 1932 / "The natives use tea made from the bark as a cure for malaria. Vern. n. Quinina"

Jamaica / W. Harris 12411 / 1916 / "Major Bitters"

1965　*P. locuples* Honduras / P. H. Allen et al. 6048 / 1951 / "Bark reported to be used to cure fevers" / "'Quina Roja'"

1966　*P. quaternaria* El Salvador / S. Calderón 1369 / 1922 / "Fr. is edible" / "'Aceitunito'"

1967　*P. pentandra* Bahama Is. / G. R. Proctor 9026 / 1954 / "'Snake stick'"

1968　*Alvaradoa amorphoides* var. *typica* Mexico / J. Bequaert 112 / 1929 / "Leaves used for a cure" / "Maya n. 'Bel-Sinik-che'"

1969　*Picrodendron macrocarpum* W.I.: Cayman Is. / G. R. Proctor 15248 / 1956 / "Fr. said to be edible" / "'Cherry'"

BURSERACEAE

1970　*Protium heptaphyllum* Brazil / R. Froes 1902 / 1932 / "Resin used as a cure for headaches" / "'Minesia'"

Brazil / F. C. Hoehne 657 / 1937 / "Resina mui aromatica e medicinal"

1971　*P. neglectum* var. *tenuifolium* Colombia / S. J. Record 12 / 30 / "Tea used for medicinal purposes" / "'Vera blanca'"

1972　*Protium* indet. Brazil / B. A. Krukoff 7666 / 1935 / "Used in Curare"

1973　*P. schlechteri* New Guinea / M. S. Clemens 10689 / 39 / "Fruit green, eaten by natives"

1974　*Tetragastris balsamifera* Haiti / E. L. Ekman 440 / 1917 / "'Bois cochon' hait."

1975　*T. altissima* Brazil / B. A. Krukoff 1563 / 1931 / "...fruit pulp edible" / "'Almesca'"

1976　*Canarium album* Hainan / S. K. Lau 1850 / 1933 / "Fruit edible" / "'Wu Lam'"

1977　*C. pimela* Hainan / F. A. McClure 767 / 1929 / "Fruits eaten" / "'Blaam'"

1978　*C. denticulatum* N. Borneo / Md. Puasa B.N.B. For. Dept. 1731 / 1932 / "Fruit edible when ripe" / "'Pininasan' (Ked.)"

1979　*C. odontophyllum* N. Borneo / Agama & Valera B.N.B. For. Dept. 9444 / 1938 / "The seeds are edible" / "'loambayau' (Mal.)"

1980　*C. hirsutum* var. *leeuwenii* Solomon Is. / S. F. Kajewski 1931 / 30 / "Nuts are eaten by the natives" / "'Kim'"

1981　*C. indicum* New Guinea / K. Mair NGF 1819 / 1945 / "Seed is a common native food" / "'Galep'"

1982　*C. oleosum* New Guinea / A. J. Hart 4560 / 52 / "Gum used in conjunction with oil (coconut) as a hair oil and skin lotion. Colourless exudation from sap possessing aromatic odour" / "'Manoi' (Tubusereua)"

1983   **C. sylvestre** New Guinea / P. v. Royen 3205 / 1954 / "seeds . . . edible" / " 'naasi' (Biak)"

1984   **C. solomonense** Solomon Is. / L. J. Brass 2847 / 1932 / "The seeds have a very pleasant almond-like flavour & are largely used by the natives for making 'puddings'" / " 'Gatoga'"

Solomon Is. / Walker & White B.S.I.P. 93 / 1945 / "Seed edible and highly prized as a food" / " 'Adoa' (M.)"

1985   **C. asperum** Philippine Is. / C. O. Frake 883 / 1958 / "Roots drunk during puerperium" / " 'Gempad' (Sub.)"

1986   **C. hirsutum** Philippine Is. / M. D. Sulit 3151 / 1949 / "According to Bukid. decoction of roots is good for stomach trouble" / " 'Kadabudabu' (Bukid.)"

1987   **C. harveyi** New Hebrides Is. / S. F. Kajewski 548 / 28 / "Nuts eaten by both whites and natives. Kernel very oily" / " 'Nungi-nuts'"

1988   **C. vitiense** Fiji Is. / A. C. Smith 996 / 1934 / "Fruit edible"

1989   **Dacryodes rostrata** N. Borneo / Matusop NBFD 7429 / 1937 / "Fruit edible, oily" / " 'Kamabyau' (Malay)"

1990   **Santiria laevigata** Sarawak / Anderson & Tahir KLNBSA 13101 / 61 / "Fruit edible" / " 'Siladah'"

1991   **Bursera fagaroides** Mexico / H. S. Gentry 2260 / 1936 / "Gum used for toothache" / " 'Torote prieta,' Mex."

1992   **B. grandifolia** Mexico / H. S. Gentry 4822 / 1939 / " . . . bark decocted for fevers" / " 'Palo mulato'"

1993   **B. penicillata** Mexico / H. S. Gentry 1585 / 1935 / "Tree with peeling bark. Medic. herbage for catarrh, resin for toothache and other ailments" / " 'Torote Copal'"

1994   **B. sarcopoda** Mexico / Y. Mexia 8901 / 1937 / "Said to scare away scorpions" / " 'Tecomaca'"

1995   **B. simaruba** Brit. Honduras / W. A. Schipp 799 / 1931 / "Natives collect light amber gum for medicinal use"

El Salvador / P. C. Standley 19636 / 1921–22 / "Juice has strong odor. Powdered fr. used for stomachache" / " 'Jiote'"

El Salvador / P. C. Standley 22302 / 1922 / "Sap with strong odor. Seeds remedy for rheumatism" / " 'Jiote,' 'Palo de jiote'"

1996   **B. trifoliata** Mexico / G. B. Hinton et al. 7781 / 35 / "Poisonous. Quincanchire"

1997   **Bursera** indet. Mexico / P. C. Standley 2201 / 1936 / "A tea said to be made from the bark as a refreshing drink" / " 'Torote negra'"

1998   **B. graveolens** Cuba / R. A. Howard 5457 / 1941 / " . . . very aromatic and called 'sassafras'"

1999   *B. tomentosa* Venezuela / H. Pittier 7927 / 1918 / "Med. An odoriferous clear resin is extracted" / "'Bálasamo'"

2000   *Garuga floribunda* New Guinea / L. J. Brass 542 / 1925 / "Edible. Kagi Kagi (Motu)"

2001   *Bursuracea* indet. Siam / Royal For. Dept. 4533 / '49 / "Edible fruits, sour taste" / "Ma-faen"

MELIACEAE

2002   *Cedrela* indet. China / Lingnan Univ. Herb. 140.4156 / 1924 / "Used in mushroom culture" / "'You chui'"

2003   *C. sinensis* China / W. T. Tsang 23064 / 1933 / "Fruit edible" / "'Chun Tsei Shue'"

China / E. H. Wilson 585 / 07 / "Shoots eaten"

2004   *Xylocarpus granatum* Fiji Is. / O. Degener 15035 / 1941 / "Chew leaf or pound bark in water & drink as a remedy for thrush" / "'Ndambi' (Serua)"

2005   *Cipadessa baccifera* China / F. A. McClure 20563 / 1937 / "Tea brewed from the fruits is prescribed for stomachache" / "'Wu Chu Yu'"

2006   *Vavaea* indet. N. Borneo / Orolfo BNB For. Dept. 5222 / 1935 / "Incense wood used by the natives for rubbing their dead before burial" / "'Chendana,' 'Sendana' (Bajau)"

2007   *V. amicorum* Philippine Is. / C. O. Frake 826 / 1958 / "Bark applied to internal pains" / "'Diak'"

Fiji Is. / Degener & Ordonez 14158 / 1941 / "'False sandalwood' (Serua, Thevua)"

2008   *V. australiana* Australia: Darwin / J. A. Gilruth NHNSW 25231 / 1914 / "White-ant resisting"

2009   *Melia azedarach* Mexico / Bro. Abbon 241 / 1911 / "'Canelo'"

2010   *M. dubia* Philippine Is. / C. O. Frake 778 / 1958 / "Leaves applied for internal pain" / "'belu' (Sub.)"

2011   *Sandoricum borneense* Borneo / Telado BNBFD 1936 / 1932 / "Edible sour fruit" / "'Santol kepas' (Ked.)"

2012   *S. koetjape* Philippine Is. / E. Fénix 20 / 1938 / "Medicinal"

2013   *Dysoxylum cauliflorum* Malay Peninsula: Selangor / A. H. Millard et al. 1622 / 59 / "Poisonous fruits, aril said to be especially toxic"

2014   *D. decandrum* New Guinea / E. E. Henty NGF 10501 / 58 / "Decoction of bark said to act as an emetic"

Philippine Is. / C. O. Frake 855 / 1958 / "Bark scraped and applied to abdomen to prevent pregnancy" / "'Lingu gunsili' (Sub.)"

2015 *D. cumingianum* Philippine Is. / M. D. Sulit 1430 / 1947 / "Wood burns even fresh" / "'Tara-tara' (Tag.)"

2016 *D. hornei* Fiji Is. / O. Degener 15308 / 1941 / "Extract of leaves used medicinally" / "'Viviniura' (Serua)"

2017 *D. lenticellare* Fiji Is. / A. R. Wagatabu 2626 / 1942 / "Fruits said to be poisonous—marked knees & swellings" / "'Riri'"

2018 *Chisocheton* indet. Malay Peninsula: Selangor / A. H. Millard et al. 1513 / 59 / "Poisonous" / "'Kayu Kelaweh' (Temuan)"

2019 *Chisocheton* indet. Philippine Is. / C. O. Frake 750 / 1958 / "Roots boiled & drunk for splenomegaly" / "'Melibusug' (Sub.)"

2020 *Aphanamixis tripetala* Philippine Is. / L. E. Ebalo 252 / 1939 / ". . . bark used in the fermentation of 'tubu'"

2021 *Amoora* indet. New Britain / A. Floyd 6564 / 54 / "Latex in bark causes severe skin irritation when cutting or sawing this tree" / "'La porubolo' (W. Nak.)"

2022 *Aglaia lancifoliata* China / W. T. Tsang 22333 / 1933 / "Fr. . . . edible" / "'Ma Lau Kwo Shue'"

2023 *A. odorata* Hainan / S. K. Lau 0782 / 1933 / "Fruit red; Lai people use the skin for shampoo"

2024 *A. glomerata* Philippine Is. / C. O. Frake 639 / 1958 / "Bark applied to pain in xiphoid process" / "'Malionad' (Sub.)"
Philippine Is. / C. O. Frake 868 / 1958 / "Bark applied for anasarca" / "'denden' (Sub.)"

2025 *A. llanosiana* Philippine Is. / C. O. Frake 818 / 1958 / "Leaves applied to broken arm" / "'begabu' (Sub.)"

2026 *A. rimosa* Philippine Is. / M. D. Sulit 1116 / 1946 / "Birds picked fruit & seed swallowed" / "'Balubar' (Tag.)"

2027 *Aglaia* indet. Philippine Is. / A. L. Zwickey 487 / 1938 / "Bark chewed for loose teeth" / "'Buga' (Lan.)"

2028 *Aglaia* indet. Philippine Is. / P. Añonuevo 44 / 1950 / "Edible" / "'Sorowan' (Manobo)"

2029 *Aglaia* indet. Philippine Is. / M. D. Sulit 3943 / 1950 / "The pulp of the seeds is eaten by Tagb. children. Tastes sweet like 'Lansones.' Worthwhile cultivating"

2030 *A. saltatorum* Fiji Is. / A. C. Smith 1476 / 1934 / "Flowers used to scent coconut oil" / "'Langakali'"

2031 *Guarea caoba* Nicaragua / Shank & Molina R. 4840 / 1951 / "'Pronto Alivio'"

2032 *G. palmeri* El Salvador / P. C. Standley 22160 / 1922 / "'Culebra,' 'Quita-calzón'"

2033 **Walsura robusta** Hainan / C. I. Lei 597 / 1933 / " . . . fr. red edible" / "'Kot Sit Shu'"

2034 **W. piscidia** INDIA: Travancore / E. W. Erlanson 5337 / 1933 / "Yields fish poison"

2035 **W. villamilii** N. Borneo / F. Melegrito 91 / 1923 / "edible"

2036 **W. multijuga** Philippine Is. / C. Frake 515 / 1958 / "Bark boiled and held in mouth for mumps" / "'Drul' (Sub.)"

Philippine Is. / C. Frake 429 / 1958 / "Bark scraped and eaten for poisoning" / "'glagu'"

2037 **Trichilia hirta** Guatemala / J. A. Steyermark 51111 / 1942 / "Reputed to be used for women's hair, take the fruit, place in water and the hair is supposed to grow longer" / " 'cacohuito'"

2038 **T. oblonga** Brazil / Y. Mexia 5598 / 1931 / "Tea of leaves used for pleurisy" / "'Pimenteira'"

MALPIGHIACEAE

2039 **Mascagnia macroptera** Mexico / W. C. Leavenworth 50 / 1940 / "'Naranjillo'"

Mexico / W. P. Hewitt 108 / 1946 / "Milky juice said to be purgative" / "'Mata Nene'"

2040 **M. septentrionalis** Mexico / G. Thurber 901 / 1851 / "The root *said to be* a specific in syphilis" / "'Gallineta'"

2041 **Tetrapteris mucronata** Colombia / R. E. Schultes 12107 / 1951 / "Said to be employed by Karapana tribe in preparing yajé"

2042 **Dinemandra glaberrima** Chile / E. E. Gigoux s.n. / 1885 / "'Té de barro, colorado'"

Chile / E. M. Johnston 4858 / 1925 / "'Te colorado'"

2043 **Gaudichaudia schiedeana** Guatemala / P. C. Standley 71488 / 1939 / "Hairs are very irritant" / "'Koko'"

2044 **Ryssopteris timoriensis** Philippine Is. / C. O. Frake 645 / 1958 / "Leaves applied for internal pain" / "'Glektan ulangan'" Sub.

2045 **Banisteria leiocarpa** Peru / C. Vargas (Herb. Univ. Cuzco-Perú 2044) / 1948 / "'Huillca bejuco'"

2046 **Banisteriopsis rosea** El Salvador / P. C. Standley 20653 / 1922 / "'Florecita de pensamiento'"

2047 **B. caapi** Colombia / G. Klug 1934 / 1931 / "'Yagé'"

2048 **B. rusbyana** Colombia / G. Klug 1971 / 1931 / "'Chagro panga,' 'oco yagé'"

2049 **B. inebriana** Colombia / G. Klug 1964 / 1931 / "'Yagé del monte'"

2050 **Cabi paraensis** Brazil / Ducke 1641 / 1944 / "'Cabi'"

2051 *Heteropteris beecheyana* Mexico / Y. Mexia 8933 / 1937 / "'Bejuco de Margarita'"

2052 *H. macrostachya* Panama / I. M. Johnston 1377 / 1946 / *"Toxic O.P."*

2053 *H. longifolia* W.I.: Dominica / J. S. Beard 660 / 1946 / "'Liane cacao'"

2054 *H. platyptera* var. *martinicensis* W.I.: Martinique / M. Stehlé (Herb.) 6032 / 1943 / "'Liane caco'"

2055 *Stigmaphyllon ellipticum* El Salvador / P. C. Standley 22158 / 1922 / "'Chinaca,' 'Bejuco de chinaca,' 'Flor de Jesús amarilla'"

2056 *S. humboldtianum* Colombia / Killip & Smith 16325 / 1926 / "(Fruits used as darts by children)" / "'Agalita de piña'"

2057 *S. fulgens* Brit. Guiana / W. A. Archer 2463 / 1934 / "(For 'Yagé' test)" / "'Kairia' (Ar.)"

2058 *Dinemagonum gayonum* Chile / R. Wagenknecht (Exped.) 18463 / 1939 / "'Sanalo-todo'"

Chile / R. Wagenknecht (Exped.) 18537 / 1940 / "'Sanalotodo'"

2059 *Galphimia gracilis* Mexico / C. & E. Seler 660 / 1888 / "'Yerva de piojo'"

2060 *G. glauca* Mexico / H. S. Gentry 1372 / 1935 / "Medic. roots remedy for urine obstruction. Cook in water and drunk hot" / "'Garbansilla,' Mex. 'Leychí,' W."

2061 *G. humboldtiana* Mexico / M. Urbinae (121 Barcena) / 1886 / "'Palo de muerto'"

2062 *Malpighia linearis* W.I.: Barbuda / J. S. Beard 377 / 1944 / "lower leaf surface with fine irritant hairs" / "'Cow itch bush'"

2063 *M. polytricha* Bahama Is. / A. E. Wight 251 / 1905 / "Appressed hairs on lower side of leaves become detached and enter the skin; somewhat painful"

2064 *M. suberosa* Cuba / J. G. Jack 6022 / 1928 / "Look out for prickles on leaves"

2065 *Malpighia* indet. Republica Dominicana / Dr. Jiménez (Herb.) 1831 / 1949 / "'Arbol de la dicha'"

2066 *Bunchosia nitida* var. *grenadensis* W.I.: St. Kitts / J. S. Beard 303 / 1944 / "'Wild coffee'"

2067 *B. glandulifera* Brazil / W. A. Archer 7655 / 1942 / "fruit . . . edible" / "'café do amazonas'"

2068 *Byrsonima cotinifolia* Mexico / Dr. Ghiesbreght 639 / 1864–70 / "Le fruit est recherché par les habitans qui le mange comme sous le nom de 7 Ex nance"

2069 *B. crassifolia* Brit. Guiana / A. C. Smith 3344 / 1938 / "Inner bark pounded into pulp and applied to wounds and abrasions of cattle and humans, either as a poultice or liquid; said to be very effective in healing. (Macusi and Wapisiana Indians)"

2070 *B. lancifolia* Brazil / Y. Mexia 5547 / 1931 / "Root scraped, boiled and extract taken for rheumatism" / " 'Cajusin' "

2071 *B. orbignyana* Bolivia / J. Steinbach 6979 / 1925 / "Cascara . . . es astringente" / " 'Coleradillo de la panpa' "

2072 *Malpighiacea* indet. Bolivia / J. Steinbach 5155 / 1920 / "Cascara amarga!" / " 'Coloradillo del monte' "

2073 *Malpighiacea* indet. Bolivia / J. Steinbach 5112 / 1920 / "Cascara muy amarga. . . . 'Coloradillo de la Pampa' "

TRIGONIACEAE

2074 *Trigonia paniculata* Brazil / Y. Mexia 4516 / 1930 / "Leaves much used for tea, fresh or dried" / "Bajadin"

VOCHYSIACEAE

2075 *Vochysia lomatophylla* Peru / F. Woytkowski 6021 / 1960 / " . . . perhaps used by Campa tribe as contraceptive?" / " 'Sacha-alfaro' "

2076 *V. rufa* Brazil / Y. Mexia 5538 / 1931 / " 'Pau doce' "

POLYGALACEAE

2077 *Polygala adenophora* Brit. Honduras / W. A. Schipp 602 / 1930 / " . . . roots have odor of menthol when crushed"

2078 *Polygala* indet. Mexico / Y. Mexia 9195 / 1938 / " . . . root has licorice odor"

2079 *P. pulchella* Brazil / Y. Mexia 4030 / 1929 / " . . . root has odor of wintergreen"

2080 *P. fallax* China / W. T. Tsang 21505 / 1932 / " . . . roots used to make medicine" / "To Tin Wong Muk"

2081 *P. paniculata* Neth. New Guinea / van Royen & Schleumer 5656 / 1961 / "Roots with a strong smell of wintergreen"

2082 *Securidaca tavoyana* Brit. N. Borneo / Bakar B.N.B. For. Dept. 3435 / 1933 / " . . . stem and bark a native substitute for soap" / "Langer (Brunei) Bongkug (Murat)"

DICHAPETALACEAE

2083 *Dichapetalum gelonioides* Philippine Is. / C. O. Frake 555 / 1958 / "Roots for snake bite" / "Glemuan Sub."

Philippine Is. / C. O. Frake 677 / 1958 / "Roots boiled for puerperium" / "Salab Sub."

Philippine Is. / C. O. Frake 788 / 1958 / "Leaves applied for charlie-horse and amenorrhea" / "bengnay-belagen Sub."

2084   *D. cymosum* S. Africa: Transvaal / R. Marloth 9518 / 1920 / "very poisonous"

EUPHORBIACEAE

2085   *Savia sessiliflora* Venezuela / H. Pittier 7687 / 1917 / "'Guayabo cereso'"

2086   *Securinega suffruticosa* China? / J. C. Liu (Peking Union Med. Coll., Dept. of Pharm. L.808) / 1927

2087   *Fluggea virosa* Philippine Is. / M. Adduru 46 / 1917 / "roots boiled for curing teeth gum" / "'Malunit'" Ib.

2088   *Phyllanthus acuminatus* El Salvador / S. Calderón 409 / 1922 / "'Pimientillo'"

     Mexico / E. C. Stewart 159 / 1935 / "'Payúl'"

     Colombia / H. García B. s.n. / 1942 / "'Barbasquillo'"

     Colombia / G. Klug 1750 / 1930 / "'Barbasco simaron'"

     Peru / Stork, Horton & Vargas 10487 / 1939 / "'Anil'"

     Peru / Y. Mexia 6253 / 1931 / "Crushed leaves used as fish poison" / "'Yutúle'"

2089   *P. acidus* El Salvador / S. Calderón 237 / 1921 / "'Guinda'" "'Grosella'"

     El Salvador / P. C. Standley 19422 / 1921 / "'Guinda'" / "...fr.... very acid, little eaten"

     Philippine Is. / A. L. Zwickey 244 / 1938 / "...frts. used in making vinegar..." / "'Eba'"

2090   *P. elsiae* Mexico / Y. Mexia 1005 / 1926 / "'Pimientillo'"

     El Salvador / P. C. Standley 22234 / 1922 / "'Guinda'"

2091   *P. micrandrus* Guatemala / J. A. Steyermark 51756 / 1942 / "'Chabinté'"

2092   *P. tequilensis* Mexico / J. T. Howell 10448 / 1932 / "'Manzanilla'"

2093   *P. mimosoides* W.I.: Dominica / J. S. Beard 645 / 1946 / "'Tamarind grand bois'"

2094   *P. niruri* Argentina / T. M. Pedersen 4817 / 1958 / "'Santa Maria'"

     Brazil / R. Froes 1829 / 1932 / "Used for fever and kidney ailments" / "'Guebra pedra'"

     Marianas Is. / F. R. Fosberg 24969 / 1946 / "'maigo lalo' used with Euphorbia hirta as vaginal astringent, also to relieve rectal itching. Leaves mashed and inserted"

     Marianas Is. / F. R. Fosberg 25380 / 1946 / "'maigo lalo' used medicinally, plant boiled in water and drunk for dysentery"

2095   *P. piscatorum* Venezuela / L. Williams 15865 / 1942 / "...las hojas machacadas son empleadas para embarbascar;...a veces cultivado en los conucos" / "N.v.: 'Púperi' (Baniba); 'Barbasco'"

Venezuela / L. Williams 13392 / 1940 / "Lvs. used as fish poison" / "'Barbasco caicareño'"

2096 *P. pseudo-conami* Colombia / G. Klug 1757 / 1930 / "Used as fish poison"

Colombia / F. J. Hermann 11233 / 1944 / "Rotenone 0.0%"

2097 *P. salviaefolius* Venezuela / F. Tamayo 2395 / 1942 / " . . . lo emplean para madurar 'curas'" / "N.v.: Yuquero"

2098 *Phyllanthus* indet. Brit. Guiana / J. S. de la Cruz 962 / 1921 / "'Conaparoo,' a fish poison . . . "

2099 *P. juglandifolius* Ecuador / A. S. Hitchcock 20567 / 1923 / "'Jobo'"

2100 *P. brasiliensis* Brazil / R. Froes 12416/160 / 1941 / "Used as fish-poison; leaves are mixed with clay" / "'Tarira-Cunanby'"

Brazil / B. A. Krukoff's 4th Exped. to Braz. Amaz. 5211 / no year / "On an Indian plantation. (Used as 'fish poison')"

2101 *P. rosellus* Brazil / Williams & Assis 7223 / 1945 / "Make a tea of this for inflammation of the kidneys. Drink it. 'Quebra pedra'"

2102 *P. tenellus* Brazil / A. Gehrt 3004 / 1920 / "'Quebra pedra,' 'Herva pombinha'"

2103 *P. chacoensis* Argentina / Eyerdam & Beetle 22983 / 1938 / " . . . fr. . . . edible"

2104 *P. anceps* Hainan / F. A. McClure 8890 / 1922 / "drug plant (for rash); names reported, 'Pa pui ts'oi'" / "It is used to make a tea which is good for cough and to remove phlegm from the throat"

2105 *P. emblica* China / T. T. Yü 1331 / 1932 / "Fruit for medicine"

India: Punjab / W. Koelz 1751 / 1931 / "Fruit used for medicine commonly by Tibetans and other hill people" / "'Amrla,' 'Kürura'" Urdu

2106 *P. urinaria* China / W. T. Tsang 21395 / 1932 / " . . . used as medicine"

2107 *P. pulcher* Cambodia / Poilane 14288 / 1927 / " . . . il est medicinal. . . . Employé comme désinfectant par exemple après avoir touché un mort mettre à trumper dans l'eau, avec laquelle on se lavera"

2108 *Phyllanthus* indet. Solomon Is. / S. F. Kajewski 2551 / 1931 / "The bark is macerated and drunk for colds" / "'Jerima'"

2109 *P. amarus* Philippine Is. / A. L. Zwickey 145 / 1938 / " . . . frs. eaten by children . . . " / "'Kanunung' (Lan.); 'Kanulug'? (Lan.)"

POLYNESIA: Mangareva / A. Agassiz et al. 110 / 1905 / "Used as poison for catching fish"

2110 *P. reticulatus* Philippine Is. / C. O. Frake 835 / 1958 / "leaves applied to pinworms" / "'bengnay-nubu'?'" Sub.

2111 *Phyllanthus* indet. Caroline Is. / F. R. Fosberg 32066b / 1950 / "'Ukalelathib' said to be used as medicine, boiled; used as drink for sore heads or bodies. Has astringent properties, stops pain"

2112 *Phyllanthus* indet. New Hebrides Is. / S. F. Kajewski 274 / 1928 / "Leaves crushed and used by natives for fevers" / "'Narmlee'"

2113 *Glochidion eriocarpum* China / W. T. Tsang 21835 / 1933 / "Roots used for medicine"

2114 *G. hayatae* China / C. K. Shang 26 / 1926 / "Fruit . . . edible" / "'Tsin Pui tsz Shue'"

2115 *G. acuminatum* India: Assam / L. F. Ruse 114 / 1923 / "Extract from the leaves is given to cattle for stomach complaints" / "'Dieng-soh-par-maso'"

2116 *G. assamicum* India: Assam / L. F. Ruse 115 / 1923 / "extract from the leaves is given to cows for stomach complaints" / "'Dieng-soh.par-masi-rit'"

2117 *Glochidion* indet. Siam / native collector S312 (Royal For. Dept. 4512) / 48 / " . . . astringent leaves. Edible leaves" / "'Kai Mod.'"

2118 *G. laevigatum* Malaya / G. anak Umbai for A. H. Millard (K.L. 1632) / 1959 / "Phytochemical Survey of Malaya" / "Medicinal and poisonous"

2119 *G. wallichii* Malaya / G. anak Umbai for A. H. Millard (K.L. 1627) / 1959 / "Phytochemical Survey of Malaya" / "Sap medicinal" / "'Roy Koy'" Temuan

2120 *Glochidion* indet. Sumatra / C. G. G. J. van Steenis 9723 / 1937 / " . . . arillus white; edible"

2121 *G. magnificum* Dutch New Guinea / Kanehira & Hatusima 11820 / 1940 / "Myrmecophilous plant"

2122 *Glochidion* indet. Solomon Is. / S. F. Kajewski 2025 / 1930 / "Pigeons are very fond of the fruit" / "'Gook-or-oto'"

2123 *G. album* Philippine Is. / C. O. Frake 356 / 1957 / "Roots boiled and rubbed in, for indigestion" / "'Gagdal'" Sub.

2124 *G. cauliflorum* Philippine Is. / M. D. Sulit 3297 / 1949 / "Decoction of roots mixed with roots of cacao and coffee given to women for quick delivery" / "'Pabatã'" Bukid.

2125 *G. pubicarpum* Philippine Is. / C. O. Frake 539 / 1958 / "Agricultural ritual" / "'gepulu'?"

2126 *Glochidion* indet. Philippine Is. / C. O. Frake 598 / 1958 / "Medicine for sick pigs" / "'getalip'" Sub.

2127 *Glochidion* indet. Philippine Is. / C. O. Frake 685 / 1958 / "Roots boiled and drunk for rigid abdomen" / "'gebana'" Sub.

2128 *Glochidion* indet. Philippine Is. / C. O. Frake 542 / 1958 / "Leaves applied to ulcers" / "'Kendulun'"

2129 *Glochidion* indet. Philippine Is. / C. O. Frake 747 / 1958 / "Leaves applied to skin eruptions" / "'denglas'" Sub.

2130 *Glochidion* indet. Philippine Is. / C. O. Frake 763 / 1958 / "Agric. ritual; roots applied for internal pains" / "'giua'" Sub.

2131 *Glochidion* indet. Philippine Is. / C. O. Frake 813 / 1958 / "Roots boiled and drunk for stomachache" / "'gites'" Sub.

2132 *Glochidion* indet. Philippine Is. / C. O. Frake 737 / 1958 / "Roots boiled and drunk for puerperium" / "'gangyus'" Sub.

2133 *Glochidion* indet. Philippine Is. / P. Añonuevo 231 / 1952 / "Fruit eaten by birds when ripe" / "'Bano'" Bilaan

2134 *G. concolor* Tonga Is. / T. G. Yuncker 15363 / 1953 / " . . . leaves used in medicine" / "'Malolo'" Tongan

Fiji Is. / O. Degener 14950 / 1941 / "Fijians eat leaf raw for stomache trouble according to Nadarivatu native. According to Serua native, leaf eaten raw for thrush" / "'Galo'" Savatu

Fiji Is. / O. Degener 15164 / 1941 / "Chew leaf for thrush. Eat leaves to stop diarrhea and bloody stool" / "'Galo'" Serua

2135 *Breynia fruticosa* Indochina / E. Poilane 21189 / 1932 / " . . . utilisée contre le mal de dent"

2136 *Breynia* indet. Solomon Is. / S. F. Kajewski 2586 / 1931 / "The leaves are macerated in water and put inside the mouth to stop tooth-ache"

2137 *Breynia* indet. Solomon Is. / S. F. Kajewski 2476 / 1931 / "When a native has a boil the leaves of this small tree are heated with lime and applied to it" / "'Ra-rahu'"

2138 *B. rhamnoides* Philippine Is. / A. L. Zwickey 19 / 1938 / "infusion of leaves for stomachache" / "'Tagom-tagom' (Bis.); 'Katorog' (Lan.)"

2139 *Sauropus changiana* China / F. C. Chang s.n. / 1950 / "'Lung-li-yah'" [Accompanying this sheet are a half dozen or so letters concerning the identity of the collection. The correspondence indicates that F. C. Chang, head of the Chemistry Department of Lingnan University, Canton, wanted in 1949. to identify this plant, which had been used as a drug in Canton for ages and from which an antibiotic supposedly had been recently prepared. The common name is translated as Dragon's Tongue.]

2140 *S. spathulaefolius* Laos / Poilane 20622 / no year / "Utilisé contre les maux de gorge; prendre des racines, les bouillir et boire . . . "

2141 *Drypetes duquei* Colombia / J. M. Duque Jaramillo 89a / no year / "'Marfil,' 'Cafetillo serrano'"

2142 *D. perreticulata* Hainan / S. K. Lau 304 / 1932 / "Used as medicine" " . . . fruit edible"

2143 *D. vitiensis* Fiji Is. / O. Degener 15430 / 1941 / "For headache, squash juice of leaves into nose and ears" / "'Meme'" Ra

2144 *Petalostigma glabrescens* Australia: Queensland / C. T. White 7666 / 1931 / "Fruit red with a thin very bitter fleshy pericarp"

2145 *P. quadriloculare* Australia: Queensland / S. F. Kajewski 94 / 1928 / "...'berry' very bitter to the taste" / "'Bitter Bark' or 'Quinine Berry'"

2146 *Hieronyma oblonga* Mexico / Dr. Berendt s.n. / 63 / "'Cascarilla——? = 'Copalchi'"

Venezuela / J. A. Steyermark 60462 / 1944 / "Small deer (locho) eat fallen fruit" / "'Cari-yau-ki-yek'"

2147 *H. caribaea* W.I.: St. Lucia / J. S. Beard 508 / 1945 / "Fruit eaten by birds" / "'Bois d'amande'"

2148 *H. moritziana* Venezuela / J. A. Steyermark 56578 / 1944 / "'Quin-du canelo'"

2149 *Aporosa sphaerosperma (A. chinensis?)* Hainan / S. K. Lau 189 / 1932 / "Fr....edible"

Hainan / W. T. Tsang 111 / 1928 / "...fr. yellow; medicines"

2150 *A. selangorica* Malaya / G. anak Umbai for A. H. Millard (K.L. 1501) / 1959 / "Phytochemical Survey of Malaya" / "edible fruits"

2151 *Baccaurea ramiflora (B. cauliflora?)* Hainan / C. I. Lei 600 / 1933 / "fruit...edible"

China / C. W. Wang 76394 / 1936 / "fr....favourite of the natives... very sweet"

China / H. T. Tsai 61284 / 1934 / "fls. edible"

2152 *Baccaurea* indet. China / C. W. Wang 77661 / 1936 / "Fruit...Very sweet, edible"

2153 *B. cauliflora* Indo-China / W. T. Tsang 29136 / 1939 / "...fr. edible"

2154 *B. sylvestris* Indo-China / A. Chevalier or Poilane 1269 / 1920 / "...il donne des fruits comestibles"

2155 *Baccaurea* indet. Siam / T. Smitinand 716 / 51 / "Frs. said to be acid and used as flavouring" / "'Som-hooke'"

2156 *Baccaurea* indet. Indo-China / Poilane 11093 / 24 / "Les fruits...très bons..."

2157 *Baccaurea* indet. Cambodia / E. Poilane 17655 / 1930 / "Fruit comestible..."

2158 *B. bracteata* Brit. N. Borneo / Tandom (B.N.B. For. Dept. 2820) / 1933 / "Fruit edible" / "Nom. v. Tampui paya (Brunei)"

2159 *B. stipulata* Brit. N. Borneo / Apostol (B.N.B. For. Dept. 3700) / 1933 / "Fruit...edible"

2160 *B. wallichii* Brit. N. Borneo / Balajadia (B.N.B. For. Dept. 2554) / 1932 / "Fruit . . . edible" / "'Kokonau' (Dusun)"

2161 *Baccaurea* indet. Malaya / G. anak Umbai for A. H. Millard (K.L. 1591) / 59 / "Phytochemical Survey of Malaya" / "Poisonous fruits"

2162 *Baccaurea* indet. Philippine Is. / G. E. Edaño 2920 / 1951 / "The fruit is edible" / "'Lipsop'" Tag.

2163 *B. wilkesiana* Fiji Is. / O. Degener 15252 / 1941 / "Squash leaf in cold water, strain in coconut cloth and drink as blood purifier" / "'Kutu'" Serua

2164 *Antidesma delicatulum* China / W. T. Tsang 28309 / 1937 / "fr. . . . edible"

2165 *A. fordii* China / W. T. Tsang 23135 / 1933 / "fr. . . . edible"
China / W. T. Tsang 26561 / 1936 / "fr. . . . edible"
Indo-China / Poilane 10522 / 1924 / " . . . les fruits sont mangés par les oiseaux"

2166 *A. ghaesembilla* China / W. T. Tsang 23840 / 1934 / "fr. . . . edible"
Indo-China / Poilane 6267 / 1923 / " . . . les fruits sont comestibles"
Brit. N. Borneo / Tandom (B.N.B. For. Dept. 2828) / 1933 / "Edible fruit"
New Guinea: Papua / L. J. Brass 5746 / 1934 / " . . . sub-acid fruit edible and pleasantly-flavored"
Philippine Is. / P. Añonuevo 52 / 1950 / "Fruit when ripe is eaten raw"

2167 *A. gracile* China / W. T. Tsang 21087 / 1932 / "fruit . . . edible"

2168 *A. henryi* Hainan / F. A. McClure (Canton Christian Coll. 8980) / 1922 / " . . . the Lois bruise the leaves, mix with sour wine, heat and apply to bruise to clear up the blood there" / "'Ng ut ch'a'"

2169 *A. japonicum* Ryukyu Is. / Walker, Sonohara, Tawada & Amano 6199 / 1951 / "Fruits edible"

2170 *A. bunius (A. thorelianum?)* Indo-China / Poilane 10901 / 24 / "En fruits ils sont comestibles" / "bois tendre attaqué par les termites"

2171 *A. thorelianum* Indo-China / Poilane 10126 / 1924 / " . . . attaqué par les termites . . . "

2172 *Antidesma* indet. Siam / native collector S452 (Royal For. Dept. 2841) / 49 / "Frs. edible, acid"

2173 *Antidesma* indet. Indo-China / F. Evrard 2492 / 1925 / " . . . fruits acides comestibles"

2174 *Antidesma* indet. Cambodia / E. Poilane 15417 / ca. 1929 / "Les Cambodgiens employent ses feuilles pour faire des soupes . . . "

2175   *A. bangueyense* Brit. N. Borneo / Valera (B.N.B. For. Dept. 4734) / 1935 / "Fruit edible" / "'Adadsay'" Dusun.

2176   *A. trunciflorum* Brit. N. Borneo / Enggoh (B.N.B. For. Dept. 10194) / 1939 / "Fruit edible" / "'Jarupis' (Malay)"

2177   *A. velutinosum* Malaya / G. anak Umbai for A. H. Millard (K.L. 1461) / 1959 / "Phytochemical Survey of Malaya" / "stupefying"

2178   *A. praegrandifolium* Solomon Is. / S. F. Kajewski 2593 / 1931 / "The leaves are macerated and drunk for constipation" / "'Tamburaviti'"

2179   *Antidesma* indet. Solomon Is. / S. F. Kajewski 2569 / 1931 / "When a boy has a sore shoulder the bark is taken and rubbed on it" / "'Tamb-raviti'"

2180   *Antidesma* indet. Solomon Is. / S. F. Kajewski 2398 / 1930 / "The natives eat the fruit of this tree when ripe" / "'Lassau'"

2181   *A. phanerophlebium* Philippine Is. / C. O. Frake 374 / 1957 / "Roots applied for stomachache" / "'pagakpatulangan'" Sub.

2182   *A. pentandrum* Philippine Is. / D. R. Mendoza 936 / 1949 / "edible fruit"

2183   *Antidesma* indet. Philippine Is. / M. D. Sulit 3346 / 1949 / "Fr. edible" / "'Tiga'" Bukid.

2184   *Antidesma* indet. Philippine Is. / C. O. Frake 76 / 1950 / "Leaves applied to ulcers" / "'Tegas'" Sub.

2185   *Antidesma* indet. Philippine Is. / C. O. Frake 816 / 1958 / "edible fruit—Leaves applied to lumbar pains" / "'bengnay menubu'?"

2186   *Antidesma* indet. Philippine Is. / R. B. Fox 38 / 1950 / "edible fruit —sweet and sour" / "'Agasip'" Tagb.

2187   *Antidesma* indet. Philippine Is. / P. Añonuevo 42 / 1950 / "Use to color teeth. The bark is burned and the ash is rubbed on the teeth" / "'Bongay'" Manobo

2188   *Antidesma* indet. Philippine Is. / P. Añonuevo 167 / 1950 / "Use for fish poisoning" / "'Kabasihasi'" Manobo

2189   *Antidesma* indet. Philippine Is. / A. L. Zwickey 699 / 1938 / "... frts. or leaf tea as tonic after childbirth" / "'Matelok' (Lan.)"

2190   *Uapaca nitida* S. Africa / F. A. Rogers 8601 / 1909 / "Fruit edible" / "'musorkoloy'" Chinganga

2191   *Bischofia javanica* Hainan / S. K. Lau 367 / 1932 / "fruit...edible"

      Fiji Is. / O. Degener 15218 / 1941 / "For tonsilitis eat leaf and drink the juice" / "'Tongogongo'" Serua

      Fiji Is. / O. Degener 14935 / 1941 / "When Fijians are injured by urticaceous plant, they rub affected part with inner bark of koka as cure" / "'Koka'" Savatu

New Hebrides Is. / S. F. Kajewski 310 / 1928 / "Bark used by boiling in salt water and applied to cuts" / "'No-ghor'"

2192 *Bischofia* indet. Solomon Is. / S. F. Kajewski 2638 / 1931 / "The red sap is taken and put on sores" / "'Haw-ha'"

2193 *Cleistanthus hirsutulus* Malaya / G. anak Umbai for A. H. Millard (K.L. 1473) / 1959 / "Phytochemical Survey of Malaya" / "stupefying"

2194 *Cleistanthus* indet. Philippine Is. / C. O. Frake 819 / 1958 / "Leaves applied for asthma" / "'ganga'" Sub.

2195 *Cleistanthus* indet. Philippine Is. / Sulit & Conklin 5064 / 1953 / "Decoction of root—medicine for dysentery" / "'Gamot-masaklod' (Kayo)" Hanunoo Mang.

2196 *Bridelia balansae* Hainan / F. A. McClure (Christian Coll. 8101) / 1921 / "fruits edible"

2197 *B. squamosa* India / J. Fernandes 28 / 1949 / "Fruits edible when ripe, else very astringent" / "'Asana'"

India / J. Fernandes 260 / 1949 / "The fruit is eaten, though it is not too tasty" / "'Asana'"

2198 *Bridelia* indet. Siam / native collector (Royal For. Dept. of Siam 5918) DE.102 / 48 / "A medicinal plant" / "'Sam-tsa'"

2199 *B. glauca* Philippine Is. / C. O. Frake 660 / 1958 / "Agric. ritual" / "'Gamungan'" Sub.

2200 *Daphniphyllum* indet. China / C. K. Shang 40 / 1926 / "Medicine (leaf)" / "'Ap Keuk Shue'"

2201 *D. luzonense* Philippine Is. / W. Beyer (Herb. 6852) / 1948 / "Fruit and leaves boiled and decoction is used to treat stomach ache" / "'Hala Hala'" Ifugao

2202 *Croton glandulosus* var. *pubentissimus* U.S.: Tex. / V. L. Cory 45641 / 1944 / "A croton fed upon by livestock"

2203 *C. lindheimerianus* U.S.: Tex. / V. L. Cory 34810 / 1940 / No notes with this collection, but a letter is attached from V. L. Cory to L. Croizat, suggesting that livestock had been eating *Croton* (probably this species) from Sutton County, Texas, a circumstance which Cory thought unusual. Cory said the only other *Croton* he had ever known livestock to eat was *C. monanthogynus*. The letter is dated July 28, 1942.

2204 *C. monanthogynus* U.S.: Tex. / A. Traverse 210 / 1956 / "odor of soap"

2205 *C. punctatus* U.S.: Tex. / A. Traverse 206 / 1956 / "...foliage has soapy aromatic odor"

2206 *C. neo-mexicanus* U.S.: Tex. / P. Goodrum 154 / 1940 / "Aromatic when freshly picked" "Food for doves" / "'Sageweed'"

2207 *C. eluteria* El Salvador / S. Calderón 519 / 1922 / "'Sasafrás'"

2208 *C. guatemalensis* El Salvador / M. C. Carlson 205 / 1946 / "'Copal-chillo'"

Guatemala / J. A. Steyermark 51062 / 1942 / "Bark used as remedy for chills" / "'Copalchi'"

Guatemala / J. A. Steyermark 46467 / 1942 / "'hoja amarga'"

Guatemala / J. A. Steyermark 51187 / 1942 / "Reputed to be used in curing rheumatism; boil the plant and then bathe in the juice of this for treating the rheumatism" / "'Copalchi'"

2209 *C. glabellus* Panama / Cooper & Slater 165 / ca. 1927 / "'Copachi'"

Honduras / P. C. Standley 52746 / 1927–28 / "'Cascarilian'"

Honduras / P. C. Standley 55964 / 1928 / "...aromatic" / "'Lián'"

2210 *C. reflexifolius* El Salvador / P. C. Standley 19882 / 1922 / "Bark a remedy for colds" / "'Copalchí'"

2211 *C. pyramidalis* Guatemala / J. A. Steyermark 49214 / 1942 / "'hedi-ondilla de montaña'"

Brit. Honduras / P. H. Gentle 3324 / 1940 / "'Barra blanca,' 'wild spice'"

2212 *C. draco* Mexico / R. Cárdenas 343 / 1910 / "...tenga algun tanino, cura las ulseras y sirve para vilmas" / "'Sangregrado'"

2213 *C. watsonii* Yucatan / J. Bequaert 114 / 1929 / "Indians burn it to drive mosquitoes away. 'Tan-che'"

2214 *C. jucundus* Mexico / T. S. Brandegee s.n. / 1904 / "(fragrant)" / "'Yerba Buena'"

2215 *C. repens* El Salvador / S. Calderón 966 / 1922 / "'Tostoncillo'"

Honduras / S. F. Blake 7429 / 1919 / "Used for stomache trouble" / "'Chacotole'?"

Guatemala / P. C. Standley 74927 / 1940 / "A remedy for dysentery" / "'Tostoncillo'"

2216 *C. xalapensis* Guatemala / H. Pittier 8590 / 1919 / "'Sangre drago'"

Guatemala / J. A. Steyermark 51334 / 1942 / "Used in connection with tooth-ache and sore gums" / "'Lloro sangre'"

Guatemala / J. A. Steyermark 51097 / 1942 / "Sap reputed to be used in treating wounds and for dressings" / "'drago'"

2217 *C. payaquensis* El Salvador / P. C. Standley 20941 / 1922 / "'Friega-plato'"

2218 *C. heterochrous* Honduras / A. Molina R. 1797 / 1948 / "'Pela naríz'"

2219 *C. fruticulosus* Mexico / G. B. Hinton et al. 16894 / 49 / "Leaves pungent"

2220 *C. flavescens?* Mexico / Leavenworth & Hoogstraal 1725 / 1941 / "...leaves aromatic"

2221 **C. cladotrichus** Mexico / E. Langlassé 228 / 1898 / "...employée par les indigènes pour guérir les conpures"? / "'Gordo lobo'"

2222 **C. pottsii (C. corymbulosus)** Mexico / G. Thurber 706 / 32 or 12 / "Used as tea" / "'Bruni tea'"

2223 **C. ciliato-glandulosus** Mexico / R. L. Dressler 989 / 1949 / "...said to irritate the eyes" / "'Trucha'"

Honduras / Yuncker, Dawson & Youse 5656 / 36 / "Crushed leaves more or less aromatic"

Guatemala / W. C. Shannon 3659 / 1892 / "'Yerba mala'"

Mexico / P. C. Standley 1023 / 1934 / "...avoided by the natives; 'malo por los manos.' I carelessly rubbed my eyes after handling, causing them to water and smart painfully"

Mexico / P. C. Standley 1053 / 1934 / "...avoided by natives" / "'Malo por los ojos'"

Guatemala / J. A. Steyermark 46443 / 1942 / "Leaves reputed, when contacting eyes, to cause blindness or inflammation of eyes"

Honduras / P. C. Standley 53897 / 1927 / "...shrub...with aromatic odor"

Honduras / P. C. Standley 55951 / 1928 / "'Ciega ojo'"

2224 **C. dioica** Mexico / J. G. Schaffner 871 / 1876 / "'Yerba del zorrillo'"

Mexico / E. K. Balls 4508 / 1938 / "Strong unpleasant smell"

2225 **C. lotorius** Guatemala / J. A. Steyermark 51332 / 1942 / "Used in connection with baths" / "'Sanolotoda'"

2226 **C. pagi-veteris** Guatemala / J. A. Steyermark 51328 / 1942 / "'Copalchi'"

2227 **C. pseudochiya** Mexico / C. L. & A. A. Lundell 7719 / 1938 / "'Icche' (Chili tree)"

2228 **C. tonduzii** Costa Rica / A. F. Skutch 4027 / 1939 / "'Copalchí'"

2229 **C. cubanus** Cuba / R. Combs 120 / 95 / "...aromatic"

2230 **C. astroites** Virgin Is. / W. C. Fishlock 213 / 1919 / "'White Maron'"

W.I.: Montserrat / J. S. Beard 423 / 1944 / "...leaves aromatic" / "'Black balsam'"

2231 **C. lucidus** Cuba / R. Combs 28 / 95 / "...aromatic shrub"

2232 **C. flavens** W.I.: Grenadines / P. Beard 1199 / 1945 / "leaves aromatic"

Virgin Is. / W. C. Fishlock 258 / 1919 / "'Red Maron'"

Virgin Is. / W. C. Fishlock 257 / 1918 / "'White Maron' used medicinally for bladder trouble"

W.I.: Curaçao / Curran & Haman 213 / 1917 / "'Wild salie'"

Venezuela / H. Pittier 10200 / 1922 / "...odorous"

2233   *C. spiralis* W.I.: San Domingo / J. J. Jiménez 803 / 1945 / " . . . with the odor of *Lippia dulcis* Trev."

2234   *C. discolor* Virgin Is. / J. S. Beard 321 / 1944 / "grazed by cattle" / "'Mehicky'"

2235   *C. linearis* Bahama Is. / G. R. Proctor 8836 / 1954 / "'Sarsaparilla'"
W.I.: Cayman Is. / G. R. Proctor 15058 / 1956 / "'Rosemary'"
Bahama Is. / O. Degener 19023 / 1946 / "'Marigold'; used for women's diseases"

2236   *C. poitaei* Haiti / E. C. Leonard 4141 / 1920 / " . . . sage-like odor"

2237   *C. guildingii* W.I.: St. Lucia / P. Beard 1131 / 1945 / "leaves aromatic"

2238   *C. populifolius* W.I.: Trinidad / W. E. Broadway 6386 / 1926 / "Cut portions odoriferous"
Venezuela / H. Pittier 7642 / 1917 / "Alcoholic maceration of bark against rheumatism" / "'Carcanapire'—Med."

2239   *C. bixoides* W.I.: St. Lucia / J. S. Beard 476 / 1945 / "leaves aromatic"

2240   *C. wilsonii* Jamaica / G. R. Proctor 16183 / 1957 / "Shrub with aromatic leaves" / "'Wild camphor'"

2241   *C. wullschlagelianus* W.I.: St. Lucia / J. S. Beard 520 / 1945 / " . . . leaves aromatic"

2242   *Croton* indet. W.I.: St. Lucia / P. Beard 1002 / 1945 / "leaves aromatic"

2243   *C. ferrugineus* Colombia / Bro. Daniel 2280 / 1940 / "'Drago'"

2244   *C. gossipifolius* Venezuela / H. Pittier 7532 / 1917 / "'Sangre-drago' . . . reddish, gummy sap"
Venezuela / L. Williams 12931 / 1940 / " . . . bark . . . exudes a deep reddish resin, whence the local name, which is used for cuts" / "'Sangrito'"

2245   *C. lechleri* Colombia / G. Klug 1831 / 1930 / "'Sangre de drago'"

2246   *C. magdalensis* Colombia / Bro. Daniel 2617 / 1940 / "'Drago'"

2247   *C. polycarpus* Colombia / J. Cuatrecasas 8057 / 1940 / "'Drago'"

2248   *C. rhamnifolius* Venezuela / H. Pittier 13101 / 1929 / "'Amargoso'"
Venezuela / L. Williams 12852 / 1940 / "'Carcanapire'"

2249   *C. scaber* Venezuela / L. Williams 11122 / 1939 / "'Sarasara'"

2250   *Croton* indet. Venezuela / H. Pittier 10483 / 1922 / " . . . aromatic and medicinal"

2251   *Croton* indet. Venezuela / Curran & Haman 984 / 1917 / "'Pardillo'"

2252   *C. palanostigma* Peru / Y. Mexia 6512 / 1932 / "Thick reddish juice used medicinally" / "'Sangre de drago'"

2253 *Croton* indet. Peru / J. F. Macbride 2882 / 1923 / "Frs. with strong aromatic-lemon odor"

2254 *C. peltophorus* Bolivia / J. Steinbach 5126 / 1920 / " 'Cardenillo' " / "Venenoso"

2255 *C. cajucara* Brazil / R. Froes 1701 / 1932 / " . . . used for fevers" / " 'Santa Maria' "

2256 *Croton* indet. Brazil / J. E. Leite 3753 / 1946 / "Nome vulgar: 'Sangue d'agua' "

2257 *Croton* indet. Brazil / H. M. Curran 568 / 1918 / " 'Angelico' "

2258 *C. hibiscifolius* Argentina / A. P. Rodrigo 923 / 1936 / " 'Iburá caaberá' " " 'Sangre de drago' "

2259 *C. lobatus* Uruguay / Rosengurtt B-3355 / 1940 / " . . . hedionda al refregar entre los dedos"

Argentina, Uruguay or Paraguay / T. Meyer 2.222 / 1937 / "planta de olor fétido"

2260 *C. sarcopetalus* Argentina / Schreiter 1052 / 1919 / " 'Curajero' "

2261 *Croton* indet. Argentina / A. P. Rodrigo 953 / 1936 / " 'Iburá caaberá' " " 'Sangre de drago' "

2262 *Croton* indet. Argentina / J. E. Montes 1916 / 46 / " 'Falso Arazá' "

2263 *C. cathayensis* China / W. T. Tsang 27775 / 1937 / " . . . used for medicine"

China / W. T. Tsang 28017 / 1937 / " . . . used for medicine"

2264 *C. crassifolius* Hainan / S. K. Lau 470 / 1932 / "Used to heal wounds by mixing the roots with wine"

2265 *C. tiglium* Hainan / S. K. Lau 20 / 1932 / "Natives . . . use the crushed fruit as fish bait"

Hainan / C. I. Lei 333 / 1933 / " . . . fruit, poisonous" / "This plant has poisonous substance. Water immersed with crushed barks or leaves may kill fish. Even only the bark or fruits may poison fish"

Hainan / F. A. McClure 20111 / 1932 / "Bruised leaves used to stupefy fish. (At Lik Faan Ts'uen) (Ngai Dist.) The Lois claim that only the seeds are used for this purpose"

Hainan / F. A. McClure 112 / 1922 / "seeds used to stupefy fish in streams and ponds so they can be caught easily; the plant is half cultivated and half wild"

China / H. H. Hu 123 / 1920 / "poisoning fish" / " 'Pa Tau' "?

Indochina: Tonkin / B. Balansa 4375 / 1891 / "Au dire des Annamites ses graines seraient un violent poison"

Laos / Poilane 13517 / before 1929 / "Fruit très toxique, employé pour tuer les vers dans les plaies animaux (le piler et appliquer le jus)"

Indochina / M. Poilane 26076 / before 1937 / "Purgatif: prendre 5 où 6 feuilles, les piler, les faire sécher au soleil mélanger avec fruit de tamargier et boire avec de l'eau"

Philippine Is. / E. D. Merrill 433 / 1902 / "Fruit used by natives to poison fish"

Philippine Is. / R. S. Williams 2751 / 1905 / "Fruit used to catch fish"

Philippine Is. / Ahern's collector 241 / 1904 / " . . . used in the practice of medicine and for poisoning fish. The source of Croton oil (*Oleum tiglium*). T., *Macaisa, Tuba*"

Philippine Is. / E. D. Merrill 308 / 1913 / "The croton oil plant is commonly cultivated about dwellings throughout the Philippines, its chief use being to stupefy fish. Certainly a purposely introduced plant in the Philippines"

2266   *C. delpyi* Cambodia / Hahn 133 / before 1897 / "Trâ-poung: employé contre la gale, effusion chaude"?

2267   *C. robustus* Burma / C. E. ? Parkinson 14930 / 32 / "Burmese medicinal plant" / "'Thit-yin'"

2268   *C. siamensis* Cambodia / Poilane 14671 / 1928 / "Feuilles seraient chiquées avec le bétel—"

2269   *Croton* indet. Malaya / G. anak Umbai for A. H. Millard (K.L. 1555) / 59 / "Phytochemical Survey of Malaya" / "Poisonous"

2270   *C. pampangensis* Philippine Is. / M. D. Sulit 4570 / 1952 / "Roots obtained put them on fire and let smoke inhale by hunting dogs. Smell of dogs become sharp on detecting wild pigs or deer" / "'Salimaong-kahoi'" Tag.

2271   *Croton* indet. Philippine Is. / H. C. Conklin 1162 / 1958 / "roots very bitter" / "'pit pita'" Han.

2272   *Croton* indet. Philippine Is. / C. Frake 346 / 1957 / "Leaves pounded as fish poison" / "'Puti? ganasi'" Sub.

2273   *Croton* indet. Philippine Is. / E. Maliwanag 184 / 1941 / " . . . fruits poisonous" / "'Camaysa'" Mang.

2274   *C. phebalioides* Australia: Queensland / Webb & White (Herb. Aust.) 1139 / 46 / " . . . stem and leaves with a very strong 'cascarilla' odour when cut or crushed. Local name 'Snuff Bush'"

2275   *C. verreauxii* Australia: Queensland / C. T. White 12573 / 1944 / " . . . bark with faint aromatic smell"

Australia: Queensland / C. T. White 11401 / 1937 / " . . . bark with a strong cascarilla scent when cut"

2276   *Eremocarpus setigerus* Africa: Piguetberg / A. J. Toerien 105 / 1936 / "collected as possible menace 'noxious weed' to be"

2277   *Sumbaviopsis albicans* Laos / Poilane 20220 / before 1933 / "Tiges & feuilles contre fièvre"

2278   *Chiropetalum griseum* Bolivia / J. Steinbach 5069 / 1920 / "contacto un poco irritante"

2279   *Caperonia palustris* Brit. Guiana / A. S. Hitchcock 16701 / 1919 / "Has stinging hairs"

2280   *Claoxylon indicum* Hainan / F. A. McClure 7624 / 1921 / "drug plant" / "'Siu liu pang,' 'Pak t'ung shu'"

2281   *Claoxylon* indet. Solomon Is. / S. F. Kajewski 1604 / 1930 / "The young leaves are gathered and eaten by the natives after being boiled like cabbage"

2282   *Claoxylon* indet. Solomon Is. / S. F. Kajewski 2506 / 1931 / "When a native is mad the leaves of this tree are boiled and given to him to drink" / "'Comma'"

2283   *Claoxylon* indet. Solomon Is. / S. F. Kajewski 2549 / 1931 / "The natives say that the sap of this tree is poisonous, while the bark is taken and rubbed on the abdomen for pains there" / "'Pa Pa'"

2284   *Claoxylon* indet. Solomon Is. / S. F. Kajewski 2036 / 1930 / "Young leaves are boiled and eaten as a vegetable" / "'Molletu'"

2285   *Claoxylon* indet. Solomon Is. / S. F. Kajewski 2376 / 1930 / "The leaves are boiled as a vegetable and eaten by the natives" / "'Soula'"

2286   *Claoxylon* indet. Solomon Is. / S. F. Kajewski 1882 / 1930 / "The young leaves are eaten by the natives after being boiled" / "'Te-con-ark'"

2287   *Claoxylon* indet. Solomon Is. / S. F. Kajewski 1824 / 1930 / "The young leaves are eaten by the natives after being boiled" / "'Mol-letu'"

2288   *Claoxylon* indet. Solomon Is. / S. F. Kajewski 1650 / 1930 / "Leaves used by the natives for medicine for headaches"

2289   *Claoxylon* indet. Philippine Is. / C. O. Frake 647 / 1958 / "Leaves applied to stomach for dysentery" / "'Tingkugas'"

2290   *Claoxylon* indet. Philippine Is. / C. O. Frake 511 / 1958 / "Carabas medicine" / "'kulen bia'" Sub.

2291   *Claoxylon* indet. Philippine Is. / G. E. Edaño 2015 / 1950 / "Leaves effective medicine for itchy body" / "'Halingaton'" Vis.

2292   *Mercurialis annua* Bermuda Is. / F. S. Collins 224 / 1913 / "'Stink-weed' (of white inhabitants) 'Mockery' (of negro people)"

2293   *Mallotus apelta* China / W. T. Tsang 21312 / 1932 / " . . . used as medicine"

        Hainan / F. A. McClure 7538 / 1921 / " . . . drug plant; mix leaves with *Noh mai* (glutinous rice) and use as salve for itchy rash . . ." / "'Pak pui ip'"

2294 *M. furetianus* Hainan / Tsang, Tang & Fung 116 / 1929 / " . . . used as medicine"

2295 *M. lianus* China / H. H. Hu 127 / 1920 / "Leaves used for healing cuts and similar injuries" / "'Yi Toong-Tze'"

2296 *M. microcarpus* China / Lingnan Herb. 147.4385 / 1924 / "'Mo Muk tan' . . . (Hairy wood remedy)"

2297 *M. philippinensis* Hainan / S. K. Lau 121 / 1932 / "The fruit and leaves are bitter. Water in which they have been boiled is used as a remedy for colds"

Philippine Is. / Ahern's collector 23 / 1904 / " . . . little or not at all used by the natives, although it has medicinal properties" / "T., *Banato*; I., *Buas*"

2298 *M. poilanei* Indochina / M. Poilane 26084 / before 1937 / "Racine tonique et aphrodisiaque"

2299 *M. lancifolius* Malaya / G. anak Umbai for A. H. Millard (K.L. 1636) / 59 / "Phytochemical Survey of Malaya" / "Poisonous"

2300 *M. auriculatus* Philippine Is. / C. O. Frake 720 / 1958 / "Roots chewed with betel and applied to ulcers" / "'Sungmay'" Sub.

2301 *Mallotus* indet. Philippine Is. / C. O. Frake 543 / 1958 / "Leaves applied for spleenomegaly" / "'tiolu gletin'"

2302 *Coccoceras muticum* Malaya / G. anak Umbai for A. H. Millard (K.L. 1495) / 59 / "Phytochemical Survey of Malaya" / "Poisonous fruits . . . "

2303 *Alchornea cordatum* Colombia / G. Klug 1745 / 1930 / "'Palo de reuma,' medicinal"

2304 *A. cordifolia* Sierra Leone / G. F. S. Elliot 4179 / before 1924 / "Christmas bush, leaves used as a hot press for feet . . . "

2305 *Adelia barbinervis* El Salvador / P. C. Standley 22161 / 1922 / "'Macaguite,' 'Espino blanco' . . . Pulverized seeds put on hair to make it smooth"

2306 *Cleidion spiciflorum* Siam / Kasin 169 / 1946 / "fruit and leaves medicinally used" / "'di mee'"

Philippine Is. / L. ela Ebalo 505 / 1940 / "fts. poisonous" / "'tuba-tuba'"

2307 *Cleidion* indet. Solomon Is. / S. F. Kajewski 2305 / 1930 / "The fruit are eaten with betel nut to relieve colds" / "'Nu marrie'"

2308 *Macaranga bracteata* E. ASIA / W. T. Tsang 593 / 1928 / "medicine"

2309 *M. hypoleuca* Sumatra / W. N. & C. M. Bangham 1095 / 1932 / "Stems used to make poison for insects and crawfish"

2310 *M. winkleri* Brit. N. Borneo / B.N.B. For. Dept. 3166 / 33 / "Leaves used for medicine. Wood boiled for medicine, stomach head and bone ache"

2311   *M. triloba* Malaya / G. anak Umbai for A. H. Millard (K.L. 1529) / 59 / "Phytochemical Survey of Malaya" / "Poisonous fruits"

2312   *M. densiflora* Solomon Is. / S. F. Kajewski 2296 / 1930 / "Leaves have a strong liquorice scent when drying" / "'Aba-poru'"

2313   *M. urophylla* Solomon Is. / S. F. Kajewski 2314 / 1930 / "The underneath of the leaves have a silky feel and are used by the natives to dry themselves after swimming" / "'Su-a-mangu'"

   Solomon Is. / S. F. Kajewski 2557 / 1931 / "The leaves are macerated and applied to sore legs by the natives" / "'Pasah'"

2314   *M. hispida* Philippine Is. / Sinclair & Edaño 9527 / 1958 / "Irritating brittle hairs on twigs and petioles"

2315   *M. carolinensis* Caroline Is. / C. C. Y. Wong 139 / 1947 / "The leaves are used in wrapping poi. (Elbert says the leaves are used to cover steaming breadfruit)" / "'tuup'"

2316   *M. dioica* New Hebrides Is. / S. F. Kajewski 319 / 1928 / "Dried leaf bandaged over sore to heal it" / "'Norvo-among'"

2317   *M. vitiensis* Fiji Is. / O. Degener 15182 / 1944 / "Fijians crush leaf and drink, water added, for diarrhoea" / "'Davo'" Serua

2318   *Acalypha lindheimeri* U.S.: San Bernadino / G. Thurber 354 or 3521 / 1857? / "(Known as a purgative)"

   Mexico / R. M. Stewart 487 / 1941 / "'Yerba del Cancer'"

   Mexico / R. M. Stewart 1918 / 1941 / "'Yerba de Cancer'"

   Mexico / R. M. Stewart 571 / 1941 / "'Yerba de la Vibora'"

2319   *A. phleoides* Mexico / J. Gregg 13 / 1847 / "Yerba del Cancer—famous in decoction, as wash for sore gums & loose teeth—also for foul ulcers, cancer, etc. . . . Castañuela"

   Mexico / A. Dugès 193 / no year / "'Yerba del cáncer'"

2320   *A. vagans* Mexico / M. Bourgeau 2954 / 1866? / "'Yerba del Pastor de Puebla'"

2321   *A. hypogaea* Mexico / G. B. Hinton 10572 / 37 / "'Yerba de Cancer'"

2322   *A. hederacea* Mexico / R. Santos for R. M. Stewart 2943 / 1942 / "'Yerba del Cancer'"

   Mexico / R. M. Stewart 877 / 1941 / "'Yerba del Cancer'"

   Mexico / R. M. Stewart 301 / 1940 / "'Yerba de Cancer'"

2323   *A. arvensis* Brit. Honduras / W. A. Schipp 212 / 1929 / "'Common' weed . . . often used by the 'Caribs' for medicinal purposes; the virtues it may contain are more imaginary than otherwise"

2324   *A. matudai* Mexico / E. Matuda 4520 / 1941 / "Reported to be 'edible'"

2325   *A. septemloba* Guatemala / P. C. Standley 59156 / 1938 / "'Yerba del cancer' . . . Used medicinally"

2326 *A. cuspidata* W.I.: Curaçao / Curran & Haman 5 / 1917 / "Yeerba die Sangre"

2327 *A. boehmerioides* Indochina / Eberhardt 2279 / 1916 / "médicament contre les maux de tête"

2328 *A. novo-guineensis* Solomon Is. / S. F. Kajewski 2583 / 1931 / "When a man has sore eyes, the leaves are put into boiling water and the steam allowed to go into the eyes" / "'Hambusi'"

Solomon Is. / S. F. Kajewski 2216 / 1930 / "The leaves of this plant are heated and applied to a swelling caused by a knock. The bruise immediately subsides. Maybe the action of the heat has a lot to do with it" / "'Taccara'"

Solomon Is. / S. F. Kajewski 2490 / 1931 / "The leaves are rubbed onto wounds by natives. This seems to be fairly general among the South Seas natives" / "'Garm-bu-ve'"

2329 *A. cardiophylla* Philippine Is. / C. O. Frake 766 / 1958 / "Leaves applied for headache" / "'gebukal'" Sub.

Philippine Is. / C. O. Frake 755 / 1958 / "Bark applied to deep ulcers" / "'Selimbugaun'"

Philippine Is. / Sulit & Conklin 5156 / 1953 / "Leaves of maslakot and those of banaba and panagurirong taken together and placed near 'balatik' so that wild pigs will pass through path where 'balatik' is set" / "'Maslakot-buladlad'"

2330 *A. trukensis* Caroline Is. / C. C. Y. Wong 237 / 1947 / "The flowers are used as medicine for the eyes, the leaves are pounded, expressed in a coconut cloth, mixed with water and put in the eyes. (Elbert says the wood is used for fire by friction, breadfruit picking poles and in houses)" / "'mannou'"

Caroline Is. / C. C. Y. Wong 237A / 1947 / "The flowers are used as medicine for the eyes by pounding the flowers, expressing with a piece of coconut cloth, mixing with water and putting in the eyes . . ." / "'mannou'"

2331 *A. wilkesiana* Fiji Is. / O. Degener 15444 / 1941 / "If testicles swollen then mix leaves of this and of Panax, squash in water and drink" / "'Ruru'" Ra

2332 *Epiprinus malayanus* Malaya / Gadoh (K.L. 1483) / 59 / "Phytochemical Survey of Malaya" / "'Kayu Rengkow'" Temuan

2333 *Platygyne hexandra* Cuba / Howard, Briggs, Kamb, Land & Ritland 81 / 1950 / "Vine with stinging hairs"

2334 *Acidoton urens* Jamaica / W. Harris 8682 / 1904 / "'Cow-itch'"

Jamaica / G. R. Proctor 8629 / 1954 / " . . . with stinging hairs"

2335 *Tragia urticaefolia* U.S.: La. / R. McVaugh 8473 / 1947 / "Stinging when touched . . . "

2336   *T. nepetifolia* Mexico / G. B. Hinton 10621 / 1937 / "Poisonous to touch" / "'Quemador'"

Mexico / Wiggins & Rollins 69 / 1941 / " . . . numerous stinging hairs throughout . . . "

2337   *T. teucrifolia* Mexico / E. Seler 1062 / 1895 / "'Ortiguilla'"

2338   *T. volubilis* Costa Rica / A. Smith NY736 / 1938 / "parts of plants sting severely"

Jamaica / G. R. Proctor 7773 / 1953 / " . . . with stinging hairs"

Fr. Guiana / W. E. Broadway 51 / 1921 / " . . . stinging nettle . . . "

2339   *T. yucatanensis* Brit. Honduras / W. A. Schipp 1211 / 1933 / " . . . has same stinging properties as the common nettle"

2340   *Tragia* indet. Mexico / H. S. Gentry 5613 / 1940 / "Vine with 'hot' nettles. Reported to make large areas of the mountain impassable" / "'Tachinola'"

2341   *T. incana* Uruguay / Rosengurtt B-2506 / 1938 / "'ortiga blanca'"

2342   *T. hispida* S. India / E. W. Erlanson 5217 / 1933 / " . . . with stinging hairs"

2343   *T. hirsuta* Philippine Is. / H. C. Conklin 526 / 1953 / "Leaves = skin irritant" / "'Dalāmu'"

2344   *Plukenetia volubilis* Peru / Killip & Smith 29927 / 1929 / "'Sacha-inchik' (wild peanut); seeds roasted and eaten . . . "

2345   *Dalechampia tiliaefolia* W.I.: Trinidad / W. E. Broadway 9146 / 1933 / "with stinging hairs"

2346   *D. clausseniana* Brazil / F. C. Hoehne 62 / 1917 / "'Cipó de São José'"

2347   *D. ipomoeaefolia* Liberia / D. H. Linder 23 / 1926 / " . . . with easily detachable stinging hairs"

2348   *Pera obovata* Brazil / F. C. Hoehne 27423 / 1931 / "'Páo de Sapateiro'"

2349   *Ricinus communis* Tonga Is. / T. G. Yuncker 15,110 / 1953 / "Leaves used for wrapping shellfish"

2350   *Homonoia riparia* Siam / A. Kostermans 254 / 1946 / " . . . mashed and powdered leaves are used against skin eruptions" / "'w(u)ee'"

2351   *Aleurites moluccana* Dutch New Guinea / P. van Royen 3828 / 1954 / " . . . seeds brownish black, kernel edible" / "'nabi' (Andjai dialect)"

Sumatra / W. N. & C. M. Bangham 941 / 1932 / "Fruits edible"

E.I. / Dr. Henry s.n. / no year / "'A good Nut'"

Tonga Is. / U.S. S. Pacific Exploring Exped. Herb. s.n. / 1838–42 / "Tonga tabu"

Fiji Is. / A. C. Smith 23 / 1933 / "Oil from fruit used on hair and skin" / "'Nggerenggere'; 'Lauthe'"

Cook Is. / T. G. Yuncker 9702 / 1940 / "Nuts wrapped with bananas helps ripen green fruit. Nuts used originally for torches. Also good for new baby to eat and to bathe with it" / "'tuitui'"

Fiji Is. / O. Degener 14957 / 1941 / "Serua native says meat of nut not eaten. Fijians used oil from kernel"

2352 *Aleurites* indet. Philippine Is. / Sulit & Conklin 5173 / 1953 / "Kernel eaten by Mangyan together with bananas (Sabá)" / "'Kamiri'" Mang. (Hanunoo)

2353 *Jatropha cardiophylla* U.S.: Ariz. / S. D. McKelvey 1529 / 1930 / "'Sangre de draca'"

2354 *J. podagrica* Nicaragua / Williams & Molina R. 11956a / 1946 / "this is said to be used for malaria" / "'Panzón'"

El Salvador / P. C. Standley 19376 / 1921–22 / "Cultivated" / "'Ruibarbo'"

2355 *J. macrorhiza* Mexico / G. Thurber 354 / no year / "used by the Mexicans as a purgative"

2356 *J. curcas* El Salvador / P. C. Standley 20578 / 1922 / "'Tempate' . . . milk used for eruption on lips, also applied in front of ears for eye inflammation"

W.I.: Dominica / W. H. Hodge 3207 / 1940 / " . . . boiled lves used for tea for sick"

Colombia / O. Haught 4057 / 1944 / "Reported to be poisonous"

Bolivia / J. Steinbach 6636 / 1924 / "La recina es buena para cerrar heridas" / "20 a 30 frutas constituyen un purgante drastico" / "'Piñon'"

Ecuador / A. S. Hitchcock 20594 / 1923 / "'Physic nut,' 'Piñon'"

Tonga Is. / T. G. Yuncker 15637 / 1953 / " . . . leaves used in medicine" / "'Fiki'"

2357 *J. cinerea* Mexico / H. S. Gentry 5422 / 1940 / "Raw juice applied to sores & employed in treating chills" / "'Sangrengrado'"

2358 *J. dioica* Mexico / G. B. Hinton 16610 / 46 / "Tea used to tighten teeth, invigorate hair" / "'Sangre de Grado'"

Mexico / A. H. Schroeder 155 / 1941 / "'Drago'"

Mexico or Central America / Dr. Gregg s.n. / no year / "'Sangre de Drago'" / "'Sangregrada'"

2359 *Jatropha* indet. Mexico / G. B. Hinton 13938 / 39 / "supposed medicinal qualities"

2360 *J. multifida* Dominican Republic / R. A. & E. S. Howard 9192 / 1946 / "Foliage smells like insect repellent"

2361 *J. gossypifolia* W.I.: Dominica / W. H. & B. T. Hodge 1611 / 1940 / "'Physic nut'—medicine"

Peru / G. Klug 3974 / 1933 / "'Piñon negro'"

Brazil / Y. Mexia 4477 / 1930 / "'Mandioca brava'"

2362 *J. divaricata* Jamaica / G. L. Webster 5261 / 1954 / "'mountain nut'"

Jamaica / J. R. Perkins 1395 / 1917 / "'Wild oil nut'"

2363 *Jatropha* indet. W.I.: Dominica / D. Taylor 132 / 1942 / "'medicinier blanc'"

2364 *Jatropha* indet. W.I.: Dominica / D. Taylor 131 / 1942 / "'medicinier noir' (larger leaved)"

2365 *J. angusti* (or *J. ciliata*) Peru / J. West 3796 / 1935 / "...reputed an aphrodisiac" / "'Huanarpo machu'"

Peru / C. Vargas C. 408 / 1936 / "planta afrodisiaca" / "'Huanarpo macho'"

2366 *J. urens* Brazil / R. Froes for B. A. Krukoff 1834 / 1932 / "'Ortiga de rato'...Root used for hydropsy"

2367 *Jatropha* indet. Bolivia / J. Steinbach 5133 / 1920 / "La raíz...sirve de purgante drastico"

2368 *Cnidoscolus angustidens* U.S.: Ariz. / Kearney & Peebles 14873 / 1940 / "stinging hairs"

Mexico / Leavenworth & Hoogstraal 1416 / 1941 / "'Mala mujer'"

Mexico / S. S. White 531 / 1938 / "Spiny, stinging herb"

2369 *C. stimulosus* U.S.: Fla. / L. J. Brass 14667 / 1945 / "'Nettle'" / "hairs stinging"

U.S.: Fla. / A. Traverse 648 / 58 / "Stinging hairs painful on contact; makes red welts"

2370 *C. multilobus* Mexico / R. L. Dressler 1643 / 1954 / "Very potent sting"

2371 *C. souzae* Mexico / N. Souza Novelo s.n. / 1944 / "En veces es usada con fines medicinales; parece que es diuretica" / "'Tsah'" Maya

2372 *C. tepiquensis* Mexico / J. C. Hinton 1 / 1942 / "Has a limited use as a masticatory;...sporadically it has been exploited for foreign consumption, where it may have been destined to the substitution of rubber" / "'Chilte,' 'Chicle,' 'Copal,' 'Hule Blanco'"

2373 *C. tubulosus* Mexico / R. Q. Abbott 41 / 1936 / "'Mala mujer'"

Peru / Y. Mexia 8051a / 1936 / "...violently stinging hairs on stem" / "'Vara de Angel'"

2374 *C. urens* Panama / Woodson, Jr., Allen & Seibert 1226a / 1938 / "an irritating 'ortiga'"

W.I.: Grenada / G. R. Proctor 16936 / 1957 / "...very bad stinging hairs" / "'Devil-nettle'"

Colombia / O. Haught 3683 / 1943 / "Seeds said to be used by Indians as purgative. 'Pringamoza'"

Fr. Guiana / W. E. Broadway 248 / 1921 / " . . . stinging hairs"

2375 *Cnidoscolus* indet. Mexico / G. B. Hinton 13908 / 39 / "stinging"

2376 *C. regina* Cuba / Bro. Léon 17176 / no year / "'Sabrosa'"

2377 *C. cnicodendron* Argentina / Eyerdam & Beetle 22881 / 1938 / "Stinging hairs . . ."

2378 *C. loasoides* Argentina / T. Meyer 2.146 / 1936 / "En Corrientes tiene mucha fama como planta medicinal" / "'pug-nó'"

2379 *C. longipes* Colombia / J. Cuatrecasas 15271 / 1943 / "látex caústico; cultivada"

2380 *C. vitifolius* Argentina / Eyerdam & Beetle 22500 / 1938 / "Poisonous juice and stinging hairs"

2381 *Cnidoscolus* indet. Peru / J. F. Macbride 2865 / 1923 / " . . . pubescence . . . of younger parts stinging"

2382 *Cnidoscolus* indet. Brazil / D. B. Pickel 2822 / 1931 / "The twigs and leaves have stinging hairs"

2383 *Cnidoscolus* indet. Colombia / Barkley & Gutiérrez 1825 / 1947 / "especie muy urticante"

2384 *Micrandra glabra* Brit. Guiana / A. S. Pinkus 236 / 1939 / " . . . seed edible"

2385 *M. spruceana* Brazil / J. T. Baldwin, Jr. 3683 / 1944 / "Seeds eaten by natives; see Spruce for preparation"

Colombia / P. H. Allen 3063 / 1943 / "Seeds collected by the local Indians for food. Known as 'Wak-puh' (Tucano)"

2386 *Hevea* indet. Brazil / R. Froes 11960 / 1940 / "'Casca ardosa'"

2387 *Elateriospermum tapos* Borneo / J. & M. S. Clemens 20247 / 1929 / " . . . fruits . . . eaten by monkeys"

2388 *Manihot utilissima* El Salvador / P. C. Standley 22263 / 1922 / "'Yuca' . . . Large roots boiled as vegetable"

2389 *M. brevipedicellata* Brazil / Froes 12669/35 / 1942 / "'Maniva brava'"

2390 *M. labroyana* Brazil / Froes 12663-29 / 1942 / "'Maniva brava'"

2391 *M. palmata* Brazil / Froes 12669/35 / 1943 / "'Maniva brava'"

2392 *Microdesmis casearifolia* Malaya / G. anak Umbai for A. H. Millard (K.L. 1612) / 59 / "Phytochemical Survey of Malaya" / "Poisonous fruits. Wood employed in the manufacture of kris scabbards"

2393 *Trigonostemon howii* Indochina / M. Poilane 10220 / 1924 / "Employé contre la gale, bouillir les feuilles et se baigner"

2394 *Codiaeum variegatum* Solomon Is. / S. F. Kajewski 2286 / 1930 /
"The native women use their plant to produce abortion, but I think it is
used largely as they do many other medicines in a superstitious manner"
"Planted by natives along the roads and near their ceremonial
houses" / "'Pourtucki'"

Philippine Is. / C. O. Frake 587 / 1958 / "Drinking ritual" / "'Hali-
payan'" Sub.

2395 *Codiaeum* indet. Solomon Is. / S. F. Kajewski 1926 / 1930 / "This
plant is used as a lure to catch fish. A basket or trap is made of cane
and the plant is placed inside and weighted down with stones. It is
then lowered into a fresh water creek. The fish come in and cannot
get back. What the value of the lure is I cannot say, but it seems to
be used quite a lot" / "'Mala-monke-nume'"

2396 *Codiaeum* indet. Solomon Is. / S. F. Kajewski 2210 / 1930 / "The
roots of this tree are macerated with a rough dry fig leaf (native sand
paper) mixed with betel nut and water and then eaten for pains in the
stomach" / "'Mala-etui'"

2397 *C. luzonicum* Philippine Is. / W. G. Solheim II 5 / 1950 / "Medicine
stomachache, roots used by boiling and drinking liquid after cool"

2398 *Strophioblachia glandulosa* Indochina / M. Poilane 10407 / 1924 /
"Les jeunes feuilles seraient commestible"

2399 *Fontainea pancheri* New Hebrides Is. / S. F. Kajewski 906 / 1929 /
"The tree contains a remarkable poison throughout and is used for
poisoning fish"

2400 *Baliospermum montanum* Cambodia / M. Poilane 14519 / 1928 /
"Racines employées crues contre les maux de ventre: ça réchauffe
aussi la bouche"

2401 *Chaetocarpus globosa* Dominican Republic / E. J. Valeur 288 / 1929 /
"'Guacima Cimmaron'"

2402 *Mabea fistulifera* Brazil / Y. Mexia 4473 / 1930 / "Hollow twigs used
for pipe stems, hence the name: 'Canudo de pito'"

2403 *Sebastiana pringlei* Mexico / H. S. Gentry 2329 / 1936 / "The bark
pulverized is used in poisoning fish; report Warihio" / "'Brincador,'
Mex."

2404 *S. ramirezii* Mexico / C. V. Hartman 568 / 1891 / "The most powerful poison for fish" / "'Yerba de la flecha'"

2405 *S. pachyphylla* Peru / C. Vargas C. 1969 / 1940 / "Usam como vomi-
purgante, en mui pequeña cantidad (el latex)" / "'azucanillo'"

2406 *Sebastiana* indet. Bolivia / J. Steinbach 5052 / 1920 / "Dicen que las
ovejas freñadas que comen esta planta malparen" / "'Turujiro'"

2407 *S. chamaela* var. *asperococca* Hainan / S. K. Lau 483 / 1932 / "Fr.
edible"

2408 *Excoecaria cochinchinensis* Indochina: Tonkin / B. Balansa 4980 / 1890–91 / "Arbuste cultivé autour des pagodes"

2409 *E. cochinchinensis* var. *viridis* Indochina / Poilane 10477 / 24 / "...le latex provoque des démangaisons"

2410 *E. agallocha* Fiji Is. / A. C. Smith 1 / 1933 / "...bark used for medicine" / "'Sinu'"

2411 *Homalanthus novoguineensis* Dutch New Guinea / L. J. Brass 13268 / 1939 / "...fruit attractive to birds"

2412 *H. nutans* Solomon Is. / S. F. Kajewski 2421 / 1931 / "When a man has been infected by an evil spirit the sap of this tree is drunk to get rid of this spirit" / "'Coondou'"
Solomon Is. / S. F. Kajewski 1938 / 1930 / "The leaves of this tree are pounded up and put into the water to poison fish" / "'Tim-bar-ci'"
Solomon Is. / S. F. Kajewski 2537 / 1930 / "The milky sap of this tree is drunk for gonorrhea" / "'Nu-numbu'"
Fiji Is. / O. Degener 15134 / 1941 / "Fijians mash leaves in water and drink juice as medicine for stomach trouble" / "'Datau'" Serua

2413 *Homalanthus* indet. Solomon Is. / S. F. Kajewski 2158 / 1930 / "The natives say the Possums eat the fruit and leaves" / "'Tini'"

2414 *H. populneus* Philippine Is. / P. Añonuevo 258 / 1950 / "Fish poison" / "Pounded leaves mixed with ashes and pepper are sprayed in river" / "'Balanti'" Bis.

2415 *Pimeleodendron papuanum* Solomon Is. / S. F. Kajewski 2071 / 1930 / "Possums eat the leaves of this tree..." / "'U-nore'"

2416 *Pimeleodendron* indet. Solomon Is. / S. F. Kajewski 2464 / 1931 / "The bark of this tree is macerated with water and drunk for fever" / "'Boro-coca'"

2417 *Stillingia texana* U.S.: Tex. / R. McVaugh 8304 / 1947 / "Called locally 'Queen's Delight,' milky sap said to cause skin blisters and used to cure ringworm"

2418 *S. sylvatica* U.S.: La. / R. McVaugh 8472 / 1947 / "'Indian flea root'; dried plants said to repel fleas"

2419 *S. microsperma* Brit. Honduras / C. L. Lundell 6740 / 1936 / "...odor disagreeable, very strong"

2420 *S. acutifolia* Guatemala / P. C. Standley 80138 / 1940 / "'Yerba mala'"
Guatemala / P. C. Standley 61510 / 1938 / "'Yerba mala'...Sap milky. Said to be very poisonous"

2421 *Sapium appendiculatum* Mexico / W. P. Hewitt 266 / 1948 / "reportedly poisonous" / "'Hierba de la flecha'"
Mexico / H. S. Gentry 2345 / 1936 / "'Yerba la Flecha,' Mex....Bark used to poison fish and formerly the white sap to poison arrows"

2422 *S. biloculare* Mexico / S. S. White 2916 / 1940 / "Hierba de la flecha"

Mexico / S. S. White 401 / 1938 / "Shrub, with milky juice which is said to be poisonous" / "'Hierba de la flecha'"

Mexico / C. G. Pringle s.n. / 1884 / "'Yerba de fleche,' used by the Apache Indians to poison their arrows"

2423 *S. jamaicensis* Guatemala / J. A. Steyermark 37644 / 1940 / "'Mata-palo'"

W.I.: San Domingo / J. Schiffino 128 / 1944 / "'Daguilla'"

2424 *S. macrocarpum* El Salvador / P. C. Standley 22436 / 1922 / "'Chila-mate'... Reputed very poisonous to skin"

2425 *S. glandulosum* W.I.: Barbados / J. S. Beard 623 / 1945 / "Very caustic white latex" / "'Poison tree'"

2426 *S. gibertii* Uruguay / A. Lombardo 3048 / 1938 / "'Curupí' o 'Arbol de la leche'"

2427 *S. haematospermum* Uruguay / A. Lombardo 3047 / 1938 / "'Curupí' o 'Arbol de la leche'"

2428 *S. linearifolium* Uruguay / A. Lombardo 3344 / 1938 / "'Curupí'"

2429 *S. sebiferum* Hainan / F. A. McClure 8886 / 1922 / "drug and economic plant; roots yield a purgative; 'U kau shu'"

2430 *Sapium* indet. Indochina / E. Poilane 20923 / 1932 / "Arbre a latex blanc trés abondant, trés corosif; s'il faillinait dans les yeux, il pourrait faire perdre la vue, il brule la peau.... Utilisé pour la capturer le poisson de mer dans les mares fermés du bord de la mer, il serait sans éffect en eau douce..."

2431 *S. indicum* New Guinea / L. J. Brass 1168 / 1926 / "Used as a fish poison by the Hall Sound natives. The fruits are pounded up and mixed with grated coconut, which, saturated with the poison, is eaten by the fish with fatal results"

2432 *Grimmeodendron eglandulosum* Cuba / R. Combs 519 / 95 / "A poisonous shrub or tree along the coast"

2433 *Hippomane mancinella* Mexico / Y. Mexia 1132 / 1926 / "Said to be a violent emetic" / "'Manzanita'"

2434 *Colliguaya integerrima* Argentina / Eyerdam, Beetle & Grondona 24484 / 1939 / "Poisonous—especially for horses" / "'Duraznia'"

2435 *Gymnanthes lucida* Bahama Is. / O. Degener 19136 / 1946 / "'Crab-bush,' so-named because the land crabs eat the bark and berry; bark bitter and used for tea"

Bahama Is. / O. Degener 19134 / 1946 / "Tea from leaves said to be panacea for stomach trouble"

2436 *Hura* indet. Colombia / O. Haught 4138 / 1944 / "... caustic latex *much used as fish poison*. 'Ceibo brujo' or 'Barbasco'"

2437 **Hura crepitans** Brazil / Dr. Shattuck 208 / 1924 / "Said to be used as a remedy for leprosy" / "'Assacu'"

Venezuela / Curran & Haman 667 / 1917 / "'Ceiba purgante'"

2438 **Euphorbia eriantha** U.S.: Ariz. / R. H. Peebles 6401 / 1930 / "Roots malodorous"

2439 **E. hirta** U.S.: Mich. / O. A. Farwell 8756 / 1930 / "Probably an escape from drug imported for medicinal purposes"

Argentina / T. M. Petersen 2596 / 1954 / "'Santa Maria'"

New Caledonia / Prony 1703A / 1914 / "Employée contre la dysenterie"

2440 **E. fulva** Mexico / G. B. Hinton 8599 / 35 / "Uses: medicinal only" / "'Pega hueso'"

2441 **E. campestris** Mexico / G. B. Hinton 3994 / 33 / "Smells like a fox" / "'Yerba del Zorro'"

2442 **E. schlechtendalii** Mexico / H. S. Gentry 6749 / 1942 / "Milk used for poisoning fish . . . " / "'Palo de Leche'"

Mexico / R. L. Dressler 2330 / 1957 / "said to be very poisonous, caustic, used to poison fish"

2443 **E. colletioides** Mexico / H. S. Gentry 1034 / 1934 / "Medic. milk used for eyes" / "'Blanca Flora,' Mex. 'Bekachari,' W."

2444 **E. californica** Mexico / H. S. Gentry 7129 / 1945 / "Shrub, used as a purgative" / "'Sipehui,' 'Candelilla'"

2445 **E. sphaerorhiza** Mexico / G. B. Hinton 962 / 32 / "The sheath of the bulb is masticated like chewing gum" / "'Chicle'"

2446 **E. colorata** Mexico / H. S. Gentry 6219 / 1941 / "'Contra yerba,' 'Quanta yerba'"

2447 **E. heterophylla** El Salvador / P. C. Standley 22067 / 1922 / "Milk applied to wounds" / "'Chilamatillo,' 'Hierba del duende'"

El Salvador / S. Calderón 1099 / 1922 / "Latex very irritant"

Brit. Honduras / W. A. Schipp 463 / 1929 / " . . . used for medicinal purposes by Caribs"

Mexico / H. S. Gentry 1031 / 1934 / "Milk of this plant used for washing eyes" / "'Picachari,' Mex."

Mexico / H. S. Gentry 1673 / 1935 / "Milk for bad eyes" / "'Pikachalih,' W."

Indonesia: Amboina / Mrs. E. A. de Wiljes-Hissink 2 / 1948 / "Milky sap used instead of castor oil"

2448 **E. glomerifera** Mexico / G. B. Hinton 687 / 32 / "It is used as a purge in intestinal fever" / "'Celedonia'"

2449 **E. maculata** Mexico / H. S. Gentry 543M / 1933 / "Natives cook whole plants in water to bathe wounds" / "'La Golondrina'"

2450 *E. polycarpa* Mexico / C. Lumholtz 11 / 1910 / "(Interesting for medical uses by natives)"

2451 *E. cinerascens* Mexico / W. P. Hewitt 184 / 1947 / "When bruised, a milk exuded which is placed on boils and pimples" / "'Hierba de la Golondrina'"

2452 *E. fendleri* Mexico / G. Thurber 407 / 1887 / "Called by the Mexicans—Yerba la Golondrina—and considered a certain remedy in rattle snake and other poisonous bites—The bruised fresh herb applied to the wound or the dry plant steeped in urine" ("...wine"?)

2453 *E. densiflora* Mexico / G. B. Hinton 2998 / 32 / "Uses—to wash wounds" / "'Golondrina'"

2454 *E. neriifolia* El Salvador / P. C. Standley 19246 / 1921–22 / "Milk remedy for sore throat and lips and gonorrhoea" / "'Tuna francesa'"

2455 *Euphorbia* indet. Mexico / G. B. Hinton 1552 / 32 / "The bulb sheath is masticated like chewing gum" / "'Chicle'"

2456 *E. cotiniflora* W.I.: Tobago / W. E. Broadway 4518 / 1913 / "'Manchineel'"
Peru / Y. Mexia 6506 / 1932 / "Planted on hills of Curuhinis ants, said to expel them" / "'Yuquilla'"

2457 *E. buxifolia* Bahama Is. / O. Degener 18757 / 1946 / "...dried and used for tea"

2458 *E. peplus* Peru / C. Vargas C. 390 / 1937 / "El látex es de naturaleza caústica, muy fuerte"
Chile? / G. Looser 569A? / 1924 / "maleza de mi jardin"

2459 *E. chilensis* Argentina / A. Ragonese 43 / 1933 / "Tóxica para el ganado"

2460 *E. patagonica* Argentina / F. Pastore 95 / 1912 / "Nombre vulgar 'pichoa'"

2461 *E. copiapina* Chile / E. E. Gigoux s.n. / 1885 / "'Pichoa'"

2462 *E. collina* Chile / J. L. Morrison 16769 / 1938 / "Said to be dangerous to the eyes, but this is said of all Chilean plants with milky juice"

2463 *E. cotinoides* Brit. Guiana / J. S. De La Cruz 2363 / 1922 / "Said to be very poisonous"
Brazil / Dr. Shattuck s.n. / 1924 / "Said to be used in treating leprosy" / "'a-Na-curana'"

2464 *E. pilulifera* Venezuela / H. Pittier 7792 / 1918 / "N.v. Hierba de boca; usada para curar las boqueras"
Philippine Is. / C. O. Frake 359 / 1957 / "Roots squeezed into eye for various troubles" / "'Glegaras'" Sub.

2465   *E. hypericifolia* Venezuela / F. T. Tamayo 534 / 1939 / "'Yerba de boca'"

2466   *E. serpens* Argentina / T. Meyer 2396 / 1937 / "'Yerba de la golondrina'"

Paraguay / W. A. Archer 4784 / 36 / "General remedy" / "'Tapacu cambu'"

Paraguay / W. A. Archer 4930 / 1937 / "Refreshing beverage; also as cure for 'flora de blanca,' disorder of women. Purchased in market at Asuncion" / "'Tapacú camby'"

2467   *E. prostrata* Argentina / T. Meyer 2395 / 1937 / "'Yerba de la golondrina'"

2468   *E. caecorum* Bolivia / J. Steinbach 5172 / 1920 / "La resina de la raiz blanca-gris cerosa ... contra la nube de los ojos" / "'Golondrina'"

2469   *Euphorbia* indet. Argentina / A. Soriano 2104 / 1946 / "n.v.: pichoga purgante"

2470   *E. helioscopia* Japan / K. Ichikawa 200013 / 1924 / "poisonous"

2471   *E. thymifolia* China / W. T. Tsang 21394 / 1932 / "... used as medicine"

Caroline Is. / C. C. Y. Wong 147 / 1947 / "The plant is used as medicine for leg and backache by pounding and mixing with coconut milk and coconut water, then drunk. Another method of treatment is to mix coconut meat in a bundle with hot stones and the plant rubbing the bundle on the effected parts. The plant is used in connection with a *rong* (magic), called nukuny" / "'pukkusón'"

2472   *E. atoto* Solomon Is. / S. F. Kajewski 2243 / 1930 / "The leaves are dried and mixed with coconut oil to give a good smell" / "'Ke-ross'"

2473   *E. plumerioides* New Guinea: Papua / L. J. Brass 1123 / 1926 / "Leaves chopped up and thrown into pools of water to poison fish" / "'Ohehu,' 'New Guinea Dynamite'"

New Guinea / Pulsford 21 / 54 / "Used as a fish poison, purgative and vermifuge" / "'Ai-chup' (Urip—Arapesh)"

2474   *E. obliqua* New Hebrides Is. / S. F. Kajewski 257 / 1928 / "Sap used in conjunction with charcoal for tatooing, producing blue marks" / "'Aripatepu'"

2475   *E. genistoides* S. Africa / R. Marloth 11970 / 24 / "Poisonous!"

2476   *Pedilanthus bracteatus* Mexico / Y. Mexia 969 / 1926 / "Used by natives as violent purgative" / "'Candelilla'"

2477   *P. tithymaloides* subsp. *angustifolius* Puerto Rico / P. Sintenis 769 / 1885 / "'Ipecacuana'"

2478   *Euphorbiacea* indet. Mexico / G. B. Hinton 11984 / 38 / "The spines sting like a hornet"

2479   *Euphorbiacea* indet. Philippine Is. / C. O. Frake 607 / 1958 / "medicine for cough, black urine, etc." / "'belekbut'" Sub.

2480   *Euphorbiacea* indet. Philippine Is. / C. O. Frake 843 / 1958 / "Bark scraped and applied as compress for stomachache" / "'Kingay'"?

2481   *Euphorbiacea* indet. Philippine Is. / C. O. Frake 629 / 1958 / "Agric. ritual" / "'bayabayaletin'" Sub.

2482   *Euphorbiacea* indet. Philippine Is. / M. D. Sulit 3550 / 1949 / "Poisonous—blisters the skin"

2483   *Euphorbiacea* indet. New Hebrides Is. / S. F. Kajewski 379 / 1928 / "Fruit eaten by natives" / "'Oval-lafsi'"

BUXACEAE

2484   *Coriaria thymifolia* Mexico / Y. Mexia 8972 / 1937 / " . . . fruits poisonous"

Mexico / G. B. Hinton 15398 / 39 / "Foliage said to be deadly poison for goats"

2485   *C. sinica* China / Steward & Cheo 309 / 1933 / "Edible"

ANACARDIACEAE

2486   *Buchanania arborescens* N. Borneo / Bayak B.N.B. For. Dept. 2634 / 1932 / "Native medicine for diarrhoea" / "Selangawan (Bajau)"

Philippine Is. / C. O. Frake 870 / 1958 / "Bark applied for stomachache" / "Gimbulan Sub."

2487   *B. lucida* N. Borneo / J. A. Leano 2304 / 1932 / "Juice itchy" / "Rengas"

2488   *Buchanania* indet. New Britain / A. Floyd 6462 / 54 / "The fruit is eaten only by pigeons. Not used" / "La go-be (W. Nak.)"

2489   *Mangifera minor* Solomon Is. / S. F. Kajewski 2157 / 30 / "The bark of the tree is pounded with lime and if an infant is sick in the stomach, the mixture is placed on the affected spot" / "Fyoe. Mango"

2490   *Anacardia humile* Brazil / L. O. Williams 7480 / 1945 / "'Cajo do Matto' Eat the fruit of this"

2491   *Bouea oppositifolia* Burma / J. F. Smith 80 / 1915 / " . . . fruit . . . , eaten with relish by flying foxes"

N.W. Borneo / J. Sinclair 10534 / 1960 / "Fruit edible, orange"

2492   *Spondias radlkoferi* Honduras / P. C. Standley 54022 / 1927–28 / "Sap has aromatic odor" / "'Hobo'"

2493   *S. purpurea* Mexico / H. S. Gentry 7108 / 1944 / "Fruit cooked and eaten, leaves eaten raw" / "Ciruelo de Coyote"

2494 **S. *mombin*** W.I.: Dominica / W. H. & B. T. Hodge 3341 / 1940 /
"...tea of leaves used for internal pains" / "'monbin'"

Colombia / E. L. Little 9289 / 1945 / "...fruit edible, but said to be
bad for the throat" / "'Jobo'"

2495 **S. *pinnata*** China / C. W. Wang 77565 / 1936 / "Fruit...edible" /
"Tai name, 'Marh-kouh'"

Philippine Is. / C. O. Frake 583 / 1958 / "Medicine for asthma—bark
boiled and drunk"

2496 **S. *dulcis*** Solomon Is. / S. F. Kajewski 2408 / 1931 / "The natives
have exceptional faith in the healing properties of this tree. The bark
is macerated together with the leaves and drunk as a medicine. The
leaves are rubbed on the body" / "Eula"

Fiji Is. / O. Degener 14975 / 1941 / "Fruit edible; bark used to make
preparation for toothache; leaf used to flavor meat" / "'Wi' (Serua)"

2497 ***Allospondias lakonensis*** China / H. H. Chung 2708 / 1924 / "...fruit
...edible with a somewhat wine flavor, sold in market"

2498 ***A. chinensis*** China / W. T. Tsang 21390 / 1932 / "The fruit is sweet
and sour" / "Ching Tau Shue"

2499 ***Dracontomelon*** indet. China / C. Ford s.n. / no year / "Said to
yield Chinese 'Olives'"

2500 ***D. vitiense*** New Hebrides Is. / S. F. Kajewski 244 / 1928 / "Fruit
eaten by natives" / "nar-ah"

2501 ***Pleiogynium solandri*** Fiji Is. / O. Degener 15263 / 1941 / "For
thrush and for stomachache, Fijians drink infusion of bark in hot
water" / "Tarawau Serua"

2502 ***Poupartia axillaris*** China / H. H. Chung 3443 / 1925 / "Fruit said to
be sour, edible..."

2503 ***Lannea coromandelica*** China / W. T. Tsang 16764 / 1928 / "Fr.
green; medicines" / "Hau Pei Ma Shue"

2504 ***Pistacia mexicana*** Mexico / G. B. Hinton 10258 / 37 / "Copal"

2505 ***P. coccinia*** China / A. Henry 11913 / no year / "...leaves used for
making incense"

2506 ***P. chinensis*** China / E. H. Wilson 380 / '07 / "...shoots cooked and
eaten"

2507 ***P. integerrima*** India: Punjab / W. Koelz 1822 / 1931 / "...fruit sold
for medicine" / "Kulu name 'Kukerl, Kokerl'"

India: Punjab / R. R. Stewart 780 / 1917 / "The galls are used in
medicine"

2508 ***Microstemon*** indet. N. Borneo / Anthony N.B. For. Dept. A771 /
48 / "...fruits edible and medicinal" / "Ransa Ransa akar"

2509   *Euroschinus elegans* New Caledonia: Nouméa / I. Franc 1583A / 1913 / "Sève répandant une forte odour de fénouil"

2510   *Mauria birringo* Panama / G. White 71 / 1938 / "... 'Siguella,' is claimed to be poisonous by the natives"

2511   *M. sessiliflora* Honduras / Yuncker, Dawson & Youse 5554 / 1936 / "Fruit reported edible when ripe"

2512   *M. heterophylla* Peru / Y. Mexia 8161 / 1936 / "Produces skin eruptions in some persons" / "'Maico'"

2513   *Schinus molle* Mexico / G. B. Hinton 3125 / 33 / "... for rheumatism" / "Piro"

       Argentina / W. Lossen 38 / no year / "N. vulg. Pimiento del Diablo"

2514   *S. pearcei* Peru / J. West 8075 / 1936 / "... peppery aroma" / "'Molla' or 'Mulle'"

2515   *S. weinmanniaefolius* Argentina / T. M. Petersen 1933 / 1953 / "Not eaten by cattle"

2516   *S. velutinus* Chile / J. West 3952 / 1935 / "Foliage causes skin irritation" / "'Litre'"

2517   *Lithraea caustica* Chile / J. West 4989 / 1936 / "Causes dermatitis similar to that of Rhus diversiloba" / "'Litre'"

2518   *Cotinus coggygria* France / B. de Retz s.n. / 1932 / "'Sumac Fustet'"

2519   *Comocladia engleriana* Mexico / Y. Mexia 1163 / 1926 / "Thick irritating juice ... The natives ... often affected just by passing the tree" / "'Hinchahuevos'"

2520   *C. mollifolia* Cuba / R. Combs 712 / 96 / "... juice milky causing '*Rhus* poison'"

2521   *Metopium brownei* Brit. Honduras / P. H. Gentle 4152 / 1942 / "'...black poisonwood'"

2522   *Pseudosmodingium perniciosum* Mexico / Y. Mexia 8824 / 1937 / "Sap and wood acrid causing skin eruptions" / "'Cuajiote Colorado'"

2523   *Rhus potentillaefolia* Mexico / Miranda 2090 / 1942 / "Nombre vulgar 'Teclate manse'"

2524   *R. juglandifolia* Guatemala / A. F. Skutch 1716 / 1934 / "I attribute to this a mild case of dermatitis, similar to that caused by *R. toxicodendron*"

2525   *R. microphylla* Mexico / W. P. Hewitt 177 / 1947 / "Fruit sour, refreshing" / "Aigrillo"

       Mexico / R. M. Stewart 1860 / 1941 / "'Comida de Vibora'"

       Mexico / L. A. Kenoyer 2034 / 1947 / "Low shrub, cropped by cattle"

2526   *R. virens* Mexico / R. L. Dressler 2347 / 1957 / "'Antrisco, good for teeth and gums'"

2527   *R. chinensis* China / W. T. Tsang 28275 / 1937 / "...fr. edible" / "Im Sheung Pak Shue"
      Japan / E. Elliott 31 / '46 / "Gall of leaf is source of tannin in China ..."

2528   *R. succedanea* China / C. I. Lei 379 / 1933 / "...poisonous" / "Tsat Shu"

2529   *R. taitensis* Solomon Is. / S. F. Kajewski 2494 / 1931 / "A fire is made and green leaves placed on top and a native who is deaf puts his head into the smoke to cure deafness" / "Tuma"

2530   *Semecarpus* indet. Sumatra / W. N. & C. M. Bangham 784 / 1932 / "Natives say juice of wood makes skin itch"

2531   *Semecarpus* indet. Borneo / A. Kostermans 6038 / 1951 / "The gum contains anacardol and caused deep wounds, like burning on myself and assistant. Itching carried on for 2 weeks"

2532   *S. decipiens* Solomon Is. / S. F. Kajewski 2328 / 30 / "the sap of this tree is very caustic, burning and taking the skin right off"
      Solomon Is. / S. F. Kajewski 2560 / 31 / "The bark is macerated and applied to Tinea and eruption of the skin"

2533   *S. vitiensis* Fiji Is. / A. C. Smith 5340 / 1947 / "...with extremely poisonous oil"

AQUIFOLIACEAE

2534   *Ilex rubra* Mexico / W. P. Hewitt 173 / 1946 / "...red 'berry' said to be sweet and purgative"

2535   *I. paraguariensis* Brazil / Whitford & Silveira 114 / 1918 / "...locally called 'Herva matte'"

2536   *I. aculeolata* China / W. T. Tsang 13668 / 1934 / "...root edible"

2537   *I. chapaensis* China / W. T. Tsang 26549 / 1936 / "...fr. edible"

CELASTRACEAE

2538   *Euonymus japonicus* Japan / E. Elliott 25 / '46 / "Medicinal use. Bark is decocted"

2539   *E. yedoensis* China / W. Y. Chun 4244 / 1922 / "Fruit usually cooked"

2540   *E. lanceifolia* China / J. Hers 1004 / 1919 / "local name: chan kuei tsien 'devil-killing arrow'"

2541   *E. miyakei* China / S. K. Lau 140 / 1932 / "...fr. edible" / "Tsio gui gang (Lois)"

2542   *E. bullata* India: Assam / L. F. Ruse 121 / 1923 / "...the leaves are used by natives for binding over cuts and bruises"

2543   *Euonymus* indet. N. Siam / Royal For. Dept. 4511 / '48 / "Shrub with bitter leaves. Fruit red. Edible leaves" / "Piae Farn"

2544   *Celastrus gemmata* China / W. T. Tsang 20855 / 1932 / "The natives use it as medicine" / "Kwo Shan Fung"

2545   *C. angulata* China / J. Hers 772 / 1919 / "name: ku pi shu (bitter bark)"

2546   *C. orbiculata* China / J. Hers 828 / 1919 / "leaves edible" / "Kao lan yeh"

2547   *C. stylosus* var. *glaber* China / W. T. Tsang 2320 / 1933 / " . . . used as medicine" / "Kwo Shan Fung"

2548   *C. richii* Fiji Is. / O. Degener 15427 / 1941 / "Leaves chewed for toothache" / "'Vere' (Ra)"

  Fiji Is. / O. Degener 15330 / 1941 / "If Fijian is possessed by Devil, he drinks tea made from leaf" / "'Vere'"

  Fiji Is. / O. Degener 15352 / 1941 / "Bark or leaf crushed in water—drink liquid for stomachache" / "'Vereloa'"

2549   *Maytenus boaria* Argentina / L. R. Parodi 11641 / 1934 / "Las hojas son apetecidos por el ganado" / "Maiten"

2550   *M. pseudocasearia* Brazil / Y. Mexia 5716 / 1931 / "Tea of leaves use for dysentery" / "'Tiuzinho'"

2551   *M. viscifolia* Argentina / L. R. Parodi 10878 / 1933 / "Las hojas las comen los cabras" / "'Sombra de Toro'"

2552   *Gymnosporia* indet. Thailand / Royal For. Dept. Siam 3394 / '46 / "Seeds yield oil" / "Mhark Taek"

2553   *G. acuminata* S. Africa: Transvaal / Nat. Herb. Pretoria Agric. Office D.P.I. file M/766 M32/381 / 1934 / " . . . bark and leaves produce a gutta-percha-like substance of good quality"

2554   *Bhesa paniculata* Brit. N. Borneo / Goklin B.N.B. For. Dept. 2800 / 1933 / " . . . leaves native medicine" / "Pangel-Pangel kayu (Brunei) Rarasan tatahon (Dusun)"

2555   *Zinowiewia concinna* Mexico / G. B. Hinton 2332 / 32 / "For fodder when none better is available" / "Gloria"

2556   *Z. integerrima* El Salvador / P. H. Allen 6944 / 1958 / "'Culebro' or 'Naranjillo'"

2557   *Elaeodendron xylocarpum* Virgin Is. / W. C. Fishlock 24 / 1918 / "'Wild Nutmeg'"

2558   *E. melanocarpum* Australia: Queensland / S. F. Kajewski 1334 / 1929 / "The dried specimens have a decided odour of liquorice"

HIPPOCRATEACEAE

2559   *Pristimera celastroides* Mexico / Y. Mexia 8881 / 1937 / "Mature seeds crushed, mixed with lard and used for de-lousing" / "'Hierba del Pioso'"

El Salvador / P. C. Standley 19956 / 1922 / "Paste of seeds used to kill lice etc. on man and beasts" / "'Mata-piojo'"

2560 *Salacia impressifolia* Bolivia / J. Steinbach 6426 / 1924 / "Fruta comible, . . . abundante pulpa blanca, dulce" / "Guapomo"

2561 *S. prinoides* China / S. K. Lau 154 / 1932 / "Fr. red; edible" / "Wang koa bong (Lois)"

2562 *S. integrifolia* Philippine Is. / Sulit & Conklin 5076 / 1953 / "Pulp of fruit edible"

2563 *Tontelea attenuata* Colombia / R. E. Schultes 5385 / 1943 / "Odor of cinnamon pronounced"

2564 *T. micrantha* Brazil / Williams & Assis 7429 / 1945 / "Fruit good to eat" / "'Bacu Pari'"

STAPHYLEACEAE

2565 *Turpinia affinis* China / K. M. Feng 11317 / 1947 / " . . . fr. edible" / "(Shoei-tong-guoo)"

2566 *T. arguta* China / W. T. Tsang 21028 / 1932 / " . . . fruit . . . as boil medicine"

China / W. T. Tsang 22991 / 1933 / " . . . used as medicine" / "Ng Tsun To"

ICACINACEAE

2567 *Emmotum fagifolium* Brazil / Ducke 2012 / 1946 / " . . . fruto alaranja-do doce mas adstringente, com cheiro agradavel"

2568 *Calatola laevigata* Brit. Honduras / W. A. Schipp 1366 / 1935 / " . . . large kernels of seed edible"

2569 *Poraqueiba sericea* Brazil / Chagos 79 / 54 / "O oleo extraido e usado na alimentaçao e para fritar peixes, batatas etc." / "Mari ou Umari"

2570 *Rhyticaryum longifolium* Solomon Is. / S. F. Kajewski 2072 / 1930 / "The leaves of this tree are cooked and eaten and are decidedly good, resembling a taste between cabbages and french beans" / "Nu-marrio"

2571 *Iodes ovalis* China / S. K. Lau 357 / 1932 / "To get the sap of the wood to heal the wound of the eye" / "Vang Au (Lai)"

2572 *Polyporandra scandens* Solomon Is. / S. F. Kajewski 1932 / 1930 / "The young leaves of this plant are cooked and eaten"

2573 *Phytocrene blancoi* Philippine Is. / C. O. Frake 502 / 1958 / "Roots boiled and drunk for puerperium" / "glepay Sub."

Philippine Is. / C. O. Frake 807 / 1958 / "Roots boiled and drunk for black urine" / "gelusay Sub."

Philippine Is. / C. O. Frake 881 / 1958 / "Leaves applied for pin-worms" / "Lanbid Sub."

Philippine Is. / Sulit & Conklin 5081 / 1953 / "Water from cut stem is medicine for pink eyes (singaw)" / "Balonsaguing Mang."

Philippine Is. / L. E. Ebalo 370 / 1940 / "fts. used as poison for fishes" / "Tagbanua 'tabtang'"

SAPINDACEAE

2574 **Serjania mexicana** Mexico / Y. Mexia 9220 / 1938 / "The woody vine is cut into pieces, brushed in water and used to poison fish" / "'Barbasco' 'Duba Yin,' Zapotecan name"

2575 **S. racemosa** Mexico / R. Q. Abbott 32 / 1936 / "'(bejuco de Margarite)' used in water to cure paludismo. Bathe in it and drink it"

2576 **S. goniocarpa** Mexico / G. Martínez-Calderón 367 / 1940 / ". . . used as a fish poison" / "Bejuco de Barbasco"

2577 **Paullinia fuscescens** El Salvador / P. C. Standley 22157 / 1922 / "Aril, white eaten; seeds reputed poisonous. Stems used to kill fish" / "'Nistamal, Bejuco cuadro'"

2578 **P. alata** Peru / R. Kanehira 46 / 1927 / "Root used for fish poison" / "'Macote'"

2579 **P. yoco** Colombia / R. E. Schultes 4028 / 1942 / ". . . use expressed sap as stimulant" / "'yoco'"

2580 **Allophylus timoriensis** Philippine Is. / C. O. Frake 678 / 1958 / "Bark boiled and drunk for general malaise" / "glupag-lupog Sub."

Philippine Is. / C. M. Weber 57 / 1916 / "Fr. edible. Cherry like flavor, but astringent"

2581 **Allophylus** indet. Philippine Is. / C. O. Frake 373 / 1957 / "Bark applied to burns" / "glupay Sub."

2582 **Allophylus** indet. Philippine Is. / C. O. Frake 526 / 1958 / "Bark scraped, then applied to rigid abdomen" / "Pandelaga Sub."

2583 **A. ternatus** Caroline Is. / C. Y. C. Wong 196 / 1947 / "The leaves and wood are used as medicine for *nykuna* (pain in leg with swelling) the constituents are pounded and put in coconut cloth with a hot stone, then rubbed on the affected part" / "ngo"

2584 **Toulicia bullata** Brazil / R. Froes 1947 / 1932 / "Bark used to cure fever" / "'Tipy Assu'"

2585 **Sapindus vitiensis** Fiji Is. / O. Degener 15448 / 1941 / "For sore stomach squash leaf in water and drink liquid" / "Drengdrenga Dialect Ra"

2586 **Erioglossum rubiginosum** N. Borneo / B.N.B. For. Dept. 6732 / 1936 / "Root infusion (boiled) medicine for cough" / "Suang rason (Bejau Laud)"

2587 *Aphania* indet. Philippine Is. / C. O. Frake 38247 / 1958 / "Leaves applied for headache" / "Selimpugun Sub."

2588 *Aphania* indet. Philippine Is. / C. O. Frake 38140 / 1958 / "Root applied for puerperium" / "glutain Sub."

2589 *Otophora fruticosa* N. Borneo / B.N.B. For. Dept. 3277 / 1933 / "Fruit edible" / "Balingasan (Bajau)"

2590 *Melicocca lepidopetala* Bolivia / J. Steinbach 6568 / 1924 / "Fruta comible" / "Motoyoé"

2591 *Talisia oliviformis* Guatemala / P. C. Standley 74833 / 1940 / "Fruit edible" / "'Tapaljocote'"

2592 *T. esculenta* Bolivia / J. Steinbach 6595 / 1924 / "Fruta . . . agradable para comer y sano" / "Piton dulce"

2593 *Euphoria cinerea* N. Borneo / B.N.B. For. Dept. 10427 / 1939 / "Fruit edible" / "mata kuching (Malay) mumboh (Dusun Ninabatangan)"

2594 *E. didyma* Philippine Is. / M. D. Sulit 6959 / 1945 / "Pulp of fruit edible" / "Alupag Tag."

2595 *Pseudonephelium fumatum* Philippine Is. / H. G. Gutierrez 61-115 / 1961 / "edible aril"

2596 *Nephelium beccarianum* N. Borneo / B.N.B. For. Dept. 4852 / 1935 / " . . . fruit olive green, edible" / "Osau (Sungei) Segir (Kedayan)"

2597 *Pometia acuminata* N. Borneo / B.N.B. For. Dept. 3697 / 1938 / "Fruit purple—edible" / "Merbau perempuan (Malay)"

2598 *P. pinnata* N. Borneo / H. G. Keith 9087 / 1938 / "Fruit used for bait for fishing and much liked by wild pigs" / "Nom. vern. timpangah (Tengara)"

2599 *Guioa lasiothyrsa* Philippine Is. / R. B. Fox 50 / 1950 / " . . . fruit edible" / "Bunsikag Tagb."

2600 *Guioa* indet. Philippine Is. / C. O. Frake 739 / 1958 / "Leaves applied for internal pain" / "Genubod Sub."

2601 *Cupaniopsis leptobotrys* Fiji Is. / O. Degener 15371 / 1941 / "Bark used medicinally" / "'Matawathe' (Ra)"

2602 *Mischocarpus guillauminii* Caroline Is. / C. Y. C. Wong 277 / 1947 / "The bark is used as medicine for general feeling of debility by placing the slightly pounded bark in a wooden bowl with some water and left in it for 2 to – days. A bundle of coconut cloth is made and dipped into the liquid, then sprinkled over the patient" / "N.v. ääppo"

2603 *Paranephelium spirei* China / S. K. Lau 584 / 1932 / "The inner part of fruit may be eaten" / "Kodoe dong (Lois)"

2604 *Hypelate trifoliata* W.I.: Providenciales / G. R. Proctor 9137 / 1954 / "Wine sometimes made from fruit" / "'Burn-throat'"

2605 *Exothea paniculata* Bahama Is. / J. S. Ames s.n. / 1926 / "Bitter Bough"

2606 *Harpullia arborea* N. Borneo / B.N.B. For. Dept. 2678 / 1932 / "Poison" / "Seban (Tidong)"

Solomon Is. / S. F. Kajewski 2409 / 31 / "The bark is macerated with water then drunk to allay pains" / "Ar-raru"

2607 *Ungnadia speciosa* U.S.: Tex. / F. Lindheimer 391 / 1847 / "...fruit sweet, pleasant but emetic"

2608 *Cubilia* indet. Borneo / Kostermans 7011 / 1952 / "Seeds edible when cooked, tasty" / "'Kendulo'"

SABIACEAE

2609 *Sabia purpurea* India: Assam / L. F. Ruse 5 / 1923 / "Young leaves are cooked and eaten. Sold in bazar"

2610 *Meliosma dentata* Mexico / W. C. Leavenworth 727 / 1940 / "Fruits very attractive to squirrels"

2611 *M. philippinensis* Philippine Is. / C. O. Frake 852 / 1958 / "Leaves pounded and applied for wounds" / "Gimbing imbing Sub."

Philippine Is. / P. Añonuevo 82 / 1950 / "Bark is used to bathe itchy skin" / "Daborabo Bukidnon"

2612 *Meliosma* indet. Philippine Is. / M. D. Sulit 3273 / 1949 / "...charred bark and leaves triturated and put in water—given to person having tympanites" / "Karubu-rabu Bukid."

BALSAMINACEAE

2613 *Impatiens balsamina* Philippine Is. / C. O. Frake 474 / 1958 / "Put in nest to keep eggs from spoiling" / "Silongka? emputi Sub."

RHAMNACEAE

2614 *Zizyphus pedunculata* Mexico / Smith, Peterson & Tejeda 3551 / 1961 / "'Cholulo' soap"

2615 *Z. obtusifolia* Mexico / I. L. Wiggins 5500 / 1931 / "Taste agreeable, slightly puckering"

2616 *Z. sonorensis* Mexico / Y. Mexia 1007 / 1926 / "Fr.... Edible" / "'Frutillo'"

2618 *Z. oenoplia* India: Bombay / J. Fernandes 652 / 1949 / "Fruits eaten when ripe" / "'Kanari' (Marathi)"

2619 *Z. oxyphylla* India: Punjab / W. Koelz 1523 / 1930 / "Fr. red, pleasant, very acid"

2620   *Z. rugosa* India: Assam / L. F. Ruse 412 / 1924 / "... fruit is eaten" / "Dung-soh-lang-khrithad"

2621   *Ampelozizyphus amazonicus* Venezuela / L. Williams 14562 / 1942 / "... y se dice que los indios la raspan y preparan una infusión como remedio contra las mordeduras de serpientes, de donde viene el nombre local" / "N.v. Palo de culebra"

Brit. Guiana / A. C. Smith 2830 / 1937 / "... bark fragrant, producing lather in water; used as soap by Waiwais"

2622   *Condalia ericoides* Mexico / R. M. Stewart 462 / 1941 / "'Comida de Cuervo'"

2623   *Reynosia wrightii* Cuba / Alain, Acuña & Figueiras 549 / 1956 / "'Sacalengua'"

2624   *Karwinskia calderonii* El Salvador / P. C. Standley 19118 / 1921–22 / "Pigs eat fruit and are paralized" / "'Guiliguiste, Huile huiste'"

2625   *K. parvifolia* Mexico / H. S. Gentry 7132 / 1945 / "An infusion of the root is taken as a tonic" / "Cacachila"

2626   *K. humboldtiana* Mexico / R. L. Dressler 2339 / 1957 / "... said to be poisonous to pigs" / "'Tullidor'"

Mexico / H. S. Gentry 1161 / 1934 / "Fruit edible but reported to make one weak and to produce trembling in the young boys. Near Caramechi I saw the young cholugos (Nasua narica) eat the fruit with gusto. Medic. leaves boiled in water and the tea drunk for fevers. Warihios reported to cushion their beds with the branches when they are ill"

Mexico / H. E. Moore 1266 / 1947 / "Fruit red-purple, edible used as a 'fresca'" / "Capolincillo"

2627   *Berchemia floribunda* China / Y. W. Taam 513 / 1938 / "... fr. black edible"

2628   *B. giraldiana* China / J. Hers 242 / 1919 / "Local name: Tsing shih tiao 'the green snake's stem'"

2629   *B. lineata* China / S. K. Lau 1981 / 1933 / "... fruit edible"

2630   *Rhamnella franguloides* Japan / N. Eri 100568 / 1926 / "child eats fruit"

2631   *Sageretia theezans* India: Punjab / W. Koelz 4661 / 1933 / "... fruit eaten"

2632   *Rhamnus pubescens* Peru / Stork & Horton 10352 / 1939 / "... frts. resemble those of *Celtis occidentalis*, also having similar taste but not sweet"

2633   *R. leptophylla* China / Cheo & Yen 192 / 1936 / "Lvs. used as tea"

2634   *R. napalensis* China / W. Y. Chun 5795 / 1927 / "Fruit edible"

2635　**Rhamnus** indet. China / J. Hers 2733 / 1923 / "Shan tsiao 'mountain pepper'"

2636　**R. zeyheri** S. Africa: Transvaal / E. E. Galpin M648 / 23 / "Drupes edible"

2637　**Hovenia acerba** China / G. W. Groff 4127 / 1919 / "fruits for brewing wine" / "M'an Tsz Kwo"

China / W. T. Tsang 25315 / 1935 / "...fr. edible" / "Man Tsz Kwo Tsz"

2638　**Colubrina elliptica** Mexico / R. L. Dressler 2327 / 1957 / "'Palo de arco' medicinal, used for 'granos'"

2639　**C. glomerata** Mexico / Y. Mexia 1892 / 1927 / "Used for fevers" / "'Margarita'"

2640　**C. greggii** Mexico / Millspaugh 3919 / 1903 / "'pimienta ché'"

2641　**C. heteroneura** Mexico / H. S. Gentry 5709 / 1940 / "Pie de Venado"

2642　**C. spinosa** Panama / G. P. Cooper 365 / 1928 / "Wild coffee"

2643　**C. cubensis** Cuba / J. A. Shafer 1047 / 1909 / "...fruit black, twig odor and taste of *Betula*"

2644　**C. arborescens** W.I.: Antigua / J. S. Beard 355 / 1944 / "'Soap bush' ...leaves crushed in water said to produce soapy lather"

2645　**C. ferruginosa** Jamaica / G. R. Proctor 8773 / 1954 / "'Scrubbing bush'"

2646　**C. reclinata** Santo Domingo / Pater Fuertes 393 / 1910 / "'Palo amargo'"

Virgin Is. / W. C. Fishlock 42 / 1918 / "The bark is used in the preparation of a bitter drink"

2647　**C. asiatica** Brit. N. Borneo / Goklin 2961 / 1932 / "Leaves used for vegetable" / "Sarunai kayu (Brunei)"

Philippine Is. / M. D. Sulit 4547 / 1952 / "Young leaves mixed with fish in cooking. Good to eat. Bark scraped and used as substitute for soap" / "Kabatili Tag."

Tonga Is. / T. G. Yuncker 1524 / 1935 / "Leaves lather and may be used as soap" / "Tongan name: fiho'a"

Tonga Is. / T. G. Yuncker 15619 / 1953 / "...bark forms lather and can be used as a soap substitute" / "Tongan name: Fihoa"

2648　**Alphitonia philippinensis** Sarawak / J. & M. S. Clemens 20662 / 1929 / " ...bark with odor of wintergreen"

Philippine Is. / Sulit & Conklin 5157 / 1935 / "Bark chewed—saliva swallowed to cure cough also stomach trouble" / "Salikapo Mang. (Hanunoo)"

2649   *A. moluccana* New Guinea / van Royen & Schleumer 5848 / 1961 / "Bark with heavy smell of wintergreen oil"

2650   *Alphitonia* indet. New Guinea: Papua / R. D. Hoogland 3402 / 1953 / "Specimens after treatment with formaline with strong cyanide scent" / "Hagehreh (Orokawa language, Mumuni)"

2651   *Alphitonia* indet. Solomon Is. / S. F. Kajewski 2591 / 1931 / "Bark when cut has a sarsaparilla odour" / "Hymani"

2652   *Alphitonia neo-caledonica* New Caledonia / C. T. White 2114 / 1923 / "Bark when freshly peeled has a strong odour of sarsaparilla" / "'Pomaderris'"

2653   *Gouania polygama* Honduras / P. C. Standley 52753 / 1927–28 / "Twigs give lather when chewed and are used to clean the teeth" / "'Limpia-dientes'"

El Salvador / P. C. Standley 19133 / 1921–22 / "Used to wash clothes" / "'Jaboncillo'"

2654   *G. cordifolia* Brazil / Y. Mexia 4248 / 1930 / "Tea made of the leaves for blood purifier" / "'Amora Lisa'"

2655   *Gouania* indet. Solomon Is. / S. F. Kajewski 1898 / 1930 / "The sap of this vine is used as a medicine for sick pigs and its use is firmly believed in by the natives" / "Moi-moit-si"

2656   *G. tiliaefolia* Philippine Is. / Sulit & Conklin 5176 / 1963 / "Stem cut a foot in length. Put one end in the mouth, while other end in ear of a patient. Blow through stem so that water comes out. This is supposed to be medicinal for deaf ear" / "Tagbura Mang. (Hanunoo)"

2657   *G. richii* Fiji Is. / O. Degener 15338 / 1941 / "Pound leaves in water, strain and drink for stomachache" / "'Vereola' (Ra)"

VITACEAE

2658   *Ampelocissus acapulcensis* El Salvador / P. C. Standley 21634 / 1922 / "Fr. . . . used for vinegar" / "'Uva'"

2659   *A. ochracea* Philippine Is. / Sulit & Conklin 46281 / 1952 / "Swollen underground stem is medicinal by Mangyan" / "Bagunaw Mang."

Philippine Is. / G. E. Edaño 1686 / 1949 / "Grind bark and leaves into fine particles and apply to swollen areas or to boils" / "Lagingi Ma"

2660   *Tetrastigma planicaule* China / W. T. Tsang 24192 / 1934 / ". . . fr. edible"

2661   *T. trifoliatum* Brit. N. Borneo / A. Cuadra 1472 / 1948 / ". . . fruit red, used for medicinal purposes by natives"

2662   *T. loheri* Philippine Is. / M. D. Sulit 1470 / 1947 / "Young leaves used for condiment" / "Ayu Tag."

Philippine Is. / E. Canicosa 166 / 1948 / "Quantity of young leaves smashed with salt and used for 'souring' fish or meat"

Philippine Is. / C. O. Frake 676 / 1958 / "Stem boiled and drunk for dysentery" / "glegili ? Sub."

2663 *Ampelopsis mexicana* Mexico / G. B. Hinton 5717 / 34 / ". . . edible"

2664 *A. cantoniensis* var. *grossedentata* China / W. T. Tsang 21569 / 1932 / " . . . fr. . . . edible; leaves used to make 'Tea Cake'" / "Ngau Kin So T'ang, Tin Po Cha"

China / Steward, Chiao & Cheo 18 / 1931 / "Lvs. used as substitute for tea"

2665 *A. delavayana* China / Steward, Chiao & Cheo 134 / 1931 / "Leaves fed to pigs" / "Mo Chu T'seng"

2666 *A. humulifolia* China / J. Hers 1300 / 1919 / "fruit . . . edible" / "Niu pu tao"

2667 *A. brevipedunculata* Indochina / Poilane 1715 / 1920 / ". . . les feuilles sont employés pour faire des medicaments employé contre plais et les mordures des milles-pieds" / "Ate: Mo chua thao Khan"

2668 *Cissus rhombifolia* Honduras / P. C. Standley 52861 / 1927–28 / "Sap said to produce blisters on skin" / "'Pica-mano'"

2669 *C. sicyoides* W.I.: St. Lucia / G. R. Proctor 17813 / 1958 / "'God-mort'"

W.I.: St. Vincent / G. R. Cooley 8517 / 1962 / "'Pudding bush'"

2670 *C. hexangularis* Indochina / Poilane 993 / 1920 / ". . . fruit comestibles" / "Annte: Dǎy rát"

2671 *C. adnata* New Guinea / H. A. Brown 295 / 1953 / ". . . used . . . as a remedy against snake-bite. Leaves boiled, liquid given to patient, causes him to vomit, and so to relieve (?) his condition" / "Ganauo-gana (Kukukuku)"

2672 *C. assamica* Solomon Is. / S. F. Kajewski 2410 / 1931 / "The leaves of this vine are pounded with lime and rubbed over the faces and eyes, for soreness in the latter" / "Ao-ra-rasu"

Solomon Is. / S. F. Kajewski 2215 / 1930 / "This plant is placed inside baskets or traps to lure eels . . . highly effective" / "Permapula"

Solomon Is. / S. F. Kajewski 1811 / 1930 / "When the vine is cut the juice is drunk by natives for headaches" / "Perma-pulu"

2673 *C. antarctica* Australia: Queensland / C. T. White 12688 / 45 / ". . . berries . . . astringent with very slight irritative effect"

2674 *Cayratia* indet. Indochina / W. T. Tsang 29193 / 1939 / ". . . fr. edible"

2675 *C. japonica* Solomon Is. / S. F. Kajewski 2600 / 1931 / "When a man has constipation the leaves are rubbed on the stomach" / "Alum-bum-buckwa"

2676   *C. geniculata* Philippine Is. / A. L. Zwickey 553 / 1938 / "'Sagadu-namo' (Lan.) implying 'the monkey passes it by'"

2677   *Cayratia* indet. Philippine Is. / C. O. Frake 759 / 1958 / "Leaves applied to burns" / "galemnunuk Sub."

2678   *Leea indica* Sumatra / W. N. & C. M. Bangham 831 / 1932 / "Natives make eye medicine from it"

Solomon Is. / S. F. Kajewski 2595 / 1931 / "...when a native has a boil, the leaves are taken and rubbed on it"

2679   *L. aequata* Malaya / G. anak Umbai for A. H. Millard KL1548 / 59 / "Leaves and fruits medicinal and edible" / "Jembali Makan (Temu-an)"

2680   *Vitis arizonica* Mexico / Y. Mexia 2551 / 1929 / "Fruit said to be edible" / "'Parra cimarrona'"

2681   *Vitis* indet. India: Bombay / J. Fernandes 1036 / 1950 / "Use: The tuber very deep in the ground is ground with rice (unhusked) in water and applied to cuts" / "'Palkonae' (Marathi)"

ELAEOCARPACEAE

2682   *Elaeocarpus chinensis* China / W. T. Tsang 25728 / 1935 / "...fr. black, edible"

2683   *E. duclouxii* China / W. T. Tsang 26117 / 1936 / "...fr. black, edible"

2684   *E. lanceaefolius* China / W. T. Tsang 25016 / 1935 / "...fr. black, edible" / "Tung To Shue"

2685   *E. sylvestris* China / W. T. Tsang 21651 / 1932 / "...fruit black, edible" / "Yeung Shi Ue Shue"

2686   *E. domatiferus* N. Borneo / J. & M. S. Clemens 34460 / '33 / "Fruit green edible"

2687   *E. obtusus* Sarawak / Richards 1189 / 1932 / "...flowers heliotrope scented" / "Kelampoh"

2688   *E. pedunculatus* N. Borneo / Maidin B.N.B. For. Dept. 6702 / 1936 / "...fruit edible" / "Parius parius (Malay)"

N. Borneo / Bayak B.N.B. For. Dept. 2625 / 1932 / "...fruit green, edible; bark used in native medicine" / "Perius-Perius (Brunei) Kulambobok (Bajau)"

2689   *E. floridanus* Solomon Is. / S. F. Kajewski 2201 / 30 / "These specimens when dry have a strong almond scent"

2690   *E. sepikanus* New Guinea / Hoogland & Womersley 3239 / 1953 / "After formaline treatment a distinct cyanide-scent" / "Oreo (Oro-kawa language, Mumuni)"

2691   *E. sphaericus* Solomon Is. / S. F. Kajewski 2495 / 31 / "Fruit eaten by cockatoos and pigeons" / "Hy-cundi"

Philippine Is. / C. O. Frake 872 / 1958 / "Bark applied for spleeno-megaly" / "Genusa Sub."

2692 *E. cumingii* Philippine Is. / H. G. Gutiérrez 61-21 / 1961 / "Fr. edible"

2693 *E. glaber* Philippine Is. / R. S. Williams 2134 / 1905 / "Fruit...edible"

2694 *E. chelonimorphus* Fiji Is. / O. Degener 14878 / 1941 / "Fijians eat kernel" / "'Kambi' Dialect Sabatu"

2695 *E. grandis* Fiji Is. / Degener & Ordonez 13610 / 1940 / "...fruit blue, edible"

2696 *E. graeffii* Fiji Is. / O. Degener 15369a / 1944 / "Extract of leaf drunk for stomach trouble" / "'Ndrivi' (Ra)"

2697 *E. persicifolius* New Caledonia: Paite / A. J. Nicholson 2 / 1945 / "Pteropus feed extensively on the fruit when in season. Probably rats and many birds also feed on this fruit"

2698 *Aceratium branderhorstii* New Guinea / L. F. Brass 8341 / 1936 / "...the red acidulous fr. eaten by natives" / "Posesi"

2699 *A. dasyphyllum* New Guinea / A. Kostermans 2867 / 1948 / "...fruit...edible, sour"

2700 *A. insulare* Solomon Is. / S. F. Kajewski 2326 / 30 / "The leaves were heated and applied in fighting days to spear and arrow wounds" / "Sura-uu"

2701 *Sloanea berteriana* Santo Domingo / E. L. Ekman H12301 / 1929 / "'Cacao Cimarron'"

2702 *Sloanea* indet. New Guinea / Womersley & Millar N.G.F. 7663 / 55 / "...fruits...reputed to be edible"

2703 *Vallea stipularis* Ecuador / Y. Mexia 7660 / 1935 / "The root is used for flavoring corn dishes" / "'Perilla' 'Sacha-Capuli'"

2704 *Muntingia calabura* Mexico / G. B. Hinton 5797 / 34 / "Fruit edible. Sold in market"

GONYSTYLACEAE

2705 *Gonystylus bancanus* Philippine Is. / Sulit & Conklin 5066 / 1953 / "Bark chewed—saliva swallowed—medicine for poisonous parasites found in fish" / "Balusan Dialect (Hanunoo) Mang."

2706 *G. philippinensis* Philippine Is. / F. Canicosa 42 / 1926 / "Seed very bitter"

TILIACEAE

2707 *Corchorus olitorius* Burma / C. E. Parkinson 14784 / 32 / "Leaves eaten" / "'Pilaw'"

2708   *C. capsularis* N. Borneo / H. G. Keith 1620 / 1932 / "...worn in the hair by Murut women when a man is ill with fever" / "Babas"

2709   *Luehea speciosa* Brazil / Y. Mexia 4507 / 1930 / "'Açoita cavallo'"

2710   *Luehea* indet. Brazil / R. Froes 1953 / 1932 / "Bark reported as a cure for leprosy" / "'Acorta cavalo'"

2711   *Tilia houghii* Mexico / G. B. Hinton 4120 / 33 / "Medicinal"

2712   *Grewia abutilifolia* China / W. T. Tsang 21799 / 1933 / "Fr. . . . edible" / "Ts'o to"

2713   *G. eriocarpa* Hainan / C. I. Lei 819 / 1933 / "...fruit... edible" / "Ngou Kun Shu"

2714   *G. asiatica* Burma / F. G. Dickason 7690 / '38 / "Bark used for soap" / "'thit yaw'"

2715   *G. astropetala* Indochina / Poilane 20 / 1919 / "Les femmes indigènes chiquent l'écorce de cet arbuste avec le bétel" / "Annte.: Cay long mau teie"

2716   *G. hirsuta* Burma / P. Khant 539 / 48 / "Root is used for medicinal purposes" / "Say-kha-gyi"

2717   *G. laurifolia* Indochina / Poilane 1297 / 1920 / "...tres bon pour tous travaux non attaqué par les termites" / "Ate. Cây-bū-lôt-tiá. Moïs Aloang plui lôt"

2718   *G. sessilifolia* Indochina / W. T. Tsang 27508 / 1937 / "...fr. edible"

2719   *G. laevigata* Java / Herb. Lugd. Bat. 50 / no year / "...(kopfwasch-mittel)" / "Inl. naam ki-laki"

2720   *G. acuminata* Philippine Is. / C. O. Frake 358 / 1957 / "Agricultural ritual" / "Talatab Sub."

      Philippine Is. / C. O. Frake 713 / 1958 / "Roots applied to ulcers" / "Gelubu-nutung Sub."

2721   *G. prunifolia* Fiji Is. / O. Degener 14943 / 1941 / "Native name Vauvau Dialect Savatu, Viti Levu. Special notes: According to Serua village native, 'Vauvau' is name of other plant. According to him it is called Boko-nigata. They make tea from crushed leaf and rub decoction on sick people"

2722   *G. polygama* Australia: N. Queensland / C. T. White 8691 / 1933 / "Leaves used as a cure for diarrhoea and dysentery"

2723   *G. ferruginea* Abyssinia / Schimper 885 / no year / "Blüthen als Naschwerk der Kinder sind honigsüss. Holz verbrannt giebt abscheulichen Geruch" / "Zungea"

2724   *G. occidentalis* Rhodesia / J. Borle 329 / '22 / "Fruit edible"

2725   *G. salvifolia* S.E. Africa / Menyhart s.n. / 1891 / "Frucht essbar" / "'Mutongoro'"

2726 *Microcos paniculata* China / W. T. Tsang 26504 / 1936 / "...fr.... edible"

2727 *M. crassifolia* N. Borneo / Patrick & Kadir B.N.B. For. Dept. A1852 / 1951 / "Fruit...edible, sour taste" / "Damak-damak (Kedayan)"

2728 *Microcos* indet. New Guinea / R. Pullen 1714 / 1959 / "Bark crushed and immersed in water as a fish poison" / "Kemisai (Jal: Madang) Yebiminda (Timbunke: Sepik)"

2729 *M. stylocarpa* Philippine Is. / M. D. Sulit 3003 / 1948 / "...fruit edible" / "Karung Bis."

2730 *M. vitiensis* Fiji Is. / O. Degener 15242 / 1941 / "Crush leaves in cold water and drink for remedy against 'spit blood'" / "Nithi Serua"

2731 *Vinticena retinervis* S. Africa: Transvaal / J. Borle 8 / 26 / "Fruit brown—edible" / "Arib"

2732 *Triumfetta lappula* Peru / F. Woytkowski 5872A / 1960 / "...medicinal; leaves and flowers crushed with some water applied upon itching wounds" / "'Cavayusa'"

Peru / F. Woytkowski 5785 / 1960 / "Used to make starch dipping stalks in water 48 hours. Also to concentrate scum when cooking 'chancaca' from sugar cane"

2733 *T. procumbens* Caroline Is. / C. C. Y. Wong 167 / 1947 / "The fruits are pounded, mixed with *aryng*, coconut milk and eaten. The leaves are used for *säfein sät*, sea medicine by placing the leaves and coconut meat in coconut cloth and squeezing the mixture in a bowl, then drinking the content" / "N.v. äära"

MALVACEAE

2734 *Abutilon virgatum* Peru / Y. Mexia 04104 / 1935 / "Infusion made from the whole plant, taken for shock" / "'Algodon macho'"

2735 *Wissadula periplocifolia* Argentina / I. Morel 598 / 46 / "Planta de Bañado"

2736 *Sphaeralcea incana* Mexico / A. Lopez 9 / 1947 / "'sirve para el pelo'" / "'Yerba del Indio'"

2737 *S. cisplatina* Argentina / L. R. Parodi 8178 / 1927 / "Bañado de Flores"

2738 *Malva verticillata* Peru / Y. Mexia 04150 / 1935 / "Leaves made into poultice for suppurating wounds" / "'Malva crespa'"

Peru / R. S. Shepard 8 / 1919 / "Green—fruit of cheese plant. Local name 'Malva amargo' (bitter)"

China / W. T. Tsang 23620 / 1934 / "...edible" / "Nam Yin Tsoi"

2739 *M. sinensis* China / Cheo & Yen 213 / 1936 / "...lvs. edible"

2740   *Malvastrum lacteum* Mexico / G. B. Hinton 658 / 32 / "Boiled it is given to children for indigestion. Called 'Malvaviscum' by the Indians"

2741   *M. peruvianum* Ecuador / W. H. Camp E2442 / 1945 / "The root boiled and used as a purgative" / "'Cuchi malva' (Pig's mallow)"

2742   *M. spicatum* Australia: Queensland / C. T. White 11778 / 1941 / "Sheep said to be very fond of old dried seed heads"

2743   *Sida acuta* var. *carpinifolia* Guatemala / C. C. Deam 261 / 1905 / "Used for cough"

2744   *S. rhombifolia* Nicaragua / C. Berger s.n. / 1920 / "The entire plant including the roots is mashed and soaked in water, yielding a mucilaginous liquid used as a remedy for troubles of the bladder and urethra"
China / W. T. Tsang 23083 / 1933 / "... used as medicine" / "Wong Fa Yu"

2745   *S. cordifolia* Paraguay / W. A. Archer 4787 / 1936 / "Remedy for coughs and for inflammations" / "'malva blanca'"

2746   *Sida* indet. Bolivia / J. Steinbach 5103 / 1920 / "Crece al rededor de las estancias en abondancia; parece que sirve al ganado de tonico" / "Malva"

2747   *S. retusa* N. Borneo / Goklin B.N.B. For. Dept. 2241 / 1932 / "... leaves used for medicine" / "Gurimot (Dusun)"

2748   *Anoda triangularis* Mexico / C. V. Hartmann 555 / 1892 / "Good to eat and good for fever" / "Tu-tsji-Jar"

2749   *Urena sinuata* W.I.: Trinidad / W. E. Broadway 9356 / 1933 / "Esteemed locally as a medicinal plant for dyspeptics" / "'Cousin mahoe' 'Duck foot'"

2750   *U. lobata* Philippine Is. / C. O. Frake 395 / 1957 / "Leaves pounded and applied to ulcers" / "Selimpukut Sub."

2751   *Pavonia bangii* Bolivia / W. J. Eyerdam 24908 / 1939 / "Peculiar mintlike odor of the leaves"

2752   *P. cancellata* Brazil / B. A. Krukoff 1085 / 1931 / "... leaves boiled and used for stomach diseases" / "'Malva chanana'"

2753   *Malvaviscus drummondii* U.S.: Tex. / A. Traverse 168 / 1956 / "... edible fruits"

2754   *M. arboreus* El Salvador / P. C. Standley 22690 / 1922 / "Decoction of fls. used for sore lips; of lvs. to make hair smooth & lustrous. Fr. ... eaten" / "'Manzanito'"

2755   *M. arboreus* var. *mexicana* Mexico / R. Cardenas 375 / 1910 / "Usos vulg. se toma para la tos como remedio que apropiado el cuernesuelo y las raiz de Sta Catarina para los afectados del pulmón"

Honduras / P. C. Standley 54987 / 1928 / "Leaves used as remedy for fevers" / "'Quesillo'"

Guatemala / P. C. Standley 23857 / 1922 / "...fr. eaten" / "'Sobón'"

2756 *Hibiscus sabdariffa* Panama Canal Zone / J. M. & M. T. Greenman 5192 / 1922 / "Drink made from juice of fruit. Plant called 'sour'"

Philippine Is. / H. C. Conklin 19126 / 1953 / "Fruit eaten" / "(lubā-sa)"

2757 *H. cannabinus* Jamaica / G. R. Proctor 18379 / 1958 / "'Deccan hemp'"

INDIA: N.W.P. / A. S. Bell 75 / 1901 / "Young shoots eaten" / "Pătua, Amāri, Pătsău, Satuja"

2758 *H. rosa-sinensis* Colombia / W. A. Archer 1863 / 1931 / "'Resucito'"

2759 *H. syriacus* var. *albus plenus* China / W. T. Tsang 25606 / 1935 / "...fr. edible" / "Kai Yuk Fa"

2760 *H. surattensis* spp. *surattensis* Sumatra / W. N. & C. M. Bangham 603 / 1931 / "Leaves edible, sour, similar to oxalis"

2761 *Hibiscus* indet. New Britain / A. Floyd 6476 / 54 / "The cotton wool is used for bathing sores" / "La kateli (W. Nakanai)"

2762 *H. abelmoschus* Tonga Is. / T. G. Yuncker 15585 / 1953 / "...various parts used in preparation of medicines"

2763 *H. manihot* Fiji Is. / O. Degener 14881 / 1941 / "...an important Fijian potherb" / "'Mbele' (Sabatu)"

2764 *Abelmoschus esculentus* Mexico / G. B. Hinton 4727 / 33 / "Seed sold for coffee in the hot country" / "Cafe"

2765 *A. moschatus* Panama / H. von Wedel 726 / 1940 / "'Wild Okra'"

Colombia / A. E. Lawrance 824 / 1934 / "Seeds are toasted and ground after which the powder is drunk mixed with warm water for stomach ache etc." / "'Almistrilla'"

2766 *A. moschatus* var. *moschatus* N. Borneo / Bakar B.N.B. For. Dept. 3303 / 1933 / "...leaves used as native medicine for boils" / "Siparato (Murut)"

Philippine Is. / M. D. Sulit 1031 / 1946 / "Decoction of whole plant preferably roots is said to be medicinal for cancer in the stomach" / "Kastuli Tag."

2767 *A. manihot* ssp. *tetraphyllus* fa. *luzoniensis* Philippine Is. / M. D. Sulit 1359 / 1947 / "Decoction of whole plant, especially roots is said to be medicinal for cancer of the stomach" / "Kastuli Tag."

2768 *Thespesia populnea* Solomon Is. / S. F. Kajewski 2418 / 31 / "When a man has a cold, the bark is macerated and drunk for this purpose" / "Zeu-zealu"

Tonga Is. / T. G. Yuncker 15107 / 1953 / "...leaves and fruit used in preparing medicines" / "Tongan name: Milo"

Fiji Is. / O. Degener 15106 / 1944 / "Pound bark, add hot water and drink for thrush" / "'Mulomulo' (Serua)"

2769 *Malvacea* indet. Peru / G. Klug 2039 / 1931 / "fruit edible" / "'Wild zapote'"

BOMBACACEAE

2770 *Bombax acuminata* Mexico / P. C. Standley 1451 / 1935 / "Seeds edible of a nut-like flavor" / "'Pochote,' Mex."

2771 *Hampea stipitata* Guatemala / J. A. Steyermark 41838 / 1941 / "Bark used for mecapal—very good" / "'mano leon'"

2772 *Quararibea lomensis* Peru / G. Klug 2039 / 1931 / "'Wild zapote' fruit edible"

2773 *Durio excelsa* N. Borneo / B. N.B. For. Dept. 10370 / 1939 / "Fruit edible. Nom. vern. durian munyit (Malay) durian hantu hutan (Dusun)"

2774 *D. graveolens* N. Borneo / B.N.B. For. Dept. A37 / 1947 / "Fruit— edible. Durian mah (Malay)"

2775 *D. testudinarum* Philippine Is. / M. D. Sulit 12527 / 1950 / "Impt. plant fr. substitute for food by the Tagbanuas"

STERCULIACEAE

2776 *Melochia speciosa* Mexico / T. S. Brandegee s.n. / 1904 / "Yerba buena"

2777 *M. corchorifolia* Burma / C. E. Parkinson 15077 / 32 / "...tender leaves eaten" / "Pi-law"

2778 *M. umbellata* Solomon Is. / S. F. Kajewski 2539 / 1931 / "If a man has a sore back the leaves are macerated in water and applied to the back" / "Or-mucka"

2779 *Waltheria americana* W.I.: Grenadines / A. C. Smith 10166 / 1956 / "'Monkey-bush'"

Liberia / Dinklage 2747 / 1910 / "Als Heilpflanze benutzt"

2780 *Commersonia bartramia* Philippine Is. / C. O. Frake 840 / 1958 / "Roots boiled and drunk for diarrhea" / "Langunasa Sub."

Australia: Queensland / S. F. Kajewski 109 / 1928 / "Leaves looked upon as a valuable fodder in times of drought" / "Brown Kurrajong"

2781 *Herrania purpurea* Panama / P. H. Allen 282 / 1937 / "'Coco del Monte'"

2782 *H. cuatrecasana* Colombia / R. E. Schultes 3585 / 1949 / "...edible fruits" / "Ko-kee-oʹ-chu"

2783 *H. camargoana* Brazil / Schultes & Lopez 9747 / 1948 / "'cacao de macaco'"

2784 **Guazuma ulmifolia** Mexico / Y. Mexia 9318 / 1938 / "Fruit taken in tea for 'kidney trouble'" / "'Cuahuilote'"

Mexico / Y. Mexia 9206 / 1938 / "Fruit eaten by stock"

Mexico / P. C. Standley 1176 / 1934 / "... edible fruit, eaten by Cholugos (Nasua narica) and ... by the Warihios and Mexicans"

El Salvador / P. C. Standley 19411 / 1921–22 / "Fr. sweet, little eaten, causes constipation" / "'Caulole Tapaculo'"

W.I.: Martinique / M. Hahn 645 / 1874 / "Ecorce employée en médicine"

Venezuela / L. Williams 11575 / 39 / "Se usa para purificar el papelon y el fruto es muy estimado por el ganado caballar"

Ecuador / Stork, Eyerdam & Beetle 8983 / 1938 / "Ripe fruits are said to be eaten after soaking and cooking"

2785 **Pterospermum niveum** Philippine Is. / C. O. Frake 565 / 1958 / "Bark boiled and drunk for swelling and puerperium" / "bayug Sub."

2786 **Helicteres angustifolia** China / S. K. Lau 425 / 1932 / "Used by native people to cure stomachache by boiling with its roots" / "Hin Hai hau"

2787 *H. isora* India: Bombay / J. Fernandes 583B / 1949 / "Leaves ground are applied on wounds; twisted fruits, dry, are ground and mixed with other ingredients and used medicinally for children"

2788 *H. lanceolata* Indochina / Poilane 1320 / 1920 / "... donne des fruits comestibles"

2789 **Kleinhovia hospita** N. Borneo / H. G. Keith 9344 / 1938 / "... young leaves boiled as veg." / "kati mahar (Kedayan) tanag (Murat Bokan.)"

2790 **Sterculia lepidoto-stellata** Solomon Is. / S. F. Kajewski 2587 / 1931 / "The seeds of this tree are boiled in water and the resulting concoction drunk for colds" / "Mala Mala"

2791 **S. shillinglawii** Solomon Is. / S. F. Kajewski 2411 / 1931 / "When a man is recovering from a long illness such as fever, the bark of this tree is pounded and mixed with water and then drunk as a tonic" / "Mamala"

2792 **Sterculia** indet. New Guinea / R. Pullen 951 / 1958 / "Seed ... edible" / "Sunjak (Jal)"

2793 **Sterculia** indet. New Guinea: Papua / L. J. Brass 21913 / 1955 / "... inflorescences and young leaves eaten raw by natives"

2794 *S. ferruginea* Philippine Is. / C. O. Frake 801 / 1958 / "Bark boiled and drunk for puerperium" / "glamag-ulangan"

2795 *S. graciliflora* Philippine Is. / C. O. Frake 742 / 1958 / "Leaves applied for paralysis" / "gireniren Sub."

2796 *S. spatulata* Philippine Is. / M. D. Sulit 4367 / 1951 / "...seeds roasted and eaten" / "Balinad Bis."

2797 *Sterculia* indet. Philippine Is. / P. Añonuevo 85 / 1960 / "Fruit eaten by monkeys"

2798 *Sterculia* indet. Philippine Is. / R. B. Fox 40 / 1950 / "...seeds edible when toasted and can be drunk like coffee when boiled with water" / "Bali'nod Tagb."

2799 *Scaphium affine* N. Borneo / A. Cuadra A907 / 1948 / "Fruit used as medicinal by natives"

2800 *S. beccarianum* N. Borneo / Orolfo B.N.B. For. Dept. 4787 / 1935 / "...fruit, green; medicine for fever"

2801 *Pterocymbium tinctorum* Philippine Is. / C. O. Frake 827 / 1958 / "Roots applied to ulcer" / "telutii Sub."

2802 *Heritiera macrophylla* China / S. K. Lau 243 / 1932 / "The native people eat the skin of the fruit with Ping Long" / "Kai Tau Tsz"

DILLENIACEAE

2803 *Tetracera rotundifolia* Brazil / B. A. Krukoff 6430 / 1934 / "Young branches produce irritation in contact with skin" / "'Cipo de Fogo'"

2804 *T. fagifolia* Malaya / G. anak Umbai for A. H. Millard K.L.1456 / 59 / "Medicinal" / "Meapelas"

2805 *Tetracera* indet. Borneo / Kostermans 6681 / 1952 / "The climber contains a lot of drinkable clear water" / "Tali hampelas"

2806 *Davilla aspera* Brazil / B. A. Krukoff 1178 / 1931 / "...branches contain a drinking water" / "'Murateteau'"

2807 *D. lacunosa* Brazil / L. O. Williams 5098a / 1945 / "Give tea to animals and they are impotent"
Brazil / Williams & Assis 7030 / 1945 / "Make a tea of leaves and bathe swollen parts of body to reduce swelling" / "'Cipo Caboclo'"
Brazil / Y. Mexia 5582 / 1931 / "Infusion of leaves used to wash sores" / "'Sambaibinha'"

2808 *Curatella americana* Mexico / G. B. Hinton et al. 10205 / 1937 / "Medicinal" / "Rasca vieja"

Mexico / Y. Mexia 1315 / 1926 / "Believed to be a specific for syphilis" / "Rasca la Vieja"

2809 **Doliocarpus** indet. Brazil / Ducke 501 / 1937 / "Caulis aquam potabilem praebet" / "cipo d'agua"

2810 **Dillenia turbinata** China / F. A. McClure 754 / 1929 / "Fruits edible" / "Baan Tui"

2811 **D. obovata** Siam / A. Kostermans 374 / 1946 / "...fruit is eaten in curries"

2812 **Actinidia cordata** China / W. T. Tsang 23833 / 1934 / "...fr....edible" / "Kau Hop Tang"

2813 **Saurauia conzattii** Mexico / R. E. Schultes 695 / 1939 / "Fruits edible" / "Chinantec name: ma-gwa-ni; Spanish name: mameyito"

2814 **S. costericensis** Panama / P. H. Allen 314 / 1937 / "...fruit sweet to taste"

2815 **S. euryphylla** Brit. N. Borneo / Orolfo B.N.B. For. Dept. 2863 / 1933 / "...bark used as a native medicine for sores" / "Ambisan (Kedayan)"

2816 **S. longistyla** Brit. N. Borneo / H. G. Keith 3119 / 1933 / "Bark used for chest pains" / "Moil-Moil (Idahan)"

2817 **Saurauia** indet. Brit. N. Borneo / H. G. Keith 1622 / 1932 / "Young leaves used for drawing boils" / "Sodisod (Murút)"

2818 **S. conferta** Solomon Is. / S. F. Kajewski 1976 / 1930 / "The leaves are put into the dog's mouths before going Possum hunting, to make their scent keener" / "Kokeu-mor-mokiu"

Solomon Is. / S. F. Kajewski 2706 / 1932 / "When a man is deaf the stems are taken roasted and put into the ears" / "Heina-muller"

2819 **S. schumanniana** Solomon Is. / S. F. Kajewski 2523 / 1931 / "The fruit and buds are heated and put on to sore legs" / "Lengeng sorsor"

2820 **Saurauia** indet. New Britain / K. J. White 10495 / 59 / "Ripe flowers used to heal sores" / "Manokato"

2821 **S. glabrifolia** Philippine Is. / C. O. Frake 754 / 1958 / "Young leaves applied for headache" / "mengulen Sub."

2822 **Saurauia** indet. Philippine Is. / A. L. Zwickey 175 / 1938 / "'Karimog a carabao' (Lan.), implying the leaves are eaten by carabao"

2823 **Saurauia** indet. Philippine Is. / P. Añonuevo 135 / 1950 / "Medicinal —scrape the bark and squeeze the juice. Juice is then applied on the wounds" / "Tungao-tungao Manobo"

2824 **Saurauia** indet. Philippine Is. / C. O. Frake 751 / 1958 / "Young leaves applied for internal pains" / "tumangkul Sub."

2825 **Saurauia** indet. Philippine Is. / P. Añonuevo 18 / 1950 / "Fruit edible when ripe" / "Alingongokog Manobo"

2826   *S. rubicunda* Fiji Is. / O. Degener 15383 / 1941 / "For eye trouble Fijians squash juice from leaf into eye"

OCHNACEAE

2827   *Ochna afzelii* Sierra Leone / G. F. S. Elliot 5350 / no year / "'Kunya-kume,' berries edible"

2828   *O. angolensis* Angola / A. C. Curtis 158 / 1923 / "Oil from leaves is used for ointment"

2829   *O. cinnabarina* Angola / L. S. Tucker 34 / 1924 / "Omia—from the berries an oil is made for eating purposes"

CARYOCARACEAE

2830   *Caryocar villosum* Brazil / B. A. Krukoff 1099 / 1931 / " . . . seeds a source of edible oil" / "'Piquia branco'"

2831   *C. glabrum* Brit. Guiana / J. S. De La Cruz 3814 / 1923 / "'Cola'"

2832   *C. tessmannii* Peru / J. M. Schunke 167 / 1935 / " . . . fruit . . . edible" / "'Almendra'"

       Colombia / Schultes & Cabrera 14999 / 1952 / "Rind of fruit used as fish poison . . . Seed eaten by natives" / "Barassna = e-hó; Kubeo = Kön"

QUIINACEAE

2833   *Quiina obovata* Surinam / G. Stahel 354 / '45 / "fruits edible" / "Hitsi Té"

2834   *Lacunaria crinata* Surinam / G. Stahel 302 / '45 / " . . . edible fruits" / "Marodite"

THEACEAE

2835   *Camellia purpuracea* China / W. T. Tsang 23335 / 1933 / " . . . poisonous" / "Ngau Ku Cha"

2836   *C. kissi* China / W. T. Tsang 24609 / 1934 / "fr. edible" / "Yua Cha Shue"

2837   *C. sinensis* var. *assamica* Indochina / W. T. Tsang 29277 / 1939 / " . . . fr. brown, edible"

2838   *Laplacea* indet. Neth. New Guinea / P. van Royen 4725 / 1954 / "The bark is used for catching fish since it paralyzes them when bits of the bark or its extract are thrown or is poured out in creeks or rivers" / "iniaili (Je dialect) baief (Gab-Gab)"

2839   *Schima wallichii* Burma / F. G. Dickason 7647 / 1938 / "Bark used to poison fish" / "'khiang' Haka Chin"

2840   *Ternstroemia pringlei* Mexico / G. B. Hinton 14070 / 38 / "Medicine for coughs"

Mexico / G. B. Hinton 310 / 32 / "Smells of cyanide"

2841 *T. tepezapote* Guatemala / J. A. Steyermark 51739 / 1942 / "Boiled bark considered effective remedy for snake-bites; drink infusion after boiling the bark" / "'huala-kuk'" / "'chucul'"

2842 *Adinandra bockiana* var. *acutifolia* China / Fan & Li 516 / 1935 / "...fr....edible"

2843 *A. glischroloma* China / Y. W. Taam 758 / 1938 / "...fr. edible" / "Du To Kwoh"

2844 *A. millettia* China / H. H. Hu 227 / 1920 / "Tea substitute"

China / W. T. Tsang 21353 / 1932 / "...fruit...edible" / "Wong Pan Ch'a Shue"

2845 *Eurya amplexifolia* China / W. T. Tsang 21116 / 1932 / "...fruit black, used as medicine" / "Tsuen Sam Ch'a"

2846 *E. chinensis* China / W. T. Tsang 25809 / 1935 / "...fr. black, edible" / "Ha Hai Ngau Shue"

2847 *E. acuminata* Philippine Is. / W. Beyer 6851 / 1948 / "Small stem used as chewing buyo to prevent whitish tongue"

2848 *Cleyera chingii* China / W. T. Tsang 20920 / 1932 / "...fruit...edible" / "Wong Pan Ch'a"

GUTTIFERAE

2849 *Kielmeyera rosea* Brazil / Y. Mexia 5802 / 1931 / "...slightly milky juice;...Infusion of leaves used for face eruptions" / "'Parerinha'"

2850 *Hypericum laricifolium* Venezuela / W. Gehriger 20 / 1930 / "...(muy abundante y la leña es usada para cocinar)" / "N.v. Huesito"

2851 *H. platyphyllum* Colombia / Killip & Smith 15616 / 1926 / "'Romero'"

2852 *H. japonicum* China / W. T. Tsang 21109 / 1932 / "...used as medicine" / "Tin Kei Wong"

2853 *H. patulum* China / Pater Siméon Ten 7 / 1915 / "...medicina contra Cadorem ulceris"

2854 *H. sampsoni* China / G. W. Groff 45 / 19 / "Licharis odor" / "Lát Liú"

2855 *Calophyllum inophyllum* New Guinea / P. van Royen 3239 / 1934 / "Fruit green when young, edible"

Solomon Is. / S. F. Kajewski 2405 / 1931 / "The natives use the sap of this tree as a treatment for boils and sores" / "Coilal"

Fiji Is. / A. C. Smith 177 / 1933 / "Oil from the fruit used as medicine and to scent coconut oil" / "'Ndilo'"

Fiji Is. / O. Degener 15420 / 1941 / "...extract of leaves used for eye trouble" / "'Ndilo' (Ra)"

Fiji Is. / Degener & Ordonez 13786 / 1940 / "... use oil from kernel to rub on themselves during races and athletic contests"

2856 **Clusia utilis** Guatemala / S. F. Blake 7859 / 1919 / "Juice of fruit ... used on cotton for toothache and to stamp clothes" / "N.v. 'quiebra-muela'"

2857 **Clusia** indet. Mexico / Y. Mexia 694 / 1926 / "Fl. white. Extremely fragrant"

2858 **C. flava** W.I.: Cayman Is. / G. R. Proctor 15141 / 1956 / "'Balsam'"

2859 **C. palmicida** W.I.: Tobago / W. E. Broadway 4601 / 1913 / "'Parrot-apple'"

2860 **C. insignis** Colombia / Schultes & Lopez 9520 / 1947 / "Flowers ... smelling like rancid butter"

2861 **C. cf. lineolata** Peru / F. Woytkowski 6019 / 1960 / "Medicinal" / "'Renaquilla'"

2862 **C. pavonii** Peru / R. Kanehira 351 / 1927 / "Resin used as a burning incense" / "'Incensio'"

2863 **Tovomita chachapoyasensis** Ecuador / J. A. Steyermark 53347 / 1943 / "... wood has resin which is bitter" / "'duco'"

2864 **Tovomita** indet. Colombia / A. E. Lawrance 427 / 1932 / "Smells like celery"

2865 **Chrysochlamys weberbaueri** Peru / Y. Mexia 8278 / 1936 / "... fruit edible" / "'Ciruela del Monte'"

2866 **Rheedia lateriflora** W.I.: Dominica / W. H. Hodge 3360 / 1940 / "... children eat the fruit" / "'bois chien'"

2867 **R. sessiliflora** Jamaica / Howard & Proctor 15031 / 1958 / "... fr. ... eaten by birds"

2868 **R. spruceana** Bolivia / M. Cárdenas 1256 / 1921 / "Fruit edible" / "'Achycharii'"

2869 **R. multiflora** China / W. Y. Chun 5903 / 1927 / "... seeds used as poultice for bruises"
China / W. T. Tsang 25711 / 1935 / "... fr. edible" / "Lam Nar Gui Shue"

2870 **Garcinia oblongifolia** China / W. T. Tsang 23960 / 1934 / "... fr. ... edible" / "Nam Na Kat Shue"
China / H. Fung 20097 / 1932 / "Leaves eaten by Loi; slightly acid in taste"

2871 **G. gaudichaudi** Cochinchina / L. Pierre 92 / 1866 / "fourniture excellente gommigut"

2872 **G. indica** India: Bombay / J. Fernandes 1521 / 1957 / "Skin and rind used to make acid drink. Dried rind used to make vinegar. A kind of butter is extracted from the ripe seeds"

2873 *G. xanthochymus* Burma / P. Khant 276 / 48 / "Fruits are eaten"

2874 *Garcinia* indet. Siam / C. R. Carpenter 105 / 1937 / "Eaten by gibbons"

2875 *G. beccarii* N. Borneo / Bayak for B.N.B. For. Dept. 2125 / 1932 / "...fruit...edible" / "Arui Arui (Dusun)"

2876 *Garcinia* indet. New Guinea: Papua / C. E. Carr 15466 / 36 / "The bark when masticated, mixed with a certain root and pounded with sugar cane juice yields a red dye employed by the natives" / "Biagi name ITILI"

2877 *G. oligophlebia* Philippine Is. / C. O. Frake 724 / 1958 / "Leaves applied for internal pains" / "Mulizlez Sub."

2878 *G. pancheri* New Hebrides Is. / S. F. Kajewski 342 / 1928 / "Eaten by natives when ripe" / "Ney-yah heven"

2879 *G. kajewskii* Australia: Queensland / Brass & White 190 / 37 / "...flowers...with an aromatic (cinnamon) scent"

2880 *G. kola* Sierra Leone boundary / G. F. S. Elliot 4841 / no year / "'Bitter Cola' used as a chew stick (root)"

2881 *Symphonia globulifera* Brit. Honduras / P. H. Gentle 2915 / 1939 / "'waika chew stick'"
Honduras / P. C. Standley 54477 / 1927–28 / "Remedy for 'el aire'"
Jamaica / G. R. Proctor 18452 / 1958 / "'Boar-wood'"
Jamaica / no collector s.n. / 1830 / "Hog-gum tree"
W.I.: Trinidad / J. S. Beard 146 / 1943 / "'Yellow mangue'"

DIPTEROCARPACEAE

2882 *Dipterocarpus insularis* Cochinchina / L. Pierre 1421 / 1874 / "Bois rouge estimé. L'arbre est exploité aussi pour son oléo résine"

2883 *D. intricatus* INDIA: L. Pierre 1689 / 1865 / "Exploité pour son oléo résine et son bois"

2884 *D. grandiflorus* N. Borneo / A. Cuadra A2473 / 49 / "...contains resin...medicine" / "Keruing Sinpor (Malay)"

2885 *Dipterocarpus* indet. Philippine Is. / C. O. Frake 828 / 1958 / "Sap applied to skin eruptions" / "dingan Sub."

2886 *Hopea forbesii* New Guinea / L. J. Brass 27799 / 1956 / "...resin gathered and sold with that of Vatica of the area" / "Walei"

2887 *Parashorea macrophylla* Sarawak / P. S. Ashton S19568 / 63 / "Ripe fruit cooked or eaten fresh, nutty to taste"

2888 *Cotylelobium malayanum* Sarawak / P. W. Richards 1816 / 1932 / "Fls....strong, sweet, vanilla-like scent" / "Resak peniau or Resak durian"

TAMARICACEAE

2889 *Tamarix chinensis* China / Cheo & Yen 232 / 1936 / "Juice for curing smallpox"

CISTACEAE

2890 *Helianthemum salicifolium* Kuwait / H. Dickson 95 / 62 / "Eaten by all animals" / "Jerait"

COCHLOSPERMACEAE

2891 *Cochlospermum vitifolium* Guatemala / J. A. Steyermark 48521 / 1942 / "Tree; bark used for asthma trouble and for treatment of kidney trouble" / "'pumpoflor'"

2892 *Amoreuxia palmatifida* Mexico / P. C. Standley 2280 / 1936 / "Natives eat the young tender fruits raw and the fleshy roots roasted" / "Sairja, Mex."

CANELLACEAE

2893 *Canella winteriana* Puerto Rico / Schubert & Winters 495 / 1954 / "Said to be used as a fish poison, and to cause eruption of the skin" / "'Marbasco' or 'Barbasco'"

VIOLACEAE

2894 *Rinorea deflexiflora* Brit. Honduras / P. H. Gentle 3222 / 1940 / "'Wild coffee'"

2895 *R. guatemalensis* Brit. Honduras / P. H. Gentle 4009 / 1942 / "'wild coffee'"

2896 *R. micrantha* Peru / J. M. Schunke 148 / 1935 / "'Limonsacha'"

2897 *Leonia glycycarpa* Peru / F. Woytkowski 5834 / 1960 / "Fruits ... (may be eaten; other collections have been toxic???)"

2898 *Hybanthus lanatus* Brazil / Y. Mexia 5710a / 1931 / "Infusion of leaves and root extensively used for 'deparative' cough and grippe" / "'Pacacomha'"

2899 *Viola diffusa* China / W. T. Tsang 22878 / 1933 / " ... used as medicine" / "Wong Fa Ue"

FLACOURTIACEAE

2900 *Oncoba echinata* Cuba / J. G. Jack 4224 / 1926 / "A source of Chalmoogra Oil"

2901 *Caloncoba brevipes* Liberia / G. P. Cooper 173 / 1929 / "Seeds used as medicine" / "'Gbo-ah'"

2902 *Mayna longifolia* var. *phasmatocarpa* Colombia / R. E. Schultes 5728 / 1944 / "Stems strong odour cyanide" / "nom. vulg. Karijona tribe—ha-pĕ-ta-kĭ Colombians—Cacito, Cacao blanco"

2903   *Carpotroche amazonica* Colombia / G. Klug 1923 / 1931 / "Bark used as caustic" / "'Nina caspi'"

2904   *Hydnocarpus kurzii* Cuba / J. G. Jack 4535 / 1926 / "'Chaulmoogra Oil Tree'"

Burma / P. Khant 201 / no year / "Fruit is used as medicines"

Burma / R. N. Parker 2247 / 1924 / "Flowers . . . with an unpleasant smell"

Philippine Is. / F. M. Salvoza 1003 / 1946 / "Medicinal"

2905   *H. woodii* Brit. N. Borneo / B.N.B. For. Dept. 10132 / 1938 / "Used medicinally" / "'karpos'"

2906   *H. anthelmintica* Philippine Is. / F. Esteves 34165 / 1955 / "Medicinal" / "'Bagarbas'"

2907   *H. sumatrana* Philippine Is. / F. M. Salvoza 1004 / 1946 / "Medicinal" / "introd.; planted"

2908   *Scaphocalyx spathacea* Malaya / Gadoh (K.L. 1408) / 59 / "Phytochemical Survey of Malaya . . . " / "'Kelabu' (Temuan)"

2909   *Pangium* indet. Brit. N. Borneo / B.N.B. For. Dept. 7698 / 1937 / "Fruit . . . edible. Fruit and leaves used medicinally for wounds" / "'kepayang' (malay), 'pangi' (Dusun Keningau)"

2910   *Scolopia buxifolia* Hainan / H. Fung 20361 / 1932 / " . . . fruit . . . bitter in taste"

2911   *S. spinosa* Brit. N. Borneo / B.N.B. For. Dept. 3020 / 1933 / " . . . leaves used in native medicine" / "'Rukam hutan bini' (Brunei) / 'Rotiom andu' (Busun)"

2912   *Dioncophyllum peltatum* Liberia / G. P. Cooper 303 / 1929 / "Vine used for medicine" / "'Goe-doo'"

2913   *Hasseltiopsis dioica (Pleuranthodendron mexicana)* Costa Rica / P. H. Allen 5240 / 1949 / "'Quebracho blanco'"

2914   *Homalium mollissimum* Hainan / S. K. Lau 433 / 1932 / "Use bark to obtain glue. Mix glue with wine as a wound healer" / "'Dai Sai'"?

2915   *Homalium* indet. China / R. C. Ching 7348 / 1928 / "Fruit . . . when sweet, edible"

2916   *H. caryophyllaceum* Brit. N. Borneo / B.N.B. For. Dept. 1549 / 1932 / "Tree with white flowers. Poison" / "'Malinoin' (Orang Sungei, K'tangan)"

2917   *H. nitens* Fiji Is. / L. Reay 18 / 1941 / "Fijians scrape bark mix it with cold water & use it as a tonic for its strengthening qualities for children & aged or weak folk" / "'Molaca'"

2918   *Xylosma ellipticum* Mexico / G. B. Hinton 3239 / 33 / "Birds like the fruit"

2919 *Xylosma* indet. Hanoi (Mission to Indo-China & Java) D. Bois 436 / 1902 / " . . . les feuilles servent a préparer une infusion médicinale. Les annanites en font grand usage"

2920 *Xylosma* indet. Brit. N. Borneo / Apostol B.N.B. For. Dept. 6745 / 1936 / "Edible" / "Cultivate"

2921 *Xylosma* aff. *samoense* Cook Is. / F. R. Fosberg 9756 / 1940 / " . . . fruit said to be edible" / " 'koka' "

2922 *Azara microphylla* Chile / Worth & Morrison 16439 / 1938 / " 'Arrayan' "

2923 *A. salicifolia* Argentina / S. Venturi 9602 / 1929 / " 'Duraznillo' "

2924 *Flacourtia parvifolia* Hainan / W. T. Tsang 425 / 1928 / " . . . fr. yellow; edible"

2925 *F. indica* Annam / Poilane 5602 / 23 / " . . . les fleurs sont très visitées par les abeilles, les feuilles grillées servent à faire des infusions que l'on donne aux femmes après l'accouchement; il serait très prise pour cet usage, Saigon en emporterait pour le même motif"

2926 *F. montana* India / J. Fernandes 68 / 1949 / " . . . fruit said to turn red and to be eaten locally" / " 'Atuk' "
India / J. Fernandes 134 / 1949 / "Fruit fleshy, round, tartish to the taste. Locally eaten" / " 'Champair' "

2927 *Flacourtia* indet. India / J. Fernandes 6 / 1949 / "Ripe fruits eaten by children"

2928 *F. rukam* Philippine Is. / M. D. Sulit 3214 / 1949 / "Wood made into pestle" / " 'Lanagon' " Bukid.

2929 *F. kirkii* N. Rhodesia / J. B. Davy 513 / 1929 / "Fruits not edible"

2930 *Olmediella betschleriana* Guatemala / P. C. Standley 58477 / 1938 / " 'Manzana,' 'Manzanote' "

2931 *Bennettiodendron brevipes* China / Lingnan (To & Ts'ang) 12623 / 1924 / " 'Mountain Cassia flower' "

2932 *Lunania parviflora* Bolivia / H. H. Rusby 849 / 1921 / "Lvs. sugary sweet when fresh, not after drying"

2933 *Tetrathylacium macrophyllum* Colombia / O. Haught 1373 / 1934 / "Ant-inhabiting tree . . . "
Colombia / G. Klug 1766 / 1930 / " . . . fr. edible"

2934 *Osmelia philippina* Philippine Is. / C. O. Frake 383 / 1857 / "Medicine for scabies (leaves)" / " 'Sanguyud' " Sub.

2935 *Ryania pyrifera* Brazil / A. Ducke 345 / 1940 / "Radixa (vel planta tota) venenosa"

2936 *R. spruceana* Venezuela / L. Williams 14819 / 1942 / " . . . se dice que las hojas machacadas son sumamente benenosas . . . " / " 'Guaríkama' (Baniba)"

2937 *Casearia aculeata* Brit. Honduras / P. H. Gentle 3825 / 1941 / "'Wild Lemon,' 'Limoncillo'"

Brit. Honduras / P. H. Gentle 3402 / 1940 / "'Bird berries female'"

2938 *C. arguta* Mexico / G. B. Hinton 10271 / 37 / "...fruit edible" / "'Bonetillo'"

Brit. Honduras / P. H. Gentle 3817 / 1941 / "'Sweet wood tree'"

Brit. Honduras / P. H. Gentle 2707 / 1939 / "'hard moho,' 'monkey plum'"

2939 *C. belizensis* Brit. Honduras / P. H. Gentle 3297 / 1940 / "'Drunken bayman wood'"

2940 *C. corymbosa* Mexico / G. B. Hinton 10438 / 37 / "'Trementinillo'"

2941 *C. dolichophylla* Mexico / G. B. Hinton 10204 / 1937 / "'Trementinillo'"

2942 *C. nitida* El Salvador / S. Calderón 986 / 1922 / "'Canjura,' 'Chilillo de la Huasteca'"

Mexico / R. L. Dressler 2050 / 1957 / "'Cafecillo'"

Guatemala / J. A. Steyermark 39754 / 1940 / "Fruit and leaves boiled and used as substitute for coffee" / "'Cafe de monte'"

2943 *C. ramiflora* El Salvador / P. C. Standley 21800 / 1922 / "'Limoncillo'"

W.I.: Trinidad / RDW 118 / 1908 / "'Wild Coffee'"

Venezuela / H. Pittier 7821 / 1918 / "N.v. Punta de ral, limoncillo"

2944 *C. sylvestris* El Salvador / S. Calderón 1119 / 1922 / "Fls. have odor of valerian"

Brazil / Reitz & Klein 3809 / 1956 / "'Cafeeiro do mato'"

Brazil / Y. Mexia 4725 / 1930 / "'Cafe du Matto'"

2945 *C. decandra* W.I.: Grenadines / R. A. Howard 10996 / 1950 / "...berry said to be edible"

Peru / Y. Mexia 6108 / 1931 / "...fruit...edible" / "'Limón-cáspi'"

2946 *C. guianensis* Jamaica / Howard & Proctor 14185 / 1955 / "'Wild coffee'"

2947 *C. hirsuta* Jamaica / G. R. Proctor 11780 / 1956 / "'Wild Coffee'"

Cuba / R. A. Howard 5679 / 1941 / "'Raspa lengua'"

2948 *Casearia* indet. W.I.: Martinique / H. & M. Stehlé 6172 / 1949? / "'Sapotillier Diable' (Poison) 'Sapote diable'"

2949 *C. javitensis* Brazil / R. Froes 12433/177 / 1942 / "'Pau de rato'"

2950 *C. resinifera* Brazil / R. Froes 12556/250 / 1942 / "Indians claim plant is very toxic" / "'Uarimacan'"

2951 *C. spinosa* Peru / Y. Mexia 6413 / 1932 / "'Espinha del demonio,' 'Supiecacha'"

2952 *Casearia* indet. Brazil / Y. Mexia 4883 / 1930 / "'Café do matto'"

2953 *Casearia* indet. Brazil / Y. Mexia 5011 / 1930 / "An infusion of leaves with salt, used in influenza, also for sick animals" / "'Contra herva'"

2954 *Casearia* indet. Brazil / B. A. Krukoff 5953 / 1934 / "Yields the poisonous glucoside 'Ryanine'"

2955 *Casearia* indet. Bolivia / J. Steinbach 6088 / 1924 / "'Cafecillo'"

2956 *Casearia* indet. Argentina / J. Steinbach 2144 / 1916 / "'Cafecillo'"

2957 *C. fuliginosa* Philippine Is. / C. O. Frake 858 / 1958 / "Roots applied to ulcers" / "'Glumbayulan'" Sub.

2958 *C. richii* Fiji Is. / O. Degener 15073 / 1941 / "Leaves chewed as remedy for thrush" / "'Galo' according to informant; not galo according to other" Serua

2959 *Peridiscus lucidus* Brazil / Ducke 304 / 1936 / "'Páo santa'"

STACHYURACEAE

2960 *Stachyurus himalaicus* China / W. T. Tsang 20674 / 1932 / "Used for medicine" / "Tung Fa"

TURNERACEAE

2961 *Turnera diffusa* Mexico / H. S. Gentry 5547 / 1940 / "Infused for tea and drunk as a tonic or general cure" / "Damiana"

PASSIFLORACEAE

2962 *Adenia zucca* Philippine Is. / Sulit & Conklin 5084 / 1955 / "Decoction of root given to person suffering from stomach trouble" / "Tabungaw-Amô Mang."

2963 *Passiflora coriacea* El Salvador / P. C. Standley 20602 / 1922 / "Lvs. with lard used as a poultice for wounds and swellings" / "'Murciélago'"

2964 *P. salvadorensis* El Salvador / P. C. Standley 19279 / 1921–22 / "Used as a purgative and for gonorrhoea and retention of urine" / "'Colzoncillo'"

2965 *P. foetida* var. *gossypiifolia* El Salvador / P. C. Standley 22476 / 1922 / "'Granadilla, Gravadilla de culebra'"

2966 *P. cuprea* Bahama Is. / O. Degener 18820 / 1946 / "Used for tea"

CARICACEAE

2967 *Jarilla chocola* Mexico / H. S. Gentry 1553 / 1935 / "Warihios eat the root raw or toasted and the fruit raw" / "Chocola Mex.   Kapiah, W."

2968 *Pileus mexicanus* Mexico / C. F. Millspaugh 96 / 1895 / "'The fruit, called Bonete, . . . eaten in a sort of custard'"

BEGONIACEAE

2969 *Begonia caroliniaefolia* Guatemala / J. A. Steyermark 49188 / 1942 / "The acid sap quenches the thirst and is reputed to be used in cutting fevers" / "'nitro de montaña'"

2970 *B. portillana* Mexico / P. C. Standley 1635 / 1935 / "Roots cooked as a purgative draught" / "Cana aigre, Mex. Chokopala, W."

2971 *B. plebeja* Guatemala / P. C. Standley 75304 / 1940 / "'Fuego; Pie de paloma'"

2972 *B. squarrosa* Mexico / G. B. Hinton 3075 / 33 / "Stalk eaten raw"

2973 *B. strigillosa* Guatemala / J. A. Steyermark 51853 / 1942 / "Reputed to be used for treating infection of intestine"

2974 *B. minor* Jamaica / Maxon & Killip 113 / 1920 / "Used in medicine for colds by natives"

2975 *Begonia* indet. W.I.: Dominica / W. H. & B. T. Hodge 3312 / 1940 / "'L'eau zay' used for stomach colds (tea of leaves)"

2976 *B. meridensis* Venezuela / J. A. Steyermark 55801 / 1944 / "Leaves eaten for quenching thirst (acid)—cattle eat also" / "'vinagrera'"

2977 *B. sucrensis* Venezuela / J. A. Steyermark 62760 / 1945 / "'tocino cochina'"

2978 *B. rossmanniae* Ecuador / R. E. Schultes 3474 / 1942 / "Plant boiled to prepare eye wash for conjunctivitis" / "Nombre kofán: 'a-ve-ne-e-chó'"

2979 *B. fimbristipula* China / W. T. Tsang 22895 / 1933 / ". . . fruit smoky; used as medicine" / "Tin Kwai"

2980 *Begonia* indet. Sumatra / W. N. & C. M. Bangham 776 / 1932 / "Leaves used as vegetable" / "'Rijang'"

2981 *B. oblongata* Philippine Is. / C. O. Frake 391 / 1957 / "Leaves eaten as antidote for *Dioscorea hispida*"

2982 *Begonia* indet. Solomon Is. / S. F. Kajewski 2518 / 1931 / "When a man is sick in the stomach the leaves are macerated in hot water and boiled" / "Mango"

CACTACEAE

2983 *Opuntia rastrera* Mexico / C. E. Smith et al. 3747 / 1961 / "Tuna de vipores"

2984 *O. brasiliensis* Bolivia / R. S. Shepard 192 / 1920 / "Fruit edible"

2985 *Opuntia* indet. Argentina / A. L. Cabrera 8415 / 1944 / "Semillos rojas, tinctoreas y medicinales" / "N.v. 'airompu'"

2986   *Cereus* indet. Mexico / Y. Mexia 785 / 1926 / "The fruits . . . constantly eaten by the natives. Birds eat many"

2987   *Peniocereus greggii* Mexico / R. M. Stewart 723 / 1941 / "'Huevo de Venado'"

2988   *Harrisia* indet. Dominican Republic / R. A. & E. S. Howard 8336 / 1946 / " . . . fruit yellow, seeds black, good eating"

2989   *Cleistocactus baumannii* Bolivia / J. Steinbach 7793 / 1927 / "Die Pulpa um die schwarzen Samen ist weiss, wohlschmeckend" / "Pitahayacita"

2990   *C. smaragdiflorus* Bolivia / M. Cárdenas 3705 / 1944 / "Edible fruit" / "'Sitiquira'"

2991   *Hylocereus napoleonsis* W.I.: St. Vincent / G. R. Cooley 8203 / 1962 / "Fruit . . . known locally as 'Mountain Peas.' Flavor not unpleasant. Strongly cathartic"

2992   *Epiphyllum* indet. Mexico / Y. Mexia 1476 / 1927 / "'Choconoxtle' Pleasant, slightly acid taste. Eagerly eaten. Used for intestinal maladies"

THYMELAEACEAE

2993   *Phaleria perrottetiana* Philippine Is. / A. L. Zwickey 121 / 1938 / "'Gitalis' (Lan.); young stems eaten"

2994   *Dicranolepis persei* Africa: Gold Coast / C. Vigne 1115 / 1928 / "Fruit . . . reported edible"

2995   *Wickstroemia chamaedaphne* China / J. C. Liu L.1168 / no year / "for fish poison, popular belief" / "Yüan Hao, Yüen Hsiao"

2996   *W. lanceolata* Philippine Is. / Sulit & Conklin 5104 / 1953 / "Roots boiled—decoction given to patient in amoebic dysentery (darag-is)" / "Gamot-pamará Dialect Mang."

2997   *W. foetida* Society Is. / L. H. MacDaniels 1282 / 1927 / "Used medic. for syphilis"

2998   *W. uva ursi* Hawaiian Is. / L. W. Bryan s.n. / 1939 / "Used as fish poison"

2999   *Daphnopsis americana* ssp. *salicifolia* Mexico / G. B. Hinton 10246 / '27 / "Manea de Torro"

3000   *D. macrocarpa* W.I.: St. Lucia / G. R. Proctor 21570 / 1960 / "'Mahaut-piment'"

3001   *D. americana* ssp. *caribaea* Puerto Rico / L. R. Holdridge 236 / 1940 / "Bark used for rope, liked locally as being bitter the rope is not eaten by animals"

      W.I.: Dominica / W. H. Hodge 466 / 1938 / "a tree whose inner bark smells like that of our wild cherry"

3002 *D. americana* ssp. *tinifolia* Jamaica / Howard & Proctor 14101 / 1954 / " . . . bark rank-smelling"

ELAEAGNACEAE

3003 *Hippophae rhamnoides* Tibet / J. F. Rock 13275 / 1925 / " . . . used as medicine for stomach ache"

Kashmir / W. Koelz 2942 / 1931 / "Fruit orange, very acid, pleasant"

LYTHRACEAE

3004 *Lythrum salicaria* EUROPE / N. C. Seringe (Herb. Helveticum) 2506 / no year / " . . . Herbe astringente, faible"

3005 *Woodfordia fruticosa* India / J. Fernandes 72 / 1949 / "Dried leaves and flowers are strung on a piece of twine and tied over the abdomen to foment the flow of urine" / "'Dhasti'"

3006 *Cuphea procumbens* Mexico / C. Lumholtz s.n. / 1903 / "'Herba de Calabera'"

3007 *C. racemosa* Colombia / Schultes & Smith 2097 / 1942 / "Diuretic" / "Nombre ingano: 'i-spa-na-nai-ambĕ'"

3008 *Pleurophora pungens* Chile / Worth & Morrison 16371 / 1938 / "Used as kidney medicine according to Sr. R. Wagenknecht" / "'Lengua de gallina'"

Chile / R. Wagenknecht 18481 / 1939 / "Used medicinally" / "'Lengua de gallina'"

3009 *Lagerstroemia pyriformis* Philippine Is. / C. O. Frake 866 / 1958 / "Agricultural ritual" / "'Bitang ulangan'" Sub.

3010 *L. speciosa* Philippine Is. / C. O. Frake 853 / 1958 / "Bark boiled and drunk for hematuria—Roots boiled and drunk during puerperium" / "'Benaba'" Sub.

Philippine Is. / C. O. Frake 885 / 1958 / "Roots drunk for jaundice" / "'Benaba ulangan'" Sub.

SONNERATIACEAE

3011 *Sonneratia acida (S. lanceolata)* New Guinea / L. J. Brass 980 / 1926 / "The feeding tree of a species of firefly which gathers in such numbers on certain individual trees that at night the whole tree is lighted with a soft greenish glow which is often reflected quite distinctly on the water"

3012 *Duabanga moluccana* Philippine Is. / H. C. Conklin 997 / 1957 / "Economic Uses: pagpandayun balūtu?" / "'talāpa'" Han.

CRYPTERONIACEAE

3013 *Crypteronia paniculata* Philippine Is. / C. O. Frake 784 / 1958 / "Bark applied for skin eruptions" / "'Hanambalan'"? Sub.

LECYTHIDACEAE

3014 *Planchonia papuana* Solomon Is. / S. F. Kajewski 2385 / 1930 / "The bark is macerated and the sap drunk to cure headaches"

3015 *Careya arborea* Burma / F. G. Dickason 7650 / 1938 / "Fruit used for fish poison" / "'ngal ti' Haka Chin.; 'pant pwe'"

3016 *Barringtonia cylindrostachya* Malaya / Umbai for A. H. Millard K.L. 1526 / 59 / "Fruits and bark a fish poison" / "Pone Tau (Temuan)"

3017 *B. asiatica* Solomon Is. / A. C. Smith 4744 / 1947 / "... used as a fish poison" / "'Vutu'"

3018 *B. racemosa* New Hebrides Is. / S. F. Kajewski 126 / 1928 / "The fruit ... is poisonous" / "Nevingen (Black)"

3019 *Gustavia augusta* Brazil / B. A. Krukoff 1072 / 1931 / "... seeds are a source of oil" / "'Genipaparana'"

3020 *Gustavia* indet. Peru / Ducke 1841 / 1945 / "... fructus edulis" / "'Chópe'"

3021 *Grias neuberthii* Peru / Y. Mexia 6115 / 1931 / "... pulp surrounding seed edible" / "'Mango sacha'"

3022 *Couroupita guianensis* Peru / Y. Mexia 6502 / 1932 / "fruit ... astringent" / "'Haya-uma'"

3023 *Lecythis elliptica* Venezuela / Curran & Haman 806 / 1917 / "Fruit reported poisonous" / "'Hoyamono'"

3024 *L. minor* Colombia / O. Haught 2220 / 1938 / "... seeds edible. 'Muy zabrosas' per local inhabitants" / "'coco-mono'"

3025 *L. pisonis* Brazil / Y. Mexia 4966 / 1930 / "... seeds ... edible ... flavor resembles Brazil-nut" / "'Sapucainha'"

3026 *Bertholletia excelsa* Venezuela / L. Williams 14586 / 1942 / "... semillas comibles ... la madura exuda una savia dulce muy buscada por el 'lame-ojo'; insecto pequeño que molesta los ojos" / "Jubia"

RHIZOPHORACEAE

3027 *Cynotroches lanceolata* Philippine Is. / C. O. Frake 774 / 1958 / "Leaves applied to ulcers" / "gelundem Sub."

3028 *Carallia diplopetala* Indo-China / W. T. Tsang 29073 / 1939 / "... fr. yellow, edible"

3029 *C. brachiata* New Guinea: Papua / L. J. Brass 7568 / 1936 / "The corky outer bark burned and powdered is used as a face and body paint by the local natives"

3030 *Bruguiera sexangula* Fiji Is. / A. C. Smith 2 / 1933 / "... bark used for medicine and dye" / "'Ndongo'"

3031 *Pellacalyx* indet. Philippine Is. / C. O. Frake 580 / 1958 / "Fishing ritual" / "tendengan Sub."

3032 *Pellacalyx* indet. Philippine Is. / C. O. Frake 785 / 1958 / "Roots applied for internal pain" / "bintulaw"

3033 *P. postulata* Philippine Is. / C. O. Frake 709 / 1958 / "Economic uses ritual" / "Sinegaw Sub."

ALANGIACEAE

3034 *Alangium handeli* Hainan / S. K. Lau 1780 / 1933 / ". . . root is used as medicine"

3035 *A. longiflorum* Philippine Is. / M. D. Sulit 1620 / 1947 / "Fr. edible when ripe" / "Malatapai Tag."

COMBRETACEAE

3036 *Combretum farinosum* Mexico / W. G. Wright 1226 / 1889 / "Called 'Panathãeya' by the natives"

3037 *C. cacoucia* Brazil / R. E. Schultes 8668 / 1947 / "Water in stem, acrid. Flowers red. Said to be poisonous" / "'Rabo de arara'"
Brit. Guiana / J. S. De La Cruz 1114 / 1921 / "Poisonous vine . . ."

3038 *C. alfredi* China / W. T. Tsang 21309 / 1932 / ". . . fr. edible" / "Sze Kwan Che T'ang"

3039 *C. wallichii* China / W. T. Tsang 23246 / 1933 / ". . . fruit . . . edible" / "So Kwan Tsz"

3040 *C. latifolium* India: Bombay Pres. / J. Fernandes 233 / 1949 / "The leaf is put over abscess to ripen it; the green leaf is simply tied over the sore" / "Vern. Hadabali"

3041 *C. pilosum* Siam / Royal For. Dept. 5696 / 44 / "Root used in fever remedy" / "Leb-mue-narng"

3042 *C. imberbe* S.W. Africa / J. Borle 13 / 26 / "Flowers used as cough medicine"

3043 *Terminalia* indet. N. Siam / Royal For. Dept. 2492 / 49 / "Natives chew barks with betle"

3044 *T. okari* New Guinea: Papua / C. E. Carr 12239 / 35 / "The kernal is most popular with the natives and ranks in flavour with the walnut" / "Okari"

3045 *T. solomonensis* Solomon Is. / S. F. Kajewski 2359 / 30 / "The natives eat the pulp of this fruit, they say it is black when ripe" / "Tauma"

3046 *Buchenavia capitata* W.I.: Dominica / W. H. & B. T. Hodge 3175 / 1940 / "'z'olivier'"

3047 *Conocarpus erectus* Bahama Is. / O. Degener 18826 / 1946 / ". . . boil bark and use lotion for prickly heat" / "'Black buttonwood'"

MYRTACEAE

3048 *Psidium chrysobalanoides* Brit. Honduras / W. A. Schipp 596 / 1930 /
" . . . fruits yellow, tarty, good for preserves"

3049 *P. salutare* Mexico / R. L. Dressler 2342 / 1957 / " 'Guayaba sanjuane-
ra,' 'G. coyote,' fruit small but very tasty"

3050 *Calyptropsidium sartorianum* Mexico / P. C. Standley 1205 / 1934 /
"Warihios grind or mash, sugar (panoche) fruit and say it is muy
bueno" / "Arellane, Mex. Chokey, W."

3051 *Decaspermum coriandri* Solomon Is. / S. F. Kajewski 2534 / 31 /
"When a man has a toothache, the fruit is put inside the mouth with
water and held there" / "Uiti-levor"

3052 *D. neo-ebudicum* New Hebrides Is. / S. F. Kajewski 283 / 28 / "Used
by the natives for enlarged spleen. Used by macerating with cold water and white sedge" / "Nywaso"

3053 *Pimenta jamaicensis* Jamaica / Howard, Proctor & Stearn 15025 /
1958 / " . . . fruit green, lemon-flavoured"

3054 *Cryptorhiza haitiensis* Dominican Republic / R. A. Howard 12446 /
1950 / " 'Canillo' . . . leaves used for tea"

3055 *Campomanesia lineatifolia* Ecuador / Eggers 14245 / 1892 / "Fructu
eduli" / "(v. Guayava de Palo)"

3056 *Myrcia citrifolia* W.I.: Dominica / W. H. & B. T. Hodge 3253 / 1940 /
"seeds edible" / " 'Kurupum' "

3057 *M. pulchra* Brazil / Y. Mexia 5067 / 1930 / " . . . red fruit, edible" /
" 'Jambro' "

3058 *Marlierea schomburgkiana* Brit. Guiana / Maguire & Fanshawe 22911
/ 1944 / " . . . fruit blue, edible, sweet"

3059 *Calyptranthes zuzygium* Jamaica / G. R. Proctor 22836 / 1962 /
" . . . fr. . . . sweet, edible"

3060 *Eugenia capuli* El Salvador / F. Choussy 39 / 1923 / " 'Guacoquito' "
Mexico / A. G. Pompa 1180 / 63 / " 'capulin agarrosa' "

3061 *E. florida* Bolivia / J. Steinbach 7244 / 1925 / "La fruta . . . un gusto
como la ceriza" / "Arrayan"

3062 *E. diantha* Brazil / Y. Mexia 5809 / 1931 / " 'Cafe de São José' "

3063 *Syzygium* aff. *flavescens* New Britain / A. Floyd 3504 / 54 / "The
leaves are boiled and eaten as a cure for stomach pains" / "La seuli
(W. Nakanai)"

3064 *S. corynocarpum* Fiji Is. / Degener & Ordonez 14219 / 1941 / "Fruit
. . . edible, somewhat like rose apple" / " 'Misi' or 'Ulala' "

3065 *S. corynantha* Fiji Is. / O. Degener 15344 / 1941 / " . . . leaves and
buds used for lung trouble" / " 'Lemba' "

Australia: New S. Wales / C. T. White 13053 / 46 / "Local name Sour Cherry"

3066 **Lysicarpus ternifolius** Australia: Queensland / S. L. Everist 1932 / 1939 / " . . . not attacked by termites" / " 'Budgeroo' or 'Blacktea-tree' "

3067 **Tristania glauca** New Caledonia: Nouméa / I. Franc 256A / 1914 / "Repand une odour ammoniacale"

3068 **Leptospermum liversidgei** Australia: Queensland / C. T. White 12257 / 1943 / " . . . leaves with a very strong citron scent when crushed"

3069 **Agonis flexuosa** Australia: W. Australia / C. T. White 5348 / 1927 / "Local name—Peppermint"

3070 **Melaleuca decora** Australia: Queensland / C. T. White 5583 / 1929 / "Paper-barked tea tree"

3071 **Myrtacea** indet. Brazil / A. Bornmüller 552 / 1905 / " 'Cereja' (Kirsche) mit essbaren aromat. (terpentinartig) Früchten"

3072 **Myrtacea** indet. Bolivia / H. H. Rusby 311 / 1921 / "Important medicine" / " 'Boldo' or 'Hoya de la Vida' "

MELASTOMACEAE

3073 **Schwackaea cupheoides** El Salvador / P. C. Standley 19620 / 1921–22 / " 'Sulfatillo' "

3074 **Tibouchina longifolia** El Salvador / P. C. Standley 19627 / 1921–22 / "Decoction held in mouth for toothache and applied for eye diseases" / " 'Talchinal' "

3075 **Tibouchina** indet. Mexico / G. B. Hinton 704 / 32 / "Has a strong odor of coconut"

3076 **Arthrostemma grandiflorum** Colombia / Schultes & Smith 2072 / 1942 / "Diuretic"

3077 **Melastoma candidum** China / S. K. Lau 753 / 1932 / " . . . fruit edible"

3078 **M. dodecandrum** China / W. T. Tsang 21024 / 1932 / " . . . roots used in medicine"

3079 **M. intermedium** China / H. H. Chung 996 / 1923 / " . . . used in medicine"

3080 **M. imbricatum** Fed. Malay States / Goping (Dr. King's coll.) 6023 / 1886 / "Seeds . . . eatable"

3081 **M. polyanthum** Solomon Is. / S. F. Kajewski 2510 / 31 / "When a man has gonorrhoea the sap of this tree is taken and drunk for it" / "Davundava"

3082 **Anplectrum viminale** Fed. Malay States / G. anak Umbai (for A. H. Millard) 1273 / 59 / "Medicinal" / "Akar Kedudok (Temuan)"

3083 *Dissochaeta celebica* Philippine Is. / C. O. Frake 721 / 1958 / "Leaves applied for pinworms" / "getungu-ulangan Sub."

3084 *Pogonanthera reflexa* Philippine Is. / C. O. Frake 772 / 1958 / "Leaves applied to wounds" / "gusip Sub."

3085 *Medinilla speciosa* Sumatra / W. N. & C. M. Bangham 779 / 1932 / "Natives cook lvs. for vegetable. Sour"

3086 *M. heterophylla* Fiji Is. / H. St.John 18181 / 1937 / "Medicinal use: when a boy wets bed, father hits penis with this, will not happen again" / "'Mimiolo' (parcel of water)"
       Fiji Is. / L. Reay 11 / 1941 / "Fijians grate stalk and use it internally as an aperient" / "Wadai"

3087 *M. rubescens* Solomon Is. / S. F. Kajewski 2514 / 31 / "When a man is sick in the stomach the leaves are macerated in boiling water, the resultant juice being drunk" / "Che-hila"

3088 *Tetrazygia discolor* W.I.: Montserrat / J. S. Beard 439 / 1944 / "'Hog-weed'"

3089 *T. longicollis* Dominican Republic / R. A. & E. S. Howard 8905 / 1946 / " ... berry purple, very sweet and eaten by birds"

3090 *Miconia ruficalyx* Brit. Guiana / B.G. For. Dept. 5410 / 46 / " ... fruit ... edible with faint blackberry flavour"

3091 *Clidemia capitellata* Honduras / A. Molina R. 891 / '48 / "'Huevos de Coyote'"

3092 *Bellucia costaricensis* Brit. Honduras / W. A. Schipp 319 / 1929 / " ... stamens highly perfumed"

3093 *B. grossularioides* Brit. Guiana / A. C. Smith 2918 / 1937 / " ... fruit edible"

3094 *B. weberbaueri* Brazil / B. A. Krukoff 1572 / 1931 / " ... fruit edible, sweet"

3095 *Henriettea succosa* Brit. Guiana / Maguire & Fanshawe 22882 / 1944 / " ... fruit ... sweet, edible"

3096 *Blakea pulverulenta* W.I.: Dominica / W. H. & B. T. Hodge 3476 / 1940 / " ... very fragrant"

3097 *Astronidium insulare* Solomon Is. / S. F. Kajewski 1999 / 1930 / "The possums eat the leaves of this tree" / "Tunutei"

3098 *A. victoriae* Fiji Is. / H. St.John 18230 / 1937 / " ... med. ointment with oil for headache" / "'diriniu'"

3099 *Memecylon* indet. Philippine Is. / C. O. Frake 651 / 1958 / "Roots boiled and drunk for general malaise" / "gyan-kuku? Sub."

3100 *Memecylon* indet. Philippine Is. / C. O. Frake 537 / 1958 / "Roots applied to ulcers" / "giburagat Sub."

ONAGRACEAE

3101 *Jussiaea suffruticosa* Caroline Is. / C. C. Y. Wong 137 / 1947 / "The flower stalks are crushed in the hands and rubbed on the spear point to get many fishes" / "N.v. coiro"

3102 *Epilobium denticulatum* Peru / Mr. & Mrs. F. E. Hinkley 65 / 1920 / "Medicinal uses" / "'Sagnayuyo'"

3103 *Oenothera rosea* Ecuador / W. H. Camp 2641 / 1945 / "...infusion used to 'refresh the intestines'" / "'Zchullo' or 'Zchungir'"
Peru / Macbride & Featherstone 61 / 1922 / "'Cupa sangre'"

3104 *Gaura parviflora* Australia: Queensland / C. T. White 6662 / 1930 / "Readily eaten by horses, seed eaten by cockatoos"

3105 *Fuchsia michoacanensis* El Salvador / P. C. Standley 22810 / 1922 / "...fr. black, sweet, juicy" / "'Saca-tinta'"

3106 *F. boliviana* var. *puberulenta* Bolivia / Y. Mexia o4292 / 1935 / "The fruits are eaten by natives, have sweetish taste"

3107 *Semeiandra grandiflora* Mexico / Y. Mexia 1680 / 1927 / "'Contra-hiedra'"

HALORAGACEAE

3108 *Gunnera magellanica* Chile / Y. Mexia 7969 / 1936 / "Scarlet fruit; edible"

ARALIACEAE

3109 *Plerandra solomonensis* Solomon Is. / S. F. Kajewski 2576 / 1931 / "When a man is constipated the inner sap is taken and drunk with water to cure it"

3110 *Boerlagiodendron* indet. Philippine Is. / G. E. Edaño 3130 / 1951 / "The root is grated fine, boiled in water & the infusion given to a mother who has just delivered to prevent relapse"

3111 *Meryta spathipedunculata* Solomon Is. / S. F. Kajewski 2527 / 1931 / "The bark is macerated with hot water and applied to sore legs" / "'Tangoie'"

3112 *Schefflera sciodaphyllum* Jamaica / G. L. Webster 5580 / 1954 / "...bark with carroty odor when broken"

3113 *S. odorata* Philippine Is. / A. L. Zwickey 25 / 1938 / "...heated bark scrapings in application for aching bones; perfume from leaves (?)" / "'Magonuno' (Lan.); 'Salapinin' (?) (Lan.)"

3114 *Dendropanax* indet. Bolivia / J. Steinbach 6508 / 1924 / "La ceniza tiene mucho potacio; por eso la quemar para hacer jabon"

3115 *Kissodendron australianum* var. *furfuraceum* Australia: Queensland / S. F. Kajewski 1256 / 29 / "Has faint odour resembling that of celery"

3116 *Acanthopanax gracilistylus* China / W. T. Tsang 20199 / 1932 / "For making medicine wine" / "'Ng Kwa Pe Tang'"

3117 *A. gracilistylus* var. *nodiflorus* China / W. T. Tsang 21930 / 1933 / "...flower yellow, edible; wine making" / "'Ng Kwa Pei'"

3118 *Heteropanax chinensis* China / W. T. Tsang 24584 / 1934 / "...fr. blackish red, edible"

3119 *Pentapanax willmottii* Australia: Queensland / C. T. White L o 688 / 1936 / "Celery Wood or Celery Wine"
Australia: Queensland / S. F. Kajewski 1277 / 1929 / "Leaves... when crushed reminding me of a faint mixture between Eucalyptus oil and turpentine"

3120 *Cephalaralia cephalobotrys* Australia: New S. Wales / C. T. White 12762 / 1945 / "...fruits...with a red watery sap, and somewhat aromatic flavour"

3121 *Aralia chinensis* China / T. T. Yü 19603 / 1938 / "...young shoots and leaves edible"

3122 *A. chinensis* var. *dasyphylloides* China / W. T. Tsang 22989 / 1933 / "...flower...used in medicine" / "Niu Pat Wai"

3123 *A. cordata* China / E. H. Wilson 4285 / 1910–11 / "A valuable medicine" / "Hung-Fo-Hoa"

3124 *A. decaisneana* China / W. T. Tsang 23298 / 1933 / "...used as medicine" / "Niu Pat Ki"

3125 *Delarbria collina* Solomon Is. / L. J. Brass 2679 / 1932 / "The fruit has a strong, pungent odour like betel pepper"

3126 *Mackinlaya celebica* Solomon Is. / S. F. Kajewski 1704 / 1930 / "Young leaves and shoots are eaten, after being boiled by the natives"
Philippine Is. / M. D. Sulit 9887 / 1949 / "The decoction of the leaves is used for bathing baby by natives (Bukidnons)"

3127 *Araliacea* indet. Bolivia / J. Steinbach 5605 / no year / "La madera es bien blanca. Se usa para sacar ceniza para hacer jabon"

UMBELLIFERAE

3128 *Hydrocotyle bonariensis* U.S.: Ala. / Webster & Wilbur 3537 / 1950 / "Plants have an aromatic smell"
Peru / F. W. Pennell 13148 / 1925 / "Root used as cure for toothache"

3129 *H. peruviana* Peru / Stork & Horton 10390 / 1939 / "Root has bitter odor, resembling *Menispermum*"

3130 *H. javanica* Sumatra / P. Buwalda 7064 / 1939 / "Aromatic when crushed"
Solomon Is. / S. F. Kajewski 2688 / 1931 / "The bark is macerated and used as a fish poison, the whole plant is macerated and thrown

into the water. The fish are stupified and rise to the surface" / "Ma-mahie"

Solomon Is. / S. F. Kajewski 1979 / 1930 / "The stems have an identical odor of parsley and . . . should be a valuable herb. The green leaves are rolled up in another leaf, say a banana leaf, heated and applied as a hot fomentation" / "Ky-no-care"

3131 *Centella asiatica* China / W. T. Tsang 25427 / 1935 / " . . . fr. edible" / "Lo Kong Kan Tang"

Philippine Is. / Sulit & Conklin 5161 / 1953 / "Roots together with wax of honeybees are smoked on tamed wild chickens (pañgati). When the Mangayans want to catch wild chickens (cock) they set trap and place the 'pañgati' in the center. When the wild cock hears the crowing of the 'pañgati' it comes to fight . . . then it is caught in a trap"

POLYNESIA: Tongatapu / T. G. Yuncker 15146 / 1953 / "Leaves used medicinally" / "tono"

3132 *Trachymene ericoides* Australia: Queensland / C. T. White 7087 / 1930 / "Whole plant with a 'carroty' smell"

3133 *Diposis bulbocastaneum* Chile / C. Grandjot 3520 / 1938 / "Knollen essbar"

3134 *Eremocharis longiramea* Peru / Goodspeed & Metcalf 30227 / 1942 / " . . . stem brittle, with a distinct odor suggesting sweet anise"

3135 *E. triradiata* Peru / J. West 8074 / 1936 / "Plant strongly aromatic"

3136 *Eryngium foetidum* Panama / H. Wedel 406 / 1940 / " . . . (natives use plants for stomach troubles); small pineapple-like cones . . . aromatic"

W.I.: Dominica / W. H. & B. T. Hodge 3340 / 1940 / "tea of leaves good for any sickness" / "shadowon beni"

Haiti / H. H. Bartlett 17339 / 1941 / "'culantro' odor abominable. Voodoo devil chaser"

China / W. T. Tsang 21305 / 1932 / " . . . fruit edible" / "Fan Hueng So"

3137 *E. globosum* Mexico / Y. Mexia 1596 / 1927 / "Eaten by stock"

3138 *E. heterophyllum* Mexico / W. P. Hewitt 48 / 1945 / " . . . tea used to cure kidney ailments" / "Yerba de sapo"

3139 *E. elegans* Bolivia / Y. Mexia 7801 / 1935 / "Good forage plant"

3140 *Osmorhiza occidentalis* U.S.: Ore. / A. Cronquist 7308 / 1935 / "Roots smelling strongly of licorice"

3141 *Torilis japonicus* Bermuda Is. / E. Manuel 47 / 63 / "Known to cause a skin rash on susceptible people when wet"

3142   *T. scabra* China / W. T. Tsang 23607 / 1934 / "Edible" / "Hueng Shi Tsoi"

3143   *Coriandrum sativum* China / C. Y. Chiao 2554 / 1930 / "leaves edible—seasoning"

3144   *Oreomyrrhis andicola* Argentina / T. Meyer 34214 / 1940 / "Medicinal"

3145   *Musineum* indet. U.S.: Mont. / J. W. Blankinship 23 / 90 / "Root said to be poisonous"

3146   *Arracacia atropurpuria* Mexico / G. B. Hinton 6821 / 34 / "Has strong odor of pine especially the stalk"

3147   *A. incisa* Peru / Goodspeed, Stork & Horton 11040 / 1939 / "Fleshy taproot has fragrance like that of anise"

3148   *Heteromorpha kasneri* Katanga / Quarré 1548 / 1929 / "Ombelliferi toxique"

3149   *Apium australe* Argentina / Y. Mexia 7951 / 1936 / "Stalks eaten by whites and natives, but are rather stringy"

3150   *Pituranthos triradiatus* Kuwait / H. Dickson 113 / 62 / "Eaten by camels"

3151   *Cryptotaenia canadense* U.S.: Penn. / F. R. Fosberg 15927 / 1938 / "Herbage with a celery odor when broken"
China / W. T. Tsang 21322 / 1932 / "... fruit edible" / "Ye Kan Tso"

3152   *Bunium persicum* Afghanistan / W. Koelz 11972 / 1937 / "... used as a spice"

3153   *Oenanthe javanica* China / W. T. Tsang 25076 / 1935 / "... fr. edible" / "Tiu Tsz Kwan"
China / Steward & Cheo 173 / 1933 / "Used as a vegetable"
New Guinea / K. G. Heider 12 / 1962 / "Eaten"

3154   *O. thomsonii* China / W. T. Tsang 25479 / 1935 / "... for feeding pigs" / "Shui Kan Ts'oi"

3155   *Oenanthe* indet. New Guinea / L. J. Brass 11220 / 1938 / "... used as a green vegetable"

3156   *Ligusticum* indet. China / Steward, Chiao & Cheo 472a / 1931 / "Medicinal"

3157   *Conioselinum chinense* Canada / Fernald & Smith 25930 / 1923 / "Plant not eaten by moose"

3158   *C. porteri* Mexico / W. P. Hewitt 85 / 1945 / "... roots strongly aromatic, one of the most important medicinal herbs. Tea of roots used for pneumonia, colds, ... ache" / "Matarique"

3159   *Angelica decursiva* Japan / E. Elliott 136 / 46 / "Young leaf well boiled and well washed, and eaten seasoned. Medicinal use: Root is febrifuge and stops cough" / "Nodake"

3160 *Glehnia littoralis* China / H. H. Chung 5964 / 1927 / "...leaf has a peculiar pungent odor and taste"

3161 *Pseudocymopterus montanus* Mexico / Y. Mexia 2508 / 1929 / "'Yerba del Oso'"

3162 *Pseudotaenidia montana* U.S.: Va. / F. R. Fosberg 15310 / 1938 / "strong oily odor"

3163 *Peucedanum glaucum* China / H. Wang 41716 / 1939 / "Root used as a drug"

3164 *P. medicum* China / K. P. To 2841 / 19 / "Medicinal salve" / "Tong Kwai"

3165 *Heracleum maximum* U.S.: N.M. / O. St.John 11 / 1896 / "Called wild parsnip, roots poisonous"

3166 *Daucus montanus* Chile / Worth & Morrison 6111 / 1938 / "Caraway odor"

CORNACEAE

3167 *Cornus capitata* China / R. C. Ching 21885 / 1939 / "...fr. red edible"

3168 *C. hongkongensis* China / Y. W. Taam 111 / 1937 / "...fr. yellow, edible" / "'Shan Lung Ngan'"

3169 *C. kousa* var. *chinensis* China / E. H. Wilson 223a / 08 / "...fruit red edible"

3170 *C. szechuanensis* China / E. H. Wilson 223b / 08 / "...fruits...edible"

PYROLACEAE

3171 *Chimaphila maculata* Mexico / W. P. Hewitt 129 / 1946 / "...leaves ...boiled taken as a liver remedy" / "'Hierba del Higado'"

ERICACEAE

3172 *Befaria glauca* var. *coarctata* Bolivia / Y. Mexia 04272 / 1935 / "Fruit eaten produces dizziness and discomfort, but if used for some time, the person is greatly stimulated" / "'Macha-Macha'"
Venezuela / J. A. Steyermark 58600 / 1944 / "Leaves and bark boiled in water used for colds" / "'aiýí-yek'"

3173 *Ledum glandulosum* U.S.: Ore. / W. C. Cusisk 2313 / 1899 / "Decoction of leaves used by hunters 'as tea'"

3174 *Rhododendron anthropogonoides* China / J. F. Rock 12194 / 1925 / "...leaves aromatic, odor of Eucalyptus"

3175 *R. arboreum* India: Punjab / W. Koelz 1835 / 1931 / "...fls. eaten"

3176 *R. megacalyx* Burma / F. Kingdon-Ward 20836 / '53 / "Flowers clove scented"

3177 *R. agathodaemonis* New Guinea / L. J. Brass 9271 / 1938 / "...carnation scented fls."

3178 *Leucothoë mexicana* Honduras / A. Molina R. 14000 / 48 / "'Cachimbo'"

3179 *L. mexicana* var. *pinetorium* Mexico / G. B. Hinton et al. 10159 / 37 / "Bark medicinal" / "'Nacahuite'"

3180 *Pieris japonica* Japan / E. H. Wilson 6443 / 1914 / "...fls. white, odour unpleasant plant poisonous"

3181 *Gaultheria alnifolia* Venezuela / J. A. Steyermark 62063 / 1945 / "...leaves with wintergreen-like odor" / "'pijoa'"

3182 *G. leucocarpa* f. *cumingiana* Sumatra / W. N. & C. M. Bangham 993 / 1932 / "Frt. slight wintergreen flavor"

3183 *Pernettya prostrata* var. *pentlandii* Bolivia / J. Steinbach 9583 / 1929 / "Die Frucht hat eine einschläfernde Eigenschaft. Ein zahmer Affe, welcher Beeren meiner zum eilegen bestimmten Pflanzen genascht hatte, wurde total betrunken" / "'Macha-macha'"

Bolivia / J. Steinbach 9514 / 1929 / "Frucht...soll, wenn reichlich genossen, Schwindel verursachen" / "'Macha-macha'"

3184 *P. prostrata* var. *purpurea* Colombia / Y. Mexia 7642 / 1938 / "...fruit...are poisonous; adults become violently ill, children dying from eating berries" / "'Moridera'"

3185 *P. pumila* Argentina / Y. Mexia 7924 / 1936 / "...fruit, eaten by Indians"

3186 *Arctostaphylos arbutoides* Honduras / Williams & Molina R. 14013 / 1948 / "Fruit edible" / "'Nariz de perro,' 'nariz de chucho'"

3187 *A. pungens* Mexico / H. S. Gentry 2025 / 1935 / "Warihios eat berries and report medic. Infusion made of leaves for 'tos, catarrhs, sarampeon" / "'Mansanilla,' Mex. 'Uhih,' W."

Mexico / W. P. Hewitt 1 / 1945 / "...fresh leaves ground with 50% salt and mixed with equal parts water (1 kilo to 1 kilo water) used as a purge for mules"

3188 *Vaccinium floribundum* var. *ramosissimum* Bolivia / J. Steinbach 5895 / 1921 / "Dicen que las frutas comiendo muchos embriagan" / "'Macha Macha'"

3189 *V. whitfordii* Philippine Is. / W. Beyer 6850 / 1948 / "Young leaves chewed to prevent tongue from getting white" / "'Guimu' Dialect Ifugao"

3190 *Agapetes variegata* India: Assam / L. F. Ruse 36 / 1923 / "...flowers are cooked and eaten with rice" / "'Dieng-jalanut'"

3191 *Dimorphanthera alpina* var. *pubigera* New Guinea / L. J. Brass 9363 / 1938 / "Fruit red, sweet and palatable"

3192 *Macleania ecuadorensis* Ecuador / J. A. Steyermark 53427 / 1943 / "...fruit edible, eaten as medicine en fresh state for lung trouble" / "'joyapa'"

3193 *Satyria warszewiczii* Costa Rica / R. L. Rodriguez C.293 / 1957 / "Fruit edible"

3194 *S. panurensis* Venezuela / J. A. Steyermark 62188 / 1945 / "...vine with drinkable water in stem" / "'cuspo'"

3195 *Psammisia hookeriana* Venezuela / J. A. Steyermark 55365 / 1944 / "...fruit edible" / "'coralito'"

3196 *Themistoclesia dependens* Ecuador / J. A. Steyermark 53446 / 1943 / "...fruit edible" / "'tira'"

THEOPHRASTACEAE

3197 *Clavija* aff. *mezii* Panama / W. L. Stern et al. 88 / 1959 / "For snake bite" / "'Membrillo macho'"

3198 *C.* aff. *poeppigii* Brazil / B. A. Krukoff 7665 / 1935 / "used in Curare"

3199 *Jacquinia aurantiaca* El Salvador / P. C. Standley 22327 / 1922 / "Used to kill fish" / "'Barbasco'"

MYRSINACEAE

3200 *Maesa acuminatissima* China / W. T. Tsang 22301 / 1933 / "An heal for head sores" / "'Pak Fan Shue'"

3201 *M. chisia* India: Assam / L. F. Ruse 8 / 1923 / "Leaves are given to cows as cattle fodder during shortages"

3202 *M. edulis* Solomon Is. / S. F. Kajewski 2395 / '30 / "When a man is very sick, the leaves are heated and placed on his bed, the patient then goes and lies down on them" / "'Tu-tung-isu'"

3203 *Maesa* indet. Solomon Is. / S. F. Kajewski 2521 / 31 / "Where there is a sore place it is rubbed by these leaves" / "'Camboru'"

3204 *M. laxa* Philippine Is. / C. O. Frake 823 / 1958 / "Young leaves applied for spleenomegaly" / "'glingu tabog' Sub."

3205 *Maesa* indet. Philippine Is. / L. E. Ebalo 347 / 1939 / "Vine, leaves used for seasoning fish" / "'malaw-mau' Mang."

3206 *M. parksii* Fiji Is. / O. Degener 15340 / 1941 / "Bark used medicinally. Take bark, crush, put in coconut cloth add water and drink for stomachache" / "'Mbumbumarasea' (Ra)"

3207 **M. novo-caledonica** New Caledonia / C. T. White 2079 / 1923 / "Bark said to be used as a fish poison by the natives"

New Caledonia / I. Franc 2478 / 1930 / "La sève trés abondante, fournit sur la place une boisson agréable"

3208 **Aegiceras corniculatum** China / S. K. Lau 509 / 1932 / "Fr. edible"

3209 **Ardisia amplifolia** Honduras / P. C. Standley 54781 / 1927–28 / "Fruit edible" / "'Uva de montaña'"

3210 **A. compressa** Honduras / P. C. Standley 55953 / 1928 / "Fruit edible" / "'Camaca'"

Mexico / Y. Mexia 1209 / 1926 / "Fr. edible: makes drink" / "'Gonda'"

3211 **A. paschalis** El Salvador / P. C. Standley 19967 / 1922 / "...fr. dark red, eaten, sour but of good flavor" / "'Cereza'"

3212 **A. venosa** Mexico / G. B. Hinton et al. 15410 / 39 / "Shoots are eaten"

Mexico / G. B. Hinton et al. 412 / 32 / "Sweet strong perfume"

3213 **A. bicolor** China / A. N. Steward et al. 155 / 1931 / "Roots used in Chinese medicine" / "'Pa Chao Chin Long'"

3214 **A. elegans** China / W. T. Tsang 22509 / 1933 / "Use to heal injuries" / "'Taai Lo Shan'"

3215 **A. gigantifolia** China / Tsang, Tang & Fung 142 / 1929 / "Often used as medicine" / "'Chau Ma T'oi'"

3216 **A. banceana** China / W. T. Tsang 20882 / 1932 / "Roots used as medicine" / "'Tai Loh San Shu'"

3217 **A. quinquegona** China / W. T. Tsang 25718 / 1935 / "...fr. black, edible" / "'F'o Tan Mo Shue'"

3218 **A. villosa** China / W. T. Tsang 22500 / 1933 / "Used as medicine to heal wounds" / "'Lo Shan Shue'"

3219 **A. pardalina** Philippine Is. / C. O. Frake 722 / 1958 / "Leaves applied to stomach for amenorrhea" / "'glitatun' Sub."

3220 **A. pyramidalis** Philippine Is. / C. O. Frake 842 / 1958 / "Roots boiled and drunk during puerperium" / "'Ganeban' Sub."

Philippine Is. / C. O. Frake 564 / 1958 / "Leaves applied to headache" / "'Kuluman' Sub."

Philippine Is. / R. B. Fox 5034 / 1948 / "When an Egongot has a fever, the leaves of this plant are rubbed over his feet, forehead and face, and then over his entire body" / "'Buloíoe' Dialect Egongot"

3221 **Grammadenia alpina** Venezuela / A. Jahn 970 / 1922 / "N.v.: cupis; corteza usada contre los dolores de muelas"

3222   *Conomorpha magnoliifolia* Surinam / W. A. Archer 2894 / 1934 /
"Fish poison" / "'Ayare' (Carib)"

3223   *Embelia laeta* Hainan / C. I. Lei 362 / 1933 / "...fruit red edible" /
"'Suen Kwo'"

3224   *Rapanea linearis* Hainan / S. K. Lau 354 / 1932 / "...fruit edible" /
"'So Lo Tsz'"

PRIMULACEAE

3225   *Primula egaliksensis* Newfoundland / M. L. Fernald et al. 26960 /
1924 / "...bruised root with musky odor"

3226   *P. parryi* U.S.: Utah / E. H. Graham 8387 / 1933 / "Fls.... fragrance
almost nauseating"

3227   *Lysimachia japonica* Philippine Is. / W. Beyer 6847 / 1948 / "Stem
when heated treats sore eyes"

3228   *Anagallis femina* Kuwait / H. Dickson 32 / 61 / "Eaten by sheep etc."

PLUMBAGINACEAE

3229   *Plumbago scandens* Mexico / H. S. Gentry 1047 / 1034 / "'por la gusano'" / "'Estrenina,' Mex. 'Whowechora,' W."

SAPOTACEAE

3230   *Ganua motleyana* Sarawak / J. A. R. Anderson 12900 / 60 / "Fruit
edible, used as a vegetable oil by native people" / "'Ketiau paya'"

3231   *Madhuca indica* India: Bombay Pres. / J. Fernandez 89 / 1949 /
"Fresh flowers are eaten raw; when dry they are roasted and eaten or
used for the production of alcohol. At present used as cattle food and
for power alcohol (or for illicit distillation among the jungle people)"
/ "'Mow'"

3232   *Madhuca* indet. Philippine Is. / C. O. Frake 743 / 1958 / "...bark
boiled and drunk for chest pains" / "'Sulang' Sub."

3233   *Madhuca* indet. Philippine Is. / C. O. Frake 621 / 1958 / "Roots applied for pinworms" / "'getangal' Sub."

3234   *Madhuca* indet. Philippine Is. / C. O. Frake 862 / 1958 / "Bark boiled
and bathed for burns" / "'Gwatan ulangen' Sub."

3235   *Madhuca* indet. Philippine Is. / C. O. Frake 830 / 1958 / "Sap applied for skin eruptions" / "'geniin' Sub."

3236   *M. obovata* New Hebrides Is. / S. F. Kajewski 804 / 29 / "Very lus-
cious fruit eaten by natives"

3237   *Burckella cocco* Solomon Is. / S. F. Kajewski 2399 / 31 / "This fruit
bears a delicious fruit, with a deep white flesh, similar to an apple,
but with a strong distinct flavour of its own. This is one of the few
island fruits worthy of cultivation" / "'Un-garno'"

3238 *Pouteria multiflora* W.I.: Nevis / J. S. Beard 464 / 1944 / " . . . fruit yellow, edible when ripe" / " 'Choky apple' "

3239 *Pouteria* indet. Brit. Guiana / A. C. Smith 3161 / 1938 / "Indians suck small pieces of the bark to cure colds and chest congestions" / " 'Widieko' (Wapisiana)"

3240 *Pouteria* indet. Brazil / Y. Mexia 5278 / 1930 / " 'Para tudo' "

3241 *P. endlicheri* New Caledonia: Nouméa / I. Franc 1643 / 1914 / "Fruits charnus recherchés par les oiseaux" / " 'Azau blanc' "

3242 *Sideroxylon* indet. Nyasaland / L. J. Brass 17510 / 1946 / " . . . fruit yellow, soft, edible" / " 'Pimbinyolo' (Chinyanja)"

3243 *Micropholis rugosa* Jamaica / G. R. Proctor 9955 / 1955 / " . . . fruit hard but edible" / " 'Beef apple' "

3244 *Mastichodendron angustifolium* Mexico / P. C. Standley 2931 / 1936 / "Fruit . . . eagerly sought by natives" / " 'Tempisque,' Mex."

3245 *Bumelia occidentalis* Mexico / Carter & Kellogg 2837 / 1950 / " . . . dried fruits edible" / " 'lebelama' "

3246 *B. retusa* ssp. *typica* Bahama Is. / R. A. & E. S. Howard 9995 / 1948 / " 'Wild saffron' "

3247 *Pradosia praealta* Brazil / Ducke 1663 / 1944 / " 'Casca doce' vel 'Pau doce' "

3248 *Mimusops parvifolia* Philippine Is. / E. D. Merrill 499 / 1902 / "Fruit eaten by natives" / " 'Talipopo' "

3249 *M. elengi* var. *javensis* New Hebrides Is. / S. F. Kajewski 179 / 28 / "Reputed by natives to be poisonous"

3250 *M. zeyheri* Africa / E. E. Galpin 8853 / 20 / "Fruit edible" / " 'Mopel' "

3251 *Dumoria africana* Gabun / B. A. Krukoff 123 / 30 / "Fruit edible" / " 'Douka' (N'komi)"

3252 *Manilkara bidentata* Virgin Is. / J. S. Beard 318 / 1944 / "Fruit edible" / " 'Bullet wood' "

3253 *M. jaimiqui* ssp. *emarginata* Bahama Is. / O. Degener 18825 / 1946 / " 'Sapodilla'; fruit used for glue, edible"

EBENACEAE

3254 *Euclea natalensis* Nyasaland / J. B. Davy 21392 / 1929 / "Extract of roots used for ulcers" / " 'Mtanawanjana' (Chewa), 'Nakututo' (yao)"

3255 *Diospyros sonorae* Mexico / H. S. Gentry 4777 / 1939 / "Fruit eaten eagerly by natives" / " 'Guaparin' "

3256 *D. rhombifolia* China / H. H. Chung 2071 / 1923 / " . . . fruit used to poison fish, local name = 'fish persimmon' "

STYRACACEAE

3257 **Styrax argenteus** Mexico / Y. Mexia 1504 / 1927 / "Fr. said to be red. Edible" / "'Escaramuza'"

El Salvador / P. C. Standley 2045 / 1922 / "Bark used to poison fish" / "'Estoraque'"

3258 **S. costanus** Venezuela / J. A. Steyermark 62464 / 1945 / "'olivo montañero'"

3259 **Alniphyllum fortunei** Hainan / S. K. Lau 199 / 1932 / "Cook the bark to heal cough" / "'Jia thun bau thon' (Lois)"

3260 **A. hainense** Hainan / F. A. McClure 737 / 1929 / "Drug, tea made from leaves used in care of wound" / "'Lai P'a Teng'"

3261 **Afrostyrax kamerunensis** Kamerun / J. Mildbraed 10688 / 1928 / "Rinde stark nach Knoblauch reichend"

SYMPLOCACEAE

3262 **Symplocos anomala** China / W. T. Tsang 22912A / 1933 / "...fruit good"

3263 **S. chinensis** China / W. T. Tsang 23809 / 1934 / "Used for medicine"
China / W. T. Tsang 21393 / 1932 / "...leaves used as medicine" / "'Kau Sze Shue'"

3264 **S. celastrifolia** N. Borneo / Goklin B.N.B. For. Dept. 2163 / 1932 / "Leaves used for medicinal purposes" / "'Mata kinai' (Dusun)"

3265 **S. fasciculata** Sumatra / R. S. Boeea 9974 / 1936 / "(lvs. medicine for wounds)" / "'kajoe lobaloba'"

OLEACEAE

3266 **Syringa microphylla** China / J. Hers 252 / 1919 / "Flowers used as a substitute for tea" / "'sung lo cha'"

3267 **Siphonosmanthus suavis** Bhutan / F. Ludlow et al. 16092 / 1949 / "Fragrant. Fed to cattle in winter"

3268 **Forestiera tomentosa** Mexico / A. Dugès s.n. / 1909 / "The fruit eaten by birds" / "'pico de pájaro'"

3269 **Ligustrum ibota** var. **regelianum** China / C. Y. Chiao 2648 / 1930 / "Leaves used as substitute for tea"

3270 **L. robustum** Annam / Poilane 1032 / 1920 / "...les tiges de cet arbuste sont employée pour combattre les maux de dents a cet effet, ils coupent un rameaux vert et mettre l'une des extremitées au feu l'eau on la sève qui sorte à l'autre bout, est appliqué sur la dent malade. D'autre pretend qu'on l'emploi non pas contre les maux de dent non contre les plaies"

3271 *Jasminum officinale* f. *grandiflorum* Formosa / U. Faure 736 / 1914 / "Les jeunes fleurs sont cieullirs, desechées et preparées comme les feuilles du thé"

3272 *J. lanceolarium* var. *puberulum* China / Y. W. Taam 755 / 1938 / "fl. white, ill smelling"
China / W. T. Tsang 25278 / 1935 / "fl. white, fragrant"

3273 *J. nervosum* China / W. T. Tsang 16941 / 1928 / "Medicines" / "'Tsak Ue Tam T'ang'"

3274 *J. longisepalum* S. Annam / Poilane P15 / 1919 / "Les indigènes font avec les feuilles et les fleurs une infusion qu'ils font prendre aux femmes après l'accouchment donne de petits fruits rouges non comestibles" / "'Cay-vau' Annte"

3275 *J. polyanthum* China / A. Henry 10314B / no year / "fls. . . . strong disagreeable odor"

3276 *J. didymum* Solomon Is. / S. F. Kajewski 2542 / 31 / "When a man has a headache, the bark is put on the fire and the smoke inhaled" / "'Arlu-vasu-vasu'"

3277 *Azima sarmentosa* China / S. K. Lau 3343 / 1934 / "leaves & roots edible"

3278 *Salvadora persica* Saudi Arabia / H. Dickson 7 / 61 / "Roots cut for tooth cleaning sticks" / "'Rak'"

LOGANIACEAE

3282 *Geniostoma calcicola* Tonga Is. / T. G. Yuncker 15379 / 1953 / "Bark used in medicines" / "Tongan n.: Te'epilo'amaui"

3283 *G. rupestre* Santa Cruz Is. / S. F. Kajewski 650 / 28 / "This flower has a strong unpleasant odour of carrion and by this reason attracts large quantities of flies"

3284 *G. vitiense* Tonga Is. / T. G. Yuncker 15787 / 1953 / "Bark . . . used in preparing medicine" / "Tongan n.: Te'epilo'amaui"

3285 *Spigelia humboldtiana* El Salvador / S. Calderón 1223 / 1922 / "Lombricilla"

3286 *S. anthelmia* Surinam / W. A. Archer 2911 / 1934 / "Fish poison" / "'Droegoman' (Negro English)"

3287 *Antonia ovata* Brit. Guiana / A. C. Smith 2196 / 1937 / "Used as a fish poison. Leaves stripped and placed in a hole where they are pounded to a pulp, which is used in stagnant or slowly flowing water. Said to be very effective. takes 3–4 hours to act" / "'Inacu' (Wapisiana)"
Brazil / Ducke 1372 / 1943 / "Ichthyotoxica 'Timbo'"

3288   *Strychnos brachistantha* Brit. Honduras / P. C. Gentle 3181 / 1940 / "snake seed"

3289   *S. peckii* Brit. Honduras / W. A. Schipp 121 / 1929 / "... fruits, the pulp eaten by Caribs & Maya Indians"

Brazil / B. A. Krukoff 7549 / 1935 / "Component of Curare of Tecuna Indians"

3290   *S. panamensis* Honduras / P. C. Standley 52619 / 1927–28 / "Remedy for pains" / "'Guaco'"

3291   *S. nux blanda* Burma / F. G. Dickason 7072 / 1938 / "Fruit used for fish poison" / "'kabaung'"

3292   *S. ovalifolia* Malay Peninsula: Selangor / K. M. Kochummen 93356 / 59 / "Used as a constituent for the blowpipe arrow poison" / "'Akar ipoh'"

3293   *Strychnos* indet. Philippine Is. / C. O. Frake 658 / 1958 / "Leaves applied for splenomegaly" / "'Bebekang-ulongen' (Sub.)"

3294   *Neuburgia corynocarpa* Solomon Is. / S. F. Kajewski 2010 / 30 / "Pigeons are very fond of this fruit" / "'Mukitoro'"

Solomon Is. / S. F. Kajewski 2548 / 31 / "The bark is macerated and applied to skin diseases" / "'Ba-hau'"

3295   *Fagraea racemosa* Solomon Is. / S. F. Kajewski 2336 / 30 / "The leaves of this tree are applied by natives to boils on legs. This appears to be common in the Solomons and the territory" / "'Ungara'"

3296   *F. gracilipes* Fiji Is. / O. Degener 15101 / 1941 / "Bark or leaves mashed with water, strained and drunk to purify the blood. Also a valued timber for house posts, canoes, bowls, etc." / "'Makamakandora' (Serua)"

3297   *F. schlechteri* New Caledonia: at or near St. Louis Mission / Sarlin & Buchholz 1597 / 1948 / "Said to be a medicinal plant good for rash and skin irritations. A piece of leaf in boiling water applied to skin"

3298   *Potalia amara* Venezuela / L. Williams 14352 / 1942 / "... se usa la infusion obtenida de la corteza como medicina laxante" / "'Temblador'"

3299   *Polypremum procumbens* El Salvador / Dr. A. Van Severen s.n. / 1922 / "Remedy for 'metritis'"

3300   *Buddleia sessiliflora* Mexico / H. S. Gentry 2210 / 1936 / "Has a thick, sweet, carrion odor attractive to flies" / "'Lengua Buey,' Mex."

3301   *B. americana* Peru / F. Woytkowski 6100 / 1961 / "... medicinal; erisipela" / "'Yurac-panga,' 'Ayac-manchana'"

3302   *B. ledifolia* Bolivia / L. R. Parodi 10087 / 1932 / "Flores muy perfumadas. Cultivado por los indios Aymaras"

3303 **B. asiatica** China / Tsiang & Wang 16432 / 1939 / "Plant poisonous"

Philippine Is. / C. O. Frake 38090 / 1958 / "Used in rice wine making" / "'Glentud ulangan' (Sub.)"

Philippine Is. / C. O. Frake 38029 / 1958 / "Leaves & roots applied to tumor-like growths on head glempi?" / "'mengayen'"

Philippine Is. / G. E. Edaño 17780 / 1953 / "The root is boiled, allowed to concentrate and is given as a treatment for malaria" / "Maradang. Dialect I."

3304 **B. lindleyana** China / W. T. Tsang 21193 / 1932 / " . . . fl. reddish purple, to medicine to remove toxin" / "'Yeung Bou Shue'"

3305 **B. yunnanensis** China / Tsiang & Wang 16433 / 1939 / " . . . plant poisonous"

3306 **Nicodemia madagascarensis** Argentina / A. T. Hunziker 5847 / 1944 / "Tiene fama de planta medicinal" / "'cambard'"

GENTIANACEAE

3307 **Centaurium cachanlahuen** Chile / P. Aravena 33369 / 1942 / "Used as a febrifuge, very bitter" / "'Bachanlagua'"

3308 **Gentiana detonsa** Mexico / W. P. Hewitt 88 / 1945 / "'Flor de Oso'"

3309 **G. chamuchni** Peru / J. West 8137 / 1936 / "Used medicinally by natives" / "'Shalcandino'"

3310 **G. claytonioides** Argentina / L. R. Parodi 7922 / 1927 / "Muy buscada para remedio" / "'Nencia'"

3311 **G. scabra** var. **buergeri** Japan / E. Elliott 102 / 46 / "Food use: seedlings cooked and powdered old leaves are eaten. Med. use: The dried root is used as a bitter stomachic; also used for preparing a bitter tincture" / "'Rindo'"

Japan / Ichikawa 200188 / 1924 / "Medicinal" (in Chinese characters)

3312 **Megacodon venosum** China / W. P. Fang 1043 / 1928 / "One kind of medicine"

3313 **Swertia japonica** Japan / T. Tanaka 10047/4 / 1924 / "Medicinal" (in Chinese characters)

Japan / E. Elliott 88 / '46 / "Collected during the flowering time. The whole plant is dried to be used as a stomachic & for dysentery. The whole plant is dried and decocted to cure diarrhea as well as stomach trouble and intestinal disease" / "'senburi'"

3314 **S. petiolata** Afghanistan / E. Bacon 102 / 1939 / "Used as fodder"

3315 **Swertia** indet. Burma / F. G. Dickason 7723 / 1937 / ". . . leaves bitter" / "'tlang rel'"

3316 **Halenia weddelliana** Colombia / A. Fernandez 2653 / 1954 / "Usada por los nativos para tratar enfermedades venereas" / "'Cacho de venado'"

3317   *Lisianthus frigidus* W.I.: St. Vincent / G. R. Cooley 8415 / 1962 / "Verbena odor"

3318   *Schultesia lisianthoides* El Salvador / P. C. Standley 19138 / 1921–22 / "Remedy for malaria" / "'Sulfatillo,' 'Sulfato de terra'"

3319   *S. stenophylla* El Salvador / S. Calderón 1276 / 1922 / "'Hierba de la vida'" / "'Conchalagua'"

3320   *Chelonanthus chelonoides* Peru / Y. Mexia 4153 / 1935 / "Used as a remedy for worm infested wound in cattle" / "'Tres esquinas'"

APOCYNACEAE

3321   *Allamanda cathartica* Burma / J. F. Smith 97 / 1913 / "Medicinal"

3322   *Willughbeia cochinchinensis* Cochinchina / Herb. L. Pierre 138 / no year / "Produit un caoutchouc inférieur, bon pour le pansement des plaies"

3323   *Plumeria tarapotensis* Peru / Y. Mexia 6071 / 1931 / "Root shredded, boiled & taken in tea for rheumatism" / "'Sanángo'"

3324   *P. attenuata* Brazil / Y. Mexia 4136 / 1929 / "Used as a purgative against fevers" / "'Aguniada'"

3325   *Himatanthus sucuuba* Brazil / R. Froes 1777 / 1932 / "Latex used for pulmonary illness" / "'Jauauba' or 'Sucuhuba'"

3326   *Holarrhena antidysenterica* India / J. Fernandes 1321 / 1950 / "Use: roots and bark used in medicine. Leaves used to wrap 'bidis,' or native cigarettes, thus reducing smokers' cough"

    India / J. Fernandes 1580 / 1950 / "Leaves used as wrapping tobacco in cigarettes or bidi" / "'Kuda'"

3327   *Alstonia scholaris* Burma / F. G. Dickason 6830 / 1937 / "Bark use for medicinal purposes" / "'Taungmayo'"

    New Guinea / R. D. Hoogland 5007 / 1955 / "Bark used by natives from Jal: used as medicine against various internal complaints, taken with cold water" / "Belleg (Jal), Sebab (Bembi), Tag (Usino), Bojom (Rawa)"

    Philippine Is. / M. D. Sulit 5644 / 1955 / "Bark medicinal for fever" / "Dita (Tag dialect)"

    Philippine Is. / Ahern's collector (Dept. of Interior, Bur. of Gov. Lab. 92) / 1904 / "...a decoction of the bark...used as a substitute for quinine. The bark of this tree is used in the manufacture of a poison for poisoning spears and arrows. T.V., *Dita, Ditaa*; Il., *Dalipaoen*"

3328   *A. spectabilis* New Guinea: Papua / E. Gray NGF. 7166 / 55 / "Collected for investigation as to alkaloid (reserpine) content"

3329   *A. vitiensis* Solomon Is. / S. F. Kajewski 2575 / 1931 / "Uses: the bark is put into the fire and if a native is affected with blight of the eyes, he hangs his head over the smoke" / "'Malunga'"

Fiji Is. / O. Degener 15040 / 1941 / "Fiji 'chewing gum.' Break stem and collect milk which hardens and is then ready for chewing. Put branch in fire and scrape off bark. Then put bark in coconut 'cloth' and squeeze juice into eye for sore eye. Juice is not milky" / "Buleki (Serua)"

Fiji Is. / A. C. Smith 1701 / 1934 / "Sap cooked and used for chewing" / "'Ndranga'"

3330 *A. reineckeana* Fiji Is. / A. C. Smith 1795 / 1934 / "Sap used for chewing" / "'Ndranga nggurungguru'"

3331 *A. constricta* Australia / C. J. Trist (Herb.?) 37 / 1931–32 / "'Bitter Bark'"

3332 *Cameraria beligensis* Brit. Honduras / P. H. Gentle 1294 / 1934 / "This plant is seldom collected probably because people are afraid of it, and with good reason"

3333 *Aspidosperma megalocarpon* Brit. Honduras / P. H. Gentle 2800 / 1939 / "red malady"

3334 *Aspidosperma* indet. Brit. Honduras / P. H. Gentle 2839 / 1939 / "white malady"

3335 *A. discolor* Brazil / Froes 12640 / 1942 / "Bark is used in treatment of malaria"

3336 *A. illustre* Brazil / H. M. Curran 41 / 1915 / "'Quina'"

3337 *A. nitidum* Brazil / R. Froes 12150 / 1941 / "(Used for treating malaria)" / "'Carapanauba'"

3338 *Aspidosperma* indet. Brazil / Froes 12709/82 / no year / "(Bark used in treatment of malaria)" / "'Pau de Rego'"

3339 *Catharanthus roseus* Tonga Is. / T. G. Yuncker 16259 / 1953 / "Parts used medicinally"

3340 *Geissospermum argenteum* Brit. Guiana / A. C. Smith 2825 / 1937 / "Bark sometimes used (but said not to be essential) in 'Balauitu,' Waiwai arrow-poison" / "'Uataki' (Waiwai)"

3341 *Tabernaemontana longipes* Brit. Honduras / H. O'Neill 8656 / 1936 / "Used as a source of rubber and also as a cure for beef worm" / "'Chaclachin'"

3342 *Tabernaemontana* indet. Brazil / R. Froes 12412/156 / 1941 / "Used by Indians to kill fish"

3343 *Rejoua novo-guineensis* Solomon Is. / S. F. Kajewski 1895 / 1930 / "The sap is used as a medicine, is put with cocoanut oil and rubbed on the skin to blister it" / "'Pouie-ma'"

3344 *Bonafousia undulata* Colombia / A. E. Lawrance 655 / 1933 / "White sticky sap exudes from stem when cut which stains hands black"

3345 *Macoubea guianensis* Brazil / R. Froes 1795 / 1932 / "Latex used for pulmonary diseases"

3346 *Ervatamia bufalina* Hainan / C. I. Lei 516 / 1933 / "The roots are used as medicine. One can cure sore throat by chewing the roots" / "'Shan Fan Tsiu'"

Hainan / F. A. McClure 2369 / 1922 / "drug plant" / "'Shan lat tsiu'"

Hainan / S. K. Lau 128 / 1932 / "The boiled crushed leaves are used to heal boils"

3347 *E. peduncularis* Malaya / Umbai & Millard s.n. / 1958 / "Phytochemical Survey of Malaya" / "Poisonous" / "'Pachek'"

3348 *Ervatamia* indet. Brit. N. Borneo / Random (B.N.B. For. Dept. 2832) / 1933 / "Native medicine for headache" / "'Rodok' (Bajau)"

Philippine Is. / P. Añonuevo 245 / 1950 / "Medicinal" / "The sap of the triturated leaves is applied to open wound" / "'Alibutbut' (Bis dialect)"

3349 *Pagiantha macrocarpa* Brit. N. Borneo / Goklin (B.N.B. For. Dept. 2641) / 1932 / "Fruit native medicine" / "Hinadak (Brunei), Kambau (Dusun)"

3350 *Stemmadenia donnell-smithii* El Salvador / P. C. Standley 19187 / 1921–22 / "Juice of fr. used to fasten cigar wrappers and for eye inflammation. Decoction of bark for rheumatism" / "'Cojón de puerco'"

3351 *S. tomentosa* var. *palmeri* Mexico / H. S. Gentry 1175 / 1934 / "Medic. milk used for bad eyes and 'por todo infermidades,' report Warihio" / "'Veraco,' Mex. 'Peychí,' W."

Mexico / H. S. Gentry 4950 / 1939 / "Sap regarded as medicinal" / "'Tabaca,' 'Veraco'"

Mexico / Y. Mexia 733 / 1926 / "Milky, gummy juice which when it hardens is chewed" / "'Berraco'"

3352 *Voacanga* aff. *megacarpa* Philippine Is. / C. Frake 362 / 1957 / "Source of chicle. 1) Bark drunk for internal pain. 2) Roots boiled and drunk for rigid abdomen" / "Tepalak . . . Sub."

3353 *Vallesia glabra* Mexico / H. S. Gentry 1264 / 1935 / "Medic. fruit applied to eyes when sore" / "'Sitavaro,' Mex."

3354 *Hunteria zeylanica* Malaya / A. H. Millard (K.L. 1415) / 59 / "Phytochemical Survey of Malaya" / "'Menggading' (Malay-Kedah)"

Malaya / K. M. Kochummen (Kepong 66462) / 58 / "Phytochemical Survey of Malaya" / "'Kemuning lutan'"

3355 *Alyxia pisiformis* India / Harmand 220 / 1874 ? / "Le bois est brûlé dans les pagodes et donne une fumée odorante"

3356 *A. stellata* Fiji Is. / O. Degener 15270 / 1941 / "For bloody stool take root of either this species or of the Vono Buli (large-leaved species), pound, add water, strain through coconut cloth and drink. . . ."

3357 *Rauwolfia ligustrina* Mexico / Y. Mexia 1042 / 1926 / "Fr. a berry, black when ripe. Edible"

3358 *R. tetraphylla* Mexico / G. Martínez-Calderón 90 / 1940 / "Se usar para curar heridas"

3359 *R. praecox* Bolivia / J. Steinbach 6536 / 1924 / "Sostienen los indigenes que la cáscara machacada sirve para contearastar la accion del veneno de vivoras"

3360 *R. spectabilis* Brit. N. Borneo / Madin (B.N.B. For. Dept. 1646) / 1932 / "Poison"

3361 *R. sumatrana* Philippine Is. / A. L. Zwickey 665 / 1938 / "...infusion of bark for stomachache...bark flavor of quinine..."

3362 *R. samarensis* Philippine Is. / C. O. Frake 612A / 1957 / "Medicine for malaria, bark boiled and drunk"

3363 *Ochrosia parviflora* Solomon Is. / L. J. Brass 2844 / 1932 / "Seeds said to be eaten by natives"
New Hebrides Is. / S. F. Kajewski 405 / 1928 / "Seed inside fruit eaten by natives"
Austral Is. / A. M. Stokes 49 / 1921 / "...fruit edible..."

3364 *O. oppositifolia* Cook Is. / T. G. Yuncker 9719 / 1940 / "...seeds edible"

3365 *O. miana* PAPUASIA / Prony 290 (Herb. I. Franc 1626A) / 1914 / "Fruit jaune, odorant, recherché par les oiseaux"

3366 *Bleekeria* indet. New Hebrides Is. / R. E. Burton 34 / before 1945 / "Leaves are used by natives to make an infusion to prevent fevers" / "'Narmalee'"

3367 *Kopsia lancibracteolata* Hainan / S. K. Lau 235 / 1932 / "...fruit, red, edible"

3368 *K. caudata* N. Borneo / Valera (N.B. For. Dept. 6723) / 1936 / "Stem medicinal—for skin diseases"

3369 *Cerbera manghas* Hainan / C. I. Lei 1127 / 1934 / "...fruit red, poisonous"
Hainan / S. K. Lau 313 / 1932 / "...fruit white, edible"
N. Borneo / Balajadia (B.N.B. For. Dept. 3279) / 1933 / "White latex causes blindness if it gets to the eyes"
Celebes / Eyma 3722 / 1938 / "...copious white juice. Interiorly used medicine (Ceram)"
Celebes / H. Curran 295 / 1940 / "...fruit showy but not edible"
Fiji Is. / Degener & Ordonez 13718 / 1940 / "Fijians express oil from seed"
Caroline Is. / C. Y. Wong 327 / 1948 / "The sap of the leaves are

used in making temporary tatoo (*gacow*): as the sap is dripping off the petiole, the design is made on the portion of the body of the person, then the ash from the burning of coconut husk or lamp black is used in the present day is spread on the design to make it more prominent. The design stays on for a day and can be washed off"

3370 *C. floribunda* Australia / S. F. Kajewski 1389 / 1929 / "This tree is used on the Daintree River to make butter boxes"

3371 *Thevetia peruviana* Mexico / W. P. Hewitt 24 / 1945 / "Plant reportedly poisonous when eaten. Nut carried for good luck" / " 'Hierba Fortuna' "

3372 *T. plumeriaefolia* Nicaragua / J. West 3551 / 1935 / " . . . strongly milky juice reported used in native medicine" / " 'Chilca' "

3373 *Urceola imberbis* Philippine Is. / M. D. Sulit 1322 / 1946 / "Stems gathered extensively for tying fence and fish corral"

3374 *Macrosiphonia hypoleuca* Mexico / E. Palmer 704 / 1898 / "Market of San Luis Potosí"

Mexico / G. B. Hinton 1137 / 32 / "The flower is used as a flavour in cooking 'Atole'—corn mush"

3375 *M. longiflora* Brazil / Williams & Assis 7524 / 1945 / " 'Velame' depurative, use root, cook the root and take the brew with sugar"

3376 *M. velame* Brazil / Y. Mexia 5709a / 1931 / "Medicinal; extensively used as anti-syphilitic; infusion of leaves and root" / " 'Villame branco' "

3377 *Mandevilla foliolosa* Mexico / S. S. White 2752 / 1939 / " 'Hierba del piojo' "

Mexico / R. Q. Abbott 473 / 1937 / " 'Hierba de la cucaracha' "

3378 *M. platydactyla* Mexico / G. B. Hinton 9319 / 36 / " 'Yerba cucaracha' "

3379 *M. pavonii* Peru / G. Klug 2212 / 1931 / "Huitoto Indian name: 'Iquidiá-o' "

3380 *Urechites lutea* W.I.: Cayman Is. / G. R. Proctor 15089 / 1956 / " 'Nightshade' "

3381 *Anodendron oblongifolium* Solomon Is. / S. F. Kajewski 2357 / 1930 / "The bark is used by the natives in the manufacture of fishing lines and nets"

Solomon Is. / S. F. Kajewski 2675 / 1931 / "The inner bark of the small long tendrils is used to make twine and fishing nets"

3382 *Trachelospermum gracilipes* China / Chung & Sun 526 / 1933 / "Flower—incense, white"

3383 *T. lucidum* India: Assam / L. F. Ruse 110 / 1923 / "Fruit eaten by natives"

3384   *Aganosma acuminata* Annam / Poilane 4511 / 22 / " . . . les feuilles, fleurs et fruits sont employés pour faire une boisson digestive"

        Brit. N. Borneo / Melegrito 3311 / 1933 / "Bark used for tying, leaves a native substitute for coffee or tea"

3385   *A. velutina* Philippine Is. / E. Canicosa 427 / 1949 / "Economic uses: sap"

3386   *Kibatalia blancoi* Philippine Is. / C. O. Frake 707 / 1958 / "Leaves applied for spleenomegaly"

3387   *Carruthersia macgregorii* Philippine Is. / Sulit & Conklin 5093 / 1953 / "Sap mixed with 'salagon' and applied to arrows to kill games"

3388   *Epigynum maingayi* Malaya / Ismail & Millard (K.L. 187) / 1958 / "Medicinal: Galactagogue"

3389   *Ichnocarpus frutescens* Philippine Is. / E. Canicosa 429 / 1949 / "Vine for tying fish trap"

3390   *Apocynum cannabinum* U.S.: Idaho / Rev. Spalding s.n. / no year / "Common hemp of the country used much by natives for bags and fish nets"

3391   *Strophanthus caudatus* Malaya / A. H. Millard (K.L. 177) / 1958 / "Poisonous; laticiferous; seeds glycosidal"

3392   *S. cumingii* Philippine Is. / M. D. Sulit 1550 / 1947 / "poisoning tip of arrows to kill games"

3393   *Strophanthus* indet. Philippine Is. / R. B. Fox 269 / 1949 / "ARROW POISON!"

3394   *Parsonsia helicandra* Solomon Is. / S. F. Kajewski 2230 / 1930 / "The vine is cut and the sap put on swellings of the leg"

3395   *Pottsia laxiflora* Hainan / S. K. Lau 134 / 1932 / "Used to crush it to heal a boil"

3396   *Angadenia lindeniana* Jamaica / W. T. Stearn 908 / 1956 / " . . . local man said that this is not eaten by cattle when fresh but is poisonous to cattle when dried" / " 'Nightshade' "

ASCLEPIADACEAE

3397   *Streptocaulon juventas* China / W. T. Tsang 23852 / 1934 / " . . . fl. pale yellow, edible"

3398   *Streptocaulon* indet. China / C. W. Wang 79267A / 1936 / "Edible"

3399   *S. baumii* Philippine Is. / C. O. Frake 552 / 1958 / "Agricultural ritual" / " 'Pelaing' " Sub.

        Philippine Is. / C. O. Frake 656 / 1958 / "Leaves applied for pin worms" / " 'dereparg' "? Sub.

3400   *Periploca linearifolia* Africa / A. G. Curtis 837 / 1923 / "root—smells like orris root"

3401 *Cryptostegia grandiflora* Mexico / H. S. Gentry 1663 / 1935 / "Said to be poisonous to livestock" / "'Pihuco,' Mex."

3402 *Hemidesmus indicus* INDIA / M. B. Raizada s.n. / 1932 / "The root is used as a substitute for sarsaparilla & is prescribed by native doctors either alone or in conjunction with other drugs, in the treatment of various ailments" / "Cultivated" / "'Kalisar,'... 'Indian Sarsaprilla'"

3403 *Cryptolepis infrutiana* Africa / E. H. Wilson C / 1921 / "root aromatic"

3406 *Araujia albens* Australia / C. T. White 11153 / 1934 / "...a few white moths seen caught in the throat (of the flowers)"

3407 *Asclepias verticillata* U.S.: Iowa / L. H. Pammel 564 / 1925 / "Associated with *Euphorbia corollata, E. maculata.* Considered poisonous to stock"

3408 *A. tuberosa* ssp. *tuberosa* U.S.: R.I. / J. F. Collins s.n. / 1934 / "(Butterfly-weed. Pleurisy-root)"

3409 *A. exaltata* U.S.: Mass. / E. F. Fletcher s.n. / before 1923 / "'Poke Milkweed'"

3411 *A. californica* U.S.: Calif. / Brewer 1123 / no year / "Eaten by insects"

3412 *A. curassavica* El Salvador / P. C. Standley 20780 / 1922 / "'Viborana'"

El Salvador / P. C. Standley 22087 / 1922 / "'Bivorana Sangrís'"

Mexico / V. Grant 516 / 1940 / "natives grind up flowers and put in tortillas as remedy for rabies" / "'Punchohuise'"

Mexico / E. Seler (Berlin Herb.) 3646 / no year / "'Yerva de la culebra'"

Mexico / E. Langlassé 11 / 1898 / "'Calderona'"

W.I.: Dominica / W. H. Hodge 3212 / 1940 / "roots good for fever"

Argentina / J. E. Montes 1929 / 46 / "'Caá-cambui'" / "'Yerba leche'"

Brazil / Y. Mexia 4159 / 1929 / "Poisonous to cattle" / "'Mata Rato'"

Brazil / R. Froes 1797 / 1932 / "Fruits exuding a poisonous latex" / "'Margaridinha'"

Colombia / Killip & Smith 14371 / 1926 / "'Gallu'"

Colombia / Schultes & Cabrera 17305 / 1952 / "Makuna—'Ko-há-gaw'; Barasana—'Py-ee-dĕ-́gaw'"

Solomon Is. / S. F. Kajewski 2224 / 30 / "The leaves are rubbed on the head for headaches by the natives" / "'Mataworu'"

New Hebrides Is. / S. F. Kajewski 388 / 28 / "Suspected poisonous weed" / "'Ge-vess'"

3413 *A. tuberosa* ssp. *terminalis* Mexico / C. H. & M. T. Mueller 686 / 1934 / "... tea made from this plant is said to be a cure for all ills"

3414   *A. ovata* Mexico / A. Dugès 7 / 1902 / "La racine sechée et en poudre est un sternutative violent" / "Nom. vern. Taraumara"

3415   *A. contrayerba* El Salvador / S. Calderón 1015 / 1922 / "'Ishcaco'"
Mexico / G. B. Hinton (Herb.) 9204 / 36 / "'Contra Yerba'"

3416   *A. glaucescens* El Salvador / P. C. Standley 19595 / 1921–22 / "Reputed poisonous. Lvs. put on boils to bring them to head" / "'Matacoyote,' 'Oreja de burro'"

3417   *A. hypoleuca* Mexico / S. S. White 2847 / 1940 / "'Oreja de mula'"

3418   *A. linaria* Mexico / A. Dugès s.n. / no year / "'Venenillo'"
Mexico / J. G. Schaffner 62 / 1876 / "'Venenillo'"

3419   *A. otarioides* Mexico / A. Dugès s.n. / 1893 / "La racine ou souche sechée et mise en poudre est un violent sternutatoire" / "'Taraumara'"

3420   *A. oenotheroides* El Salvador / P. C. Standley 21991 / 1922 / "'Matacoyote'"

3421   *A. brachystephana* Mexico / R. M. Stewart 389 / 1941 / "'Immortal'"

3422   *A. elata* Mexico / S. S. White 2654 / 1939 / "'Oreja de mula'"

3423   *A. campestris* Bolivia / J. Steinbach 5594 / 1921 / "Es Venenosa" / "'Urùma'"?

3425   *Asclepias* indet. Philippine Is. / E. Maliwanag 194 / 1941 / ". . . leaves applied as poultice on boil" / "'pamosangan' (Mang.)"

3426   *A. vincetoxicum* Africa / Herb. Helveticum   N. C. Seringe.-Hall. 571 / no year / "Racines employées contre les scrophules"

3427   *Calotropis gigantea* Burma / J. F. Smith 103 / 1913 / "Bark of root used as an emetic"
India: Assam / L. F. Ruse 17 / 1923 / "The leaves are heated and used on insect bites" / "'Dieng-uai'"
India: Bombay Pres. / J. Fernandes 67 / 1949 / "Milky white juice used to stem flow of blood on surface wounds"

3428   *Pycnostelma chinense* China / F. Hom 00182 / 31 / "Use as drug. If this plant is in the house there will be no entering of snakes. It has the widest use so-called 'Drug King'" / "'Liu Tiu Chuk'"

3429   *Morrenia odorata* S. America: Depto. Robles, S. del Estero (Argentina?) / R. Maldonado B. 225 / 1939 / "Con sus frutos hacen dulce muy apreciado por su propiedad de aumentar la secreción lactea" / "'doca'"
Argentina / J. B. Correa 23 / 41 / "planta lechosa—muy aromática" / "'Doca'"

3431   *Cynanchum viride* Chile / no collector s.n. / no year / "'Yerba Buena'"

3432   *Sarcostemma clausum* Honduras / A. Molina R. (Herb.?) 1788 / 1948 / "'Bejuco de pescado'"

3433 **S. cynanchoides** ssp. *hartwegii* Mexico / C. Lumholtz 17 / 1910 /
"The titmouse makes nests from the coma of the seeds. Fruits (either
raw or boiled) eaten by the Indians"

Mexico / G. Thurber s.n. / 1847 / "(Said to be poisonous)"

3434 **S. solanoides** Peru / Goodspeed & Metcalf 30229 / 1942 / "Twining
vine, soft to touch, with odor of onion when fresh . . ."

3435 **S. andinum** Peru / R. D. Metcalf 30274 / 1942 / "Used by natives for
insect bites"

3436 **Secamone micrantha** Hainan / S. K. Lau 172 / 1932 / "To boil the
leaves to heal"

3437 **Secamone** indet. Philippine Is. / P. Añonuevo 287 / 1950 or 1930 /
"Heat the fruit for a while. Pound it with a little salt and apply to
the cavity of painful tooth" / "'Lagisi'"

3438 **Gymnema** indet. China / no collector 248.6891 or 13021 / 1924 /
"Name reported: Tai ch'a yeuk . . . (Large tea drug)"

3438A **Gymnema** indet. Philippine Is. / Sulit & Conklin 5092 / 1953 /
"Roots . . . put in coconut shell containing ambers—smoke drives away
'ampakto' (bad spirits)"

3438B **Gymnema** indet. Philippine Is. / Sulit & Conklin 5068 / 1953 /
"Roots cut into small pieces, then mixed with 'binhi' (seeds for planting) to avoid destroying by rodents—hence fruitful harvest"

3439 **Gymnema** indet. Philippine Is. / Sulit & Conklin 5162 / 1953 /
"Roots cut into small pieces and mixed with palay seeds (Binhi) for
planting. The Mangyans believe that by this method they expect
fruitful harvest"

3440 **G. sylvestre** New Caledonia / H. S. McKee 2248 / no year / "Flowers
white, strong unpleasant fishy scent"

New Caledonia / R. Virot 946 / 1942 / "Fleurs à odeur fetide"

3441 **Tylophora** indet. Solomon Is. / S. F. Kajewski 2251 / 30 / "The
roots are blown along by natives in the direction of rain to make it
finish"

3442 **T. polyantha** Caroline Is. / C. C. Y. Wong 408 / 1948 / "The plant is
used in the treatment of general debility, sleepiness, and tiredness
(fely ne göf), the entire plant is taken, pounded and squeezed into a
bowl, then mixed with the water of a young coconut and the mixture
is drunk" / "'yeloy' (Rumung); 'ley' (Rul)"

3443 **Dischidia platyphylla** Philippine Is. / Sulit & Conklin 5146 / 1953 /
"The leaves are pressed to make them juicy, then applied to bagã (boil)
as topical" / "'Takdamol'" Mang. (Hanunoo)

3444 **Dischidia** indet. Philippine Is. / Sulit & Conklin 5124 / 1953 / "Roots
smashed put or mixed with 'tubã' to avoid fermentation" / "'Lubyug'"
Mang. (Hanunoo)

3444A *Dischidia* indet. Philippine Is. / H. C. Conklin 131 / 1953 / "roots fragrant"

3444B *Dischidia* indet. Philippine Is. / G. E. Edaño 1592 / 1949 / "The sap is used as a medicine for boil"

3445 *Hoya multiflorum* China / W. T. Tsang 22678 / 1933 / "Fl. white, ill-smelling"

3446 *H. ridleyi* Malaya / A. H. Millard (K.L. 1409) / 59 / "Phytochemical Survey of Malaya"

3447 *Hoya* indet. Solomon Is. / S. F. Kajewski 2278 / 30 / "The natives have a superstition that if the fruit is planted in their banana patches they will have an exceptionally good crop" / "'Tormia'"

3448 *H. pubicalyx* Philippine Is. / C. O. Frake 705 / 1958 / "Hunting ritual" / "Sasui"? Sub.

3449 *Hoya* indet. Philippine Is. / G. E. Edaño 2019 / 1950 / "latex good for wound, so with the juice of the leaves" / "'Dalicotcot'" Vis.

3449A *Hoya* indet. Philippine Is. / C. O. Frake 879 / 1958 / "Leaves applied for spleenomegaly" / "'Nalipnalig'" or "'Nalipualig'" Sub.

3449B *Hoya* indet. Philippine Is. / Sulit & Conklin 5059 / 1953 / "Sap mixed with that of dayang-dayang, boiled is used as arrow poison" / "'Laput-purú'" Mang. (Hanunoo)

3449C *Hoya* indet. Philippine Is. / P. Añonuevo 226 / 1950 / "Leaves are used as dressing for wounds" / "'Kating'" Belaan or Bilaau

3450 *H. australis* Fiji Is. / H. St.John 18100 / 1937 / "Used medicinally for child with a cough" / "'nambetiambete'"

3451 *H. vitiensis* Fiji Is. / O. Degener 14627a / 1941 / "For garlands; considered poisonous" / "'Wa(n)dra'" Sabatu

3452 *Marsdenia* indet. Mexico / H. S. Gentry 1588 / 1935 / "Warihios reported to eat the tender green fruit" / "'Tonchi'"

3452A *Marsdenia* indet. Mexico / G. B. Hinton et al. (Herb.) 5899 / 34 / "Fruit edible" / "'Cuaguayote'"

3453 *M. clausa* Dominican Republic / R. A. & E. S. Howard 8824 / 1946 / "Natives *will not* touch this plant. Said to cause instant death" / "'Curumaguey'"

3454 *M. roylei* Nepal / Polunin, Sykes & Williams 2481 / 1952 / "Copious white latex & unpleasant scent"

3455 *M. australis* Australia / G. K. Lillecrapp s.n. / 1946 / "'Native Pear'"

3456 *M. erecta* Greece / T. G. Orphanides 88 / 1849 / "Les charlatans donnent cette plante contre les morsures des chiens enragés"

3457 *Dregea volubilis* Hainan / Tsang, Tang & Fung 109 / 1929 / "Used as medicine" / "'Ya Foot'"

3458 *Matelea baldwiniana* U.S.: Mo. or Ark. / G. Engelmann s.n.? / 1835 / "corolla whitish with offensive odor"

U.S.: Mo. / O. E. Lansing, Jr. 3044 / 1911 / "Flowers ill-scented"

3459 *M. cynanchoides* U.S.: Tex. / G. J. Goodman 5846 / 1954 / "Fetid! "

3460 *M. caudata* Mexico / H. S. Gentry 1628 / 1935 / "Fruit when young is eaten by the natives (Warihios also) raw or cooked" / " 'Talayote,' Mex.; 'Talayote,' W.'

3461 *M. fruticosa* Baja Calif. / B. J. Hammerly 418 / 1941 / "Flowers purple-black with disagreeable odor"

3462 *M. petiolaris* Mexico / H. S. Gentry 1626A / 1935 / "Fruit toasted and eaten by Warihias" / " 'Mahoy piwala,' W."

3463 *M. pilosa* Mexico / J. G. Schaffner 650 / 1876 / " 'Flor del muerto negro' "

Mexico / A. Dugès 99 / 1893 / " 'Flor del Muerto'—(fleur du mort)"

3464 *M. tristifolia* Mexico / H. S. Gentry 1626 / 1935 / "Fruit toasted and eaten by Warihios" / " 'Mahoy piwala,' W."

3465 *Gonolobus barbatus* El Salvador / S. Calderón 908 / 1922 / " 'Mata-coyote' "

3466 *G. edulis* Cuba / C. F. Baker (Econ. Plants of the World) 96 / 1907 / "The juice of this fruit is said to be used in the making of refrescos by the Venezuelans"

Honduras / E. R. Mitchell 123 / 1926 / " . . . odor offensive"

3467 *G. erianthus* Mexico / Dr. Coulter 976 / probably before 1880 / " 'Flor del Muerto verde' "

3468 *Gonolobus* indet. Argentina / J. West 6142 / 1936 / "black flower, fetid smelling"

3469 *Menabea venenata* Madagascar / H. Humbert 2812 / 1924 / "Servait à empoisonner les flèches—et connue 'poison d'épreuve' utilisé dans certaines cérémonies"

3470 *Asclepiadacea* indet. Colombia / Killip & Smith 14336 / 1926 / " 'Matavivi' "

3470A *Asclepiadacea* indet. China / C. Schneider 4136 / 1914 / " . . . fl. flavis foetidis . . . "

3470B *Asclepiadacea* indet. Solomon Is. / S. F. Kajewski 2392 / 30 / "When a man has a headache, he takes a piece of this vine and fastens it round his head" / " 'A-o-popotu' "

CONVULVULACEAE

3471 *Cuscuta chinensis* China / Cheo & Yen 223 / 1936 / " . . . seed for medical use" / " . . . parasitic on Artemisia"

3472   *C. campestris* Fiji Is. / O. Degener 14934 / 1941 / "Acts as styptic; merely squash and apply to wound" / "'Wa(m)bosuthu'" Savatu

3473   *C. planiflora* EUROPE / N.C. Seringe Herb. Helveticum 1965 / no year / "Stimulante, incisive, apéritive"

3474   *Dichondra macrocalyx* Argentina / Venturi 240 / 1919 / "'Yerba Buena'"

3475   *Erycibe elliptilimba* Hainan / S. K. Lau 12 / 1932 / "The crushed roots are used as bait for fishing"

3476   *Erycibe* indet. Philippine Is. / C. O. Frake 882 / 1958 / "Leaves used for skin rash" / "'Bayang'" Sub.

3477   *Convolvulus arvensis* Afghanistan / E. Bacon 39 / 1939 / "Flower boiled, taken as laxative" / "'Peičaq'"

3478   *Merremia alata* Brazil / R. Froes 11851 / 1940 / "'Jalapa amarelo'"
Brazil / R. Froes 1764 / 1932 / "... starch from tubers used for purgative purposes" / "'Jalapa da Terra'"

3479   *M. distillatoria* Philippine Is. / R. B. Fox 33 / 1948 / "A medicine for ÚGUT, boils; merely obtain the flower of the plant, and place it on the boil" / "'BĬLÁNGƏD'" Egóngot

3480   *M. umbellata* Philippine Is. / C. Frake 472(?) (Herb. 38016) / 1958 / "Roots boiled and drunk for hematuria" / "'Leleknut'" Sub.

3481   *M. peltata* Caroline Is. / C. Y. C. Wong 288 / 1947 / "The leaves are pounded and placed in breadfruit poi to make it taste good" / "'Fitay'"

3482   *Operculina turpethum* Fiji Is. / O. Degener 15404 / 1941 / "After childbirth squash leaf in water or drink as tonic. No good as 'string'" / "'Wandamdam'" Ra

3483   *Ipomoea ancisa* Mexico / H. S. Gentry 2648 / 1936 / "Herbage decocted and drunk for stomach ailments" / "'Romeria de la Sierra'"

3484   *I. arborescens* Mexico / J. M. Greenman 267 / 1906 / "La Purga"

3485   *I. digitata* Brit. Honduras / W. A. Schipp 636 / 1930 / "... supposed to be poisonous flowers"

3486   *I. pedicellaris* Mexico / H. S. Gentry 4880 / 1939 / "Seeds employed as a purgative" / "'Mantela de Maria'"

3487   *I. quamoclit* El Salvador / P. C. Standley 20970 / 1922 / "'Condeamor'"

3488   *I. denticulata* W.I.: Dominica / W. H. & B. T. Hodge 3323 / 1940 / "'Caapi'"

3489   *I. tiliacea* W.I.: Dominica / W. H. & B. T. Hodge 3318 / 1940 / "'Caapi'"

3490 *I. crassifolia* Brazil / W. A. Archer 7678 / 1942 / "Stems and lvs. used as blood tonic" / "'Tarsta'"

3491 *Ipomoea* indet. Bolivia / J. Steinbach 7006 / 1925 / "Se le atribuye propiedad curativa contra la picada de serpente" / "'Contra vibora'"

3492 *Ipomoea* indet. Brazil / Krukoff's 5th Exped. to Braz. Amaz. 5863 / 1934 / "Roots used medicinally as a laxative" / "'Salsa'"

3493 *Ipomoea* indet. Bolivia / J. Steinbach 7077 / 1925 / "'Contraveneno'"

3494 *I. caerulea* New Britain / A. Floyd 6485 / 54 / "Young leaves boiled and put on sores and rashes" / "'La Bihela'" W. Nakanai

3495 *I. alba* Caroline Is. / C. Y. C. Wong 179 / 1947 / "The plant is used for *säfein sät* (against the bite of the sea spirit), the leaves are pounded, expressed into a bowl and mixed with coconut water, then drunk. For pain in the chest, the leaves are pounded and mixed with expressed coconut milk, then boiled and the mixture is drunk" / "'öörö-pön'"

3496 *I. congesta* Tonga Is. / T. G. Yuncker 15273 / 1953 / "Stems used medicinally" / "'Fue fakahinga'"

3497 *I. gracilis* Fiji Is. / O. Degener 15110 / 1941 / "For headache, press leaves and rub against forehead" / "'Tokatolu'" Serua

3498 *Ipomoea* indet. Cook Is. / T. G. Yuncker 10243 / 1940 / "'Tefifi tea'"

3499 *Argyreia obtusifolia* China / T. S. Pan 9 / 28 / "Medicine" / "'Pat Tuk Muk Ip'" "'Sz Chou Muk Ip'"

3500 *A. capitata* Burma / P. Khant D. R. (Herb.?) 47 / no year / "The stem is used for purgative medicine" / "'Mingoka-nwe'"

3501 *Convolvulacea* indet. Bolivia / J. Steinbach 5583 / 1924 / "Se usa contra picadas de víboras" / "'Contraveneno'"

POLEMONIACEAE

3502 *Cobaea licayensis* Ecuador / J. N. Rose 22273 / 1918 / "'Pepino del Monte'"

3502A *Polemonium delicatum* var. *typicum* U.S.: Utah / F. W. Gould 1792 / 1942 / "Plants . . . with disagreeable odor"

3502B *P. confertum* var. *eximium* U.S.: Calif. / D. D. Keck 4928 / 1938 / "very fragrant"

3502C *P. viscosum* U.S.: Mont. / Hitchcock & Muhlick 12698 / 1945 / "particularly mephitic"

3503 *Loeselia coerulea* Mexico / R. M. Stewart 1039 / 1941 / "'Gordo Lobo'"

Mexico / A. Dugès s.5 / 1900 / "Yerba de la Ventosidad, ou Moradilla"

3504 **L. greggii** Mexico / A. López 13 / 1947 / "Medicina par la calentura" / "'Yerba del Guachichile'"

3505 **L. mexicana** Mexico / Y. Mexia 9117 / 1938 / "Root emetic" / "'Espinosilla'"

HYDROPHYLLACEAE

3506 **Phacelia crenulata** U.S.: Calif. / F. W. Gould 916 / 1940 / "Flowers strongly scented"

3507 **P. palmeri** U.S.: Colo. / M. Ownbey 1498 / 1937 / "Herbage with a strong foetid odor"

3508 **P. texana** U.S.: Tex. / G. Thurber 183 / 1951 / "Plant viscid & with a strong & disagreeable odor"

3509 **Phacelia** indet. Argentina / T. Meyer 34211 / 1940 / "Medicinal—flor azul (para curar heridas)" / "'Hutina'"

3510 **Wigandia reflexa** Dominican Republic / R. A. & E. S. Howard 8586 / 1946 / "...hairs spiny & stinging(?)" / "'tabacco cimarron'"

3511 **W. caracasana** Colombia / E. Dryander 2443 / 1939 / "Herb...parece Tabacco los campasinos le usan c.T." / "'Tabaquillo'"

3512 **Nama densum** U.S.: Ore. / A. Cronquist 7159 / 1953 / "Plants with a slightly mephitic odor"

3513 **N. hispidum** Mexico / W. P. Hewitt 100 / 1946 / "Remedial herb, boiled as a tea and drunk for gassy stomach" / "'Hierba de la ventosidad'"

3514 **N. hispidum** var. **spathulatum** Mexico / G. B. Hinton 16951 / 51 / "Tremendously toxic, causing first blisters, a week later swelling, fever"

BORAGINACEAE

3515 **Cordia alba** El Salvador / P. C. Standley 19674 / 1921–22 / "...fr. ...extremely sweet, viscid, eaten" / "'Tiguilote,' 'Tiguilote negro'"
Colombia / S. J. Record 1 / 30 / "Fruit edible" / "'Uvita'"

3516 **C. alliodora** El Salvador / P. C. Standley 19717 / 1922 / "Decoction of lvs. used externally for bruises & swellings" / "'Laurel'"
El Salvador / P. C. Standley 19264 / 1921–22 / "Lvs. heated & placed on wounds" / "'Laurel'"

3517 **C. boissieri** Mexico / G. Thurber 870 / 1852 / "Fruit said to possess intoxicating properties"
Mexico / V. Grant 501 / 1940 / "Natives make a syrup from fruit for pneumonia" / "'Nacahuita'"
Mexico / C. H. Muller 2623 / 39 / "Locally fruits are used to manufacture molasses"

3518 *C. brevispicata* Mexico / R. V. Moran 6885 / 1959 / "Herbage with chicken soup mix odor"

3519 *C. collococca* El Salvador / P. C. Standley 22371 / 1922 / " ... fr. red, eaten, very sweet & somewhat astringent, very mucilaginous" / "'Manuno'"

Jamaica / W. T. Stearn 902 / 1956 / " ... ripe fruits ... with juicy not unpleasant orange pulp" / "'cherry'"

3520 *C. glabra* Nicaragua / C. F. Baker 2549 / 1903 / "With a strong resinous odor"

3521 *C. cylindrostachya* Mexico / R. V. Moran 7408 / 1959 / "Used for tea" / "'Confituria'"

El Salvador / P. C. Standley 21896 / 1922 / "Used to coagulate rubber and indigo" / "'Cuaja-tinta'"

Mexico / E. Stewart 11 / 1935 / "Maya name: Elemuy"

3522 *C. dentata* Mexico / G. B. Hinton 3755 / 1933 / "fr. edible" / "'Chirimo'"

Jamaica / W. T. Stearn 307 / 1956 / "Fruits ... with a colourless, strong, unpleasant-tasting pulp around the stone"

3523 *C. foliosa* Guatemala / J. A. Steyermark 50806 / 1942 / "Reputed to be used for inflammation and stomach ache; boil leaves and drink the infusion in treating the remedy" / "'Salviasanta'"

3524 *C. inornata* Mexico / Y. Mexia 1080 / 1926 / " ... seed hard. Edible" / "'Perlitas'"

3525 *C. morelosana* Mexico / E. Langlassé 39 / 1898 / "Fleurs en infusions contre les points de côté" / "'Cherare'"

Mexico / G. B. Hinton 3122 / 33 / "Flower boiled with fruit of cirian taken for pneumonia" / "'Chiraire'"

3526 *C. seleriana* Mexico / E. Langlassé 119 / 1898 / "'Ami de Iguan'"

3527 *C. spinescens* Honduras / P. C. Standley 53713 / 1927–28 / "'Carne asada'"

3528 *C. stenoclada* Mexico / G. Martínez-Calderón 628 / 1941 / "Fruta contiene buena goma"

3529 *C. stellifera* Brit. Honduras / W. A. Schipp 1040 / 1932 / "Flowers sweetly perfumed"

3530 *C. curassavica* Puerto Rico / H. F. Winters 553 / 55 / "Said to be used medicinally" / "'Jaraguaso'"

3531 *C. divaricata* W.I.: Martinique / R. A. Howard 11721 / 1950 / " ... foliage aromatic"

3532 *C. laevigata* W.I.: St. Lucia / J. S. Beard 495 / 1945 / " ... fruit white, glutinous, sweet" / "'Chypre'"

3533   *C. exarata* Haiti / E. L. Ekman 6701 / 1926 / "Smells like celery when fresh, from coumarine when drying"

3534   *C. gerascanthoides* Jamaica / Maxon & Killip 1441 / 1920 / "...flowers very fragrant"

3535   *C. globosa* Cuba / E. P. Killip 13944 / 1931 / "'Yierba de la sangre'"

3536   *C. obliqua* Cuba / J. G. Jack 4528 / 1926 / "'Sea coast Tea,' 'Gout Tea'"

3537   *C. sebestena* Cuba / M. Curbelo 5684 / 1921 / "'Vomitel amarillo'"
Virgin Is. / J. S. Beard 314 / 1944 / "Fruit edible white outer layer" / "'Wild nut tree'"

3538   *C. sulcata* Dominican Republic / R. A. & E. S. Howard 9746 / 1946 / "...fruit...sweet and good to the taste"

3539   *C. bicolor* Colombia / R. Romero Castañeda 1681 / 1949 / "Fruto... pulpa amarilla, comestible"

3540   *C. corymbosa* Argentina / M. Lillo 12054 / 1912 / "'Yerba Buena'"

3541   *C. ecalyculata* Brazil / A. Gehrt 5304 / 21 / "'Chá de Bugre'"
Brazil / Y. Mexia 5201 / 1930 / "Tea made of leaves for blood disorders" / "'Cutiéra'"

3542   *C. ferruginea* Colombia / Killip & Smith 16336 / 1926 / "'Mulato'"

3543   *C. lutea* Peru / J. West 3575 / 1935 / "...fruit...reported as edible and sweet" / "'Overo'"
Ecuador / R. Espinosa 1831 / 47 / "Se usa como medicinal" / "'Overal'"

3544   *C. nodosa* Bolivia / H. H. Rusby 778 / '21 / "Edible white fruit"
Peru / J. M. Schunke 10 / 1935 / "'Pacacuracaspa'"
Bolivia / J. Steinbach 5400 / 1921 / "La fruta...es dulce y fresco"

3545   *C. verbenacea* Brazil / B. Rodrigues 963 / 1943 / "E medicinal" / "'Balieira'"

3546   *C. dichotoma* China / F. A. McClure 2626 / 1922 / "Fruits used to poison fish" / "'Fung Tsang shu Kau Shu'"
Burma / F. G. Dickason 7464 / 1938 / "Leaves used for Burmese cheroots. Flowers edible. Fruits sticky—used as paste" / "'Thanapet'"

3547   *C. wallichii* India: Bombay Pres. / J. Fernandes 211 / 1949 / "Fruit is eaten"

3548   *C. griffithii* N. Borneo / Saw For. Research Inst. Kepong 71668 / 52 / "Fruit can be made into gum" / "'Kendal'"

3549   *C. rufescens* Brazil / G. Hatschbach 3443 / 1953 / "...frutos comestivels"

3550   *C. cummingiana* Solomon Is. / S. F. Kajewski 2428 / 31 / "When a native gets a burn, the charcoal is taken and rubbed on the burned place" / " 'Conga natere' "

3551 *C. subcordata* New Guinea / H. A. Brown 90 / 1952 / "The leaves are boiled and the liquid used to bathe limbs of people with muscular or rheumatic pains" / " 'Utimafu' (Elema, Iokea)"

New Hebrides Is. / S. F. Kajewski 184 / 1928 / "Natives tell me they eat fruit"

3552 *C. myxa* New Hebrides Is. / S. F. Kajewski 397 / 28 / "Bark used for grass skirts and berries used to feed pigs" / " 'Yalehoi' "

3553 *Bourreria oxyphylla* Brit. Honduras / C. L. Lundell 6439 / 1936 / " . . . fls. very fragrant"

3554 *B. pulchra* Mexico / J. M. Greenman 360 / 1906 / " . . . flowers of an unpleasant odor"

3555 *B. ovata* Bahama Is. / R. A. & E. S. Howard 10250 / 1948 / "Used as tea" / " 'Strongbark' "

Bahama Is. / D. Fairchild 2571 / 1932 / "Fruit edible" / " 'Strongbark' "

3556 *B. succulenta* W.I.: Martinique / H. & M. Stehlé 6051 / 1943 / " 'Bonbon rouge' "

3557 *Ehretia acuminata* China / F. A. McClure 20558 / 1937 / "Fruit olive green, edible" / " 'Wu-lan-tzu' "

3558 *E. dicksoni* China / Fan & Li 361 / 1935 / " . . . fr. edible"

3559 *E. longiflora* China / Y. W. Taam 2033 / 1941 / "flower pink, ill-smelling"

3560 *E. microphylla* China / S. K. Lau 137 / 1932 / "The fruit is edible" / " 'Mon Thai' (Lois)"

3561 *E. navesii* Philippine Is. / C. O. Frake 664 / 1958 / "Barks applied for athletes foot" / " 'gesising' Sub."

Philippine Is. / C. O. Frake 683 / 1958 / "Agricultural ritual. Medicine for internal pains" / " 'tingu' Sub."

Philippine Is. / C. O. Frake 692 / 1958 / "Bark used as fish poison" / " 'genengnam' Sub."

Philippine Is. / C. O. Frake 740 / 1958 / "Bark boiled and drunk for cough" / " 'matilak' Sub."

Philippine Is. / C. O. Frake 671 / 1958 / "Leaves applied to ringworm" / " 'gantul' Sub."

3562 *Cortesia cuneato* Argentina / A. R. Leal 7963/207 / 1942 / "Muy comido por las cabras" / " 'Campa' "

3563 *Coldenia hispidissima* var. *latior* U.S.: Ariz. / E. U. Clover 5262 / 1939 / "Supais boil root and drink liquid for stomach trouble" / " 'Ka-aw' "

3564   *C. canescens* Mexico / R. M. Stewart 358 / 1941 / "'Yerba del Pobre'" Mexico / L. R. Stanford et al. 137 / 1941 / "Heavily grazed by goats"

3565   *C. greggii* Mexico / W. P. Hewitt 159 / 1946 / "Tea brewed from twigs taken medicinally" / "'Regeneradora'"

3566   *Ixorhea ischudiana* Mexico / R. Schreiter 5238 / 1927 / "...flor muy fragrante (medicinal)"

3567   *Tournefortia glabra* Mexico / B. Conzatti 3842 / 1919 / "Empleada en cataplasmas a los pies contra la influenza" Guatemala / J. A. Steyermark 35434 / 1940 / "Leaves and stem with musky solanaceous odor"

3568   *T. hartwegiana* Mexico / P. C. Standley 1157 / 1934 / "Medic. for insect stings & snake bites same as with Confituria grande" / "'Confituria negra'"

3569   *T. hirsutissima* Panama / R. E. Woodson, Jr. et al. 721 / 1938 / "Flowers—very fragrant"

3570   *T. bicolor* W.I.: Dominica / W. H. & B. T. Hodge 3263 / 1940 / "...pounded leaves put on sores and boils by Caribs" / "'mey wet'"

3571   *T. staminea* Jamaica / Howard & Proctor 15049 / 1958 / "...flowers ...with strong fragrance like violets"

3572   *T. volubilis* W.I.: Cayman Is. / W. Kings 140 / 38 / "Used for female trouble" / "'Aunt Eliza Bush'"

3573   *T. brevilobata* Ecuador / W. H. Camp 2560 / 1945 / "Infusion of lvs. drunk by women a day or two after childbirth to 'clean everything out'" / "'Guarocallo pequeño'"

3574   *T. fuliginosa* Colombia / W. H. Camp 336 / 1944 / "...corolla... very fragrant" Colombia / R. E. Schultes 3273 / 1942 / "Fruit edible"

3575   *T. ramosissima* Ecuador / A. Rimbach 29 / no year / "Foliage and fruit good forage for sheep" / "'Pesso'"

3576   *T. rolloti* Colombia / O. Haught 5918 / 1947 / "Rank-smelling"

3577   *Messerschmidia gnaphaloides* Puerto Rico / P. Sintenis 527 / 1885 / "Té del mar"

3578   *Heliotropium glabriusculum* U.S.: Tex. / B. H. Warnock c550 / 1938 / "...flower with most unusual fragrant odor"

3579   *H. angiospermum* Mexico / G. B. Hinton 11981 / 38 / "'Alacran chico'"

3580   *H. angustifolium* Mexico / L. R. Stanford et al. 185 / 1941 / "Heavily grazed by goats"

3581   *H. fallax* Guatemala / J. A. Steyermark 50681 / 1942 / "'Yerba de torro'"

3582 *H. procumbens* Mexico / G. B. Hinton 773 / 32 / "'Moco de Guajo-lote'"

3583 *H. rufipilum* Guatemala / P. C. Standley 76426 / 1940 / "'Yerba de alacrán'"

Ecuador / J. A. Steyermark 52948 / 1943 / "fruit edible" / "'caspa-rosari' or 'huagracaya'"

3584 *H. arborescens* Peru / Stork & Horton 9953 / 1938 / "Agreeably fragrant, suggesting lemon"

3585 *H. argenteum* Ecuador / F. Prieto 2555 / 1945 / "Infusion of lvs. given to women who have fits and spasms during childbirth" / "'Guagra-callo blanco'"

3586 *H. sinuatum* Chile / C. Muñoz B-14 / 1935 / "Empleada como leña y forrage" / "'Palo negro' o 'heliotropo del campo'"

3587 *H. tiaridioides* Brazil / R. Ruiz C56 / 43 / "Contra molestias respiratorias" / "'Crista de galo'—'Rabo de macaco'"

3588 *H. ramosissimum* Arabia / H. Dickson 10 / 61 / "Leaves used as a cure for snake bite" / "'Ram Ram'"

3589 *Cynoglossum* indet. N. Siam / Royal For. Dept. 3820 / 1948 / "Medicinal root used in remedy of fever" / "'Yah Sarm-sib-sawng-rark'"

3590 *Hackelia brachytuba* Sikkim / R. E. Cooper 172 / 1913 / "...fls. almond scented"

3591 *Dasynotus spinocarpon* Arabia / H. Dickson 1 / 61 / "Eaten by sheep & camels" / "'Demargh al Jerboa' (Brains of Jerboa)"

3592 *Eritrichium rupestre* India: Punjab / W. Koelz 8468 / 1936 / "Use medicinal"

3593 *Plagiobothrys calandrinioides* Argentina / O. Boelcke 4438 / 1950 / "...comida por los animales"

3594 *Antiphytum heliotropioides* Mexico / L. R. Stanford et al. 170 / 1941 / "...heavily grazed by goats"

3595 *Mertensia ciliata* U.S.: Nev. / M. S. Jeppesen 184 / 1936 / "Forage value very high   grazed by C., H., S., & G."

3596 *M. fusiformis* U.S.: Colo. / C. R. Towne 98764 / 44 / "Forage value fair   grazed by C. & H."

3597 *Trigonotis peduncularis* China / W. T. Tsang 26089 / 1936 / "...used as pig food"

3598 *Macromeria guatemalensis* Guatemala / J. A. Steyermark 50069 / 1942 / "Reputed to be used to cure colds; cook the leaves and drink infusion" / "'té de monte'"

3599 *Lithospermum cobrense* Mexico / C. V. Hartman 714 / 1891 / "Flowers very fragrant"

3600   *L. discolor* Mexico / Y. Mexia 712 / 1926 / "'Yerba de la vivora'"

3601   *L. erythrorhizon* China / C. Y. Chiao 2660 / 1930 / "Tuberous roots soaked in wine for drinking purpose"

3602   *L. callosum* Kuwait / H. Dickson 72 / 61 / "Much eaten by camels" / "'Hamāt'"

3603   *Arnebia tinctoria* Kuwait / H. Dickson 34 / 61 / "Eaten by sheep etc." / "'Chahil'"

3604   *Boraginacea* indet. Cuba / Bro. Crisógono 3801 / 1946 / "'Vomitel colorado'"

3605   *Boraginacea* indet. Cuba / Bro. Leon 11450 / 1923 / "'Vomitel blanco'"

VERBENACEAE

3606   *Verbena carolina* Mexico / G. B. Hinton 2729 / 32 / "Triturated taken for malaria"

3607   *V. elegans* Mexico / H. S. Gentry 2730 / 1936 / "Decoction made of herbage for stomach troubles" / "'Moradilla,' Mex."

3608   *V. litoralis* Mexico / G. B. Hinton 3731 / 33 / "Juice from the macerated plant taken for malaria"
Bolivia / J. Steinbach 5137 / 1920 / "Medicinal. Se usa en contusiones"
Peru / Mr. & Mrs. F. E. Hinckley 64 / 1920 / "Used as a purgative" / "'Berbena'"
Colombia / Killip & Smith 21197 / 1927 / "Used for fevers"
Paraguay / W. A. Archer 4788 / 1936 / "General remedy, also used for coughs" / "'Vervena'"

3609   *Lantana horrida* U.S.: Tex. / A. Traverse 1437 / 1960 / "Resinous odor"

3610   *L. achyranthifolia* Mexico / G. B. Hinton 1765 / 32 / "Food for birds" / "'Frutilla'"

3611   *L. camara* El Salvador / P. C. Standley 19208 / 1921–22 / "...fls.... used to cause sweat in fevers etc. Root used to purify blood. Fr. eaten" / "'Cinco negritos'"

3612   *L. involucrata* Mexico / H. S. Gentry 1278 / 1933 / "Medic. herbage cooked in water to make wash for insect stings and rattlesnake bites, same as with Confituria grande" / "'Confituria blanca,' Mex."

3613   *L. polyanthus* W.I.: Dominica / W. H. Hodge 855 / 1938 / "...leaves with minty odor"

3614   *L. fucata* Brazil / Y. Mexia 5414 / 1930 / "...flower made into syrup for coughs" / "'Cambará'"

3615   *L. peduncularis* Galapagos Is. / H. K. Svenson 6 / 1930 / "Strong disagreeable odor in leaves"

3616   *Lippia alba* Mexico / G. B. Hinton 13149 / 38 / "Leaves very fragrant and their brew is good for the stomach ache" / "'Tarete'"

Cuba / F. Marie-Victorin 21 414 / 1943 / "(cultivé en guise de Menthe autour des bohios)"

3617   *L. barbata* Mexico / H. S. Gentry 3755 / 1938 / "An infusion is made of the odorous herbage for treating colds and similar ailments" / "'Poleo'"

3618   *L. dulcis* Mexico / G. B. Hinton 764 / 32 / "Used in stomach trouble of children" / "'Yerba dulce'"

3619   *L. palmeri* Mexico / H. S. Gentry 4188 / 1939 / "The odorous herbage is used as a condiment" / "'Oregano'"

3620   *L. umbellata* Mexico / E. Palmer 479 / 1896 / "...decoction used for rheumatism" / "'Yerba de Santa Gertrude'"

3621   *L. nahuire* Mexico / H. S. Gentry 5721 / 1940 / "...shrub...with licorice-like odor. Tea is made locally from the foliage" / "'Nahuire'"

3622   *L. pringlei* Mexico / P. C. Standley 1179 / 1934 / "Medic. sap for toothache (chew bark?), leaves put in hot water then coated with mentholatum and applied to bruise, sore, headache etc." / "'Matayaki,' 'Choila,' 'Chokili,' Mex."

3623   *L. scaberrima* Dominican Republic / R. A. & E. S. Howard 9677 / 1946 / "...ls. sweet smelling and eaten like mint"

3624   *L. affinis* Brazil / Y. Mexia 5625 / 1931 / "Made into tea & used for dysentery" / "'Camará'"

3625   *L. betulifolia* Brazil / Y. Mexia 6068 / 1931 / "Leaves used as a drink 'Tambú'"

3626   *L. chilensis* Chile / R. Wagenknecht 18423 / 1939 / "Used largely by Chileans as medicine"

3627   *L. deserticola* Argentina / T. Meyer 4045 / 1945 / "Medicinal: estomacal, abortiva, etc." / "'Rica-rica'"

3628   *L. grisebachiana* Argentina / P. Jorgensen 1025 / 1915 / "...por purificar la sangre"

3629   *L. hastulata* Argentina / L. R. Parodi 9709 / 1931 / "Aromatico y medicinal" / "'Rica-rica'"

Argentina / J. West 6301 / 1936 / "Used as flavoring in mate" / "'Rica-rica'"

3630   *L. turbinata* Argentina / R. S. Ruiz H 5 / '41 / "...medicinal" / "'poleo'"

3631   *L. nodiflora* Annam / Poilane P-65 / 1919 / "...cette herbe sert de nouriture au porcs" / "'Cây eau dāu.A.'"

3632   *Lippia* indet. Argentina / L. R. Parodi 11036 / 1933 / "...medicinal ..." / "'Yierba Linga'"

3633   *Stachytarpheta cayennensis* W.I.: Dominica / W. H. & B. T. Hodge 3305 / 1940 / "...tea from plant used to cure any illness" / "'vervain'"

3634   *Priva cuneata* Chile / W. J. Eyerdam 24646 / 1939 / "It has edible tubers ... "

3635   *Citharexylum affine* Mexico / Y. Mexia 584 / 1926 / "Fruit ... much eaten by birds" / "'Jalacote'"

3636   *Duranta erecta* W.I.: Antigua / J. S. Beard 343 / 1944 / "'Pigeonberry'"

3637   *Callicarpa americana* U.S.: Tex. / A. Traverse 135 / 1956 / "Leaves & twigs have musky pungent odor"

3638   *C. longifolia* China / To & Ts'ang 12480, 12556 / 1924 / "Names reported: 'Tai yan mat' (Great man); 'Fat fung shu' (Leprosy tree)"

3639   *C. reevesii* China / G. W. Groff 4050 / 1919 / "Used in the preparation of a medicine used for injuries" / "'Pok Kwat T'am'"

3640   *C. rubella* China / F. Hom 00142 / 31 / "Drug plant" / "'Kap Ts'ing'"

3641   *C. pedunculata* Solomon Is. / S. F. Kajewski 2420 / 30 / "When a small baby in arms is sick the fruit is chewed with betel nut and spat into the baby's mouth" / "'Bau'"

3642   *C. pentandra* var. *paloensis* Solomon Is. / S. F. Kajewski 2540 / 31 / "The bark is macerated with water and drunk for colds" / "'Kimberi'"

3643   *C. cana* Caroline Is. / C. C. Y. Wong 135 / 1947 / "Wood and leaf are used for headache, *Semun cukó* (chicken sickness). Wood is boiled and used mixed with coconut, then eaten. Leaves are pounded and eaten. Fruit is pounded and put in eyes" / "'ääkyn'"

3644   *C. elegans* Philippine Is. / G. E. Edaño 1691 / 1949 / "The leaves are used as medicine for headache" / "'Lilay' (Ma)"

3645   *C. formosana* Philippine Is. / I. P. Paniza 9419 / 1948 / "Insecticide" / "'Tigao' (Tag.)"

3646   *Aegiphila martinicensis* W.I.: Dominica / W. H. & B. T. Hodge 3320 / 1940 / "'grain' used in traps to lure doves"

3647   *Teysmanniodendron* indet. Borneo / Kostermans 6019 / 1951 / "Living bark ... yellow with sugar-cane smell" / "'Kaju gédang'"

3648   *T. pteropodum* Philippine Is. / M. D. Sulit 2636 / 1948 / "Ashes of wood very itchy when come in contact with skin" / "'Tikiko'"

3649   *Cornutia grandifolia* var. *normalis* Panama / W. L. Stern et al. 56 / 1959 / "...crushed lvs. with rank aroma of mint"

3650   *C. grandifolia* Panama / H. von Wedel 2173 / 1941 / "'Musciallago' (bat)"

3651 *C. odorata* Ecuador / Y. Mexia 8497 / 1936 / "...leaves strongly and pleasantly aromatic. Leaf used in infusion for rheumatic pains" / "'Ulape'"

3652 *Premna maclurei* China / S. K. Lau 145 / 1932 / "Fruit edible" / "'Sai gau ek' (Lois)"

3653 *P. microphylla* China / W. T. Tsang 21056 / 1932 / "...fr....edible" / "'Tso Woh Tse'"

3654 *P. chevalieri* Indochina: Tonkin / W. T. Tsang 29526 / 1935 / "fr.... edible"

3655 *P. integrifolia* Malay Peninsula / C. X. Furtado 37416 / 1941 / "Leaves eaten by Tamils"

3656 *P. cumingiana* Philippine Is. / C. O. Frake 793 / 1958 / "Leaves soaked and drunk for general malaise" / "'gepalay' (Sub.)"

3657 *P. odorata* Philippine Is. / M. Adduru 37 / 1917 / "Used to prevent insects from attacking tobacco & corn" / "'Lassi' (Ib.)"

3658 *P. gaudichaudii* Caroline Is. / C. Y. Wong 192 / 1947 / "The women wear the flowers on their ears to attract their lover (Elbert says the plant is used in love magic)" / "'umakau'"

Caroline Is. / C. Y. Wong 361 / 1948 / "The roots are used in the making of medicine for fever with pains under the arms, legs and on the chest (fely ni gor), the roots are scraped and placed in a coconut sheath, then squeezed into a betel nut bowl; a coconut is used for the water and mixed with the expressed sap and the mixture is then drunk.... The leaves are used in the treatment of muscle aches, the leaf is warmed over an open flame and then placed on the affected spot. The birds eat the fruits" / "'ar'"

3659 *P. taitensis* var. *rimatarensis* Fiji Is. / Degener & Ordonez 13793 / 1940 / "Used to dye hair black" / "'Rauvula'"

3660 *Vitex gaumeri* Mexico / J. Bequaert 63 / 1929 / "Leaves used as horse fodder" / "'Dachnik' (Maya) (−green ear)"

3661 *V. pyramidata* Mexico / P. C. Standley 2952 / 1936 / "Fruit edible ..." / "'Hupari' (Mex.)"

3662 *V. mollis* Mexico / Y. Mexia 540 / 1926 / "Fr. black with pleasant acid taste. Much eaten. Bark used for fever remedy"

3663 *V. cymosa* Bolivia / J. Steinbach 6428 / 1924 / "...fruta...del tamaño de un olivo, a que se parece tambien alyo en el gusto. Es fruta muy sana" / "'Zarumá'"

3664 *V. rotundifolia* Japan / K. Uno s.n. / 1950 / "seeds medicine" / "'Hamagō'"

3665 *V. negundo* var. *bicolor* Samoa / Garber & Christophersen 611 / 1921 / "Leaves rubbed on body and head to cure fever" / "'namulega'"

Fiji Is. / Degener & Ordonez 13620 / 1940 / "...leaves made into a poultice and used for abrasions"

3666 *V. doniana* Nyasaland / L. J. Brass 17074 / 1946 / "...fruit...edible" / "'Nrindimbi' (Chinjanja)"

3667 *Gmelina papuana* New Guinea / L. J. Brass 695 / 25 / "Fruit...eaten by cassowary"

3668 *Faradaya* indet. Solomon Is. / S. F. Kajewski 2543 / 31 / "This in common with other vines is used for gonorrhea, the bark is macerated with water, the resulting concoction being drunk" / "'Ala-ta-homa'"

3669 *Faradaya* indet. New Guinea / L. J. Brass 8069 / 1936 / "Fls. showy white, carnation scented"

3670 *Clerodendrum fragrans* W.I.: Dominica / W. H. & B. T. Hodge 3308 / 1940 / "Flowers are boiled for headache" / "'Moxella'"

3671 *C. cyrtophyllum* China / W. T. Tsang 24204 / 1934 / "...edible" / "'Yiu Tze Ts'oi'"

China / H. H. Chung 1715 / 1923 / "...whole plant with a nauseating smell"

3672 *C. japonicum* China / W. T. Tsang 21068 / 1932 / "...flower...ill smelling, edible, used as medicine" / "'Chau Sze Mool Lai'"

3673 *C. kwangtungense* China / W. T. Tsang 21581 / 1932 / "...fruit edible" / "'Pak Tsz Shue'"

3674 *C. trichotomum* China / H. H. Hu 558 / 1920 / "...leaves with offensive smell"

3675 *C. cochinchinensis* Indochina / Poilane 706 / 1919 / "Employé comme medicament pour les enfants" / "'Cây leo trăng. Ate'"

3676 *C. confusum* Solomon Is. / S. F. Kajewski 1925 / 30 / "The bark is stripped and the wet sappy part is applied by the natives to places which are sore" / "'Koru-kopu'"

Solomon Is. / S. F. Kajewski 2341 / 30 / "When the piccanninies have sores on their bodies and skin diseases the leaves are put in a bamboo with water and allowed to rot. They are then rubbed on the sores" / "'Kaka-fair'"

Solomon Is. / S. F. Kajewski 2502 / 31 / "The leaves are boiled in water and applied to sore legs"

3677 *C. inerme* Solomon Is. / S. F. Kajewski 2407 / 31 / "When a man has trouble with his eyes, including blindness, the leaves are placed in hot water and calico thrown over the head and dish, allowing the fumes to reach the head" / "'A-la-loi-alugi'"

3678 *C. intermedium* Philippine Is. / H. G. Gutiérrez 61-19 / 1961 / "Medicinal leaves with coconut oil applied plaster-like for headache" / "'Talinongay' (Ilk.)"

3679 *C. lanuginosum* Philippine Is. / C. O. Frake 588 / 1958 / "Medicine for splenomegaly" / "derunal (Sub.)"

Philippine Is. / P. Añonuevo 210 / 1950 / "Medicinal. Scrape the bark and apply to the forehead" / "Asni (Bil.)"

3680 *C. minabassae* Philippine Is. / C. O. Frake 499 / 1958 / "Leaves applied to stomachache" / "'papait-ulongan' (Sub.)"

3681 *C. glabrum* S. Africa: Natal / Watt & Brandwyk 1734 / 27 / ". . . used medicinally"

3682 *Congea* indet. Siam / native collector 403 (Royal For. Dept. 3810) / 1949 / "Roots used as laxative"

3683 *Verbenacea* indet. Colombia / Killip & Smith 16713 / 1927 / "(used for indigestion)" / "'Mejorana'"

LABIATAE

3684 *Ajuga decumbens* China / F. Hom A-625 / 31 / "used for drug" / "'Kai Na Ts'oi'"

3685 *Teucrium canadense* U.S.: Mo. / O. E. Lansing, Jr. 3186 / 1911 / "'Snake-weed'"

Mexico / H. E. Moore, Jr. 2940 / 1947 / "'toronjil'"

3686 *Teucrium* indet. Mexico / V. E. Grant 506 / 1940 / "Sweet aromatic odor of foliage" / "Crushed herbage to cool fever" / "'Verbena'"

3687 *T. bicolor* Chile / E. E. Gigoux s.n. / 1886 / "'Oreganillo'"

3688 *T. nudicaule* Chile / G. Geisse s.n. / 1886 / "'Pimientilla'"

3689 *T. japonicum* China / W. T. Tsang 27818 / 1937 / "fl. pink, ill-smelling"

3690 *T. stoloniferum* China / Y. W. Taam 2194 / 1941 / "flower pink, ill-smelling"

3691 *Teucrium* indet. China / C. K. Shang 42 / 1926 / "Medicine" / "'Chau Mui Ts'o'"

3692 *Tetraclea coulteri* U.S.: Tex. / R. McVaugh 8255 / 1947 / ". . . odor of plant strong, like that of burdock"

3693 *Isanthus brachiatus* U.S.: Ohio / Webb, Rood et al. 1569 / 1923 / "'False Pennyroyal'"

U.S.: Ohio / E. E. Laughlin 1026 / 1908 / "'False Pennyroyal'"

U.S.: Wis. / N. C. Fassett 17599 / 1935 / "(plants aromatic)"

3694 *Trichostema dichotomum* U.S.: Mass. / E. F. Fletcher s.n. / no year / "'Bastard Pennyroyal'"

3695 *T. lanceolatum* U.S.: Calif. / H. F. Copeland 1631 / 38 / "'Vinegar weed,' abundant and malodorous . . ."

U.S. or Canada / Cusick 1498 / 1887 / "Has a very offensive odor and is said to injure honey when worked upon by bees"

3696 *T. oblongum* U.S.: Calif. / L. Constance 2342 / 1938 / "strongly 'citronella-scented'..."

U.S.: Wash. / W. N. Suksdorf 34 / 1882 / "A plant with a disagreeable smell"

3697 *Rosmarinus officinalis* Colombia / Killip & Smith 16714 / 1927 / "(Used for rheumatism)" / "'Romero'"

Ecuador / J. N. Rose 23576 / 1918 / "'Romaro.' Used medicinally as a tea"

3698 *Haplostachys munroi* Hawaiian Is. / F. R. Fosberg 12532 / 1935 / "herbage aromatic"

3699 *Gomphostemma philippinarum* Philippine Is. / R. B. Fox 37 / 1948 / "A medicine for warts (?), BUTÍ in Tagalog, BUSÍ in Egóngot. Take the leaf of this plant and squeeze to obtain a juice. Mix this juice with a small portion of juice obtained from the citrus, KABUYAO. Then apply the mixture to the BUSÍ" / "'ANÁDJOP'" Egóngot

3700 *Scutellaria epilobiifolia* U.S.: Wis. / I. A. Lapham s.n. / no year / "'Mad-dog Skull-cap'"

3701 *S. lateriflora* U.S.: Mass. / E. F. Fletcher s.n. / no year / "'Mad-dog Skullcap'"

U.S.: Penn. / S. P. Sharples 209 / no year / "'Mad-dog Scull-cap'"

U.S.: Mich. / Mrs. C. C. Deam s.n. / 1901 / "'Mad-dog Skullcap'"

U.S.: Ohio / R. J. Webb 78 / 1895 / "'Mad-dog Skullcap'"

3702 *S. australis* U.S.: Mo. / J. R. Churchill s.n. / 1918 / "'Shepherd Nut'"

3703 *Scutellaria* indet. Ecuador / J. N. Rose 22960 / 1918 / "'Cardiaca'"

3704 *S. japonica* Japan / E. Elliott 147 / 46 / "Medicinal Use: Whole herb (decocted) is tonic; cures convulsion and stops diarrhoea" / "'Hiki-okoshi'"

3705 *Lavandula* indet. Turkey / E. K. Balls B2377 / 1935 / "Whole plant strongly camphor-scented"

3706 *Marrubium vulgare* Mexico / C. & E. Seler 1603 / 1895 / "'Amor seco'"

3707 *Agastache nepetoides* U.S.: Ohio / E. E. Laughlin 978 / 1907 / "'Catnip Giant-hyssop'"

3708 *A. urticifolia* U.S.: Utah / R. K. Vickery, Jr. 2452 / 1959 / "'Horse Mint'"

3709 *A. cana* Mexico / S. S. White 3486 / 1940 / "Plant with licorice odor"

3710 *A. mexicana* Mexico / G. B. Hinton 15030 / 39 / "Smells of anis. Medicinal"

3711 *A. rugosa* China / W. T. Tsang 25314 / 1935 / "...fr. edible"

3712 *Lophanthus rugosus* China / Steward, Chiao & Cheo 715 / 1935 / "Aromatic herb" / "Used for yeast cakes and Chinese medicine"

China / F. A. McClure 20577 / 1937 / "Crushed leaves and stems very aromatic; used as flavoring in food"

3713 *Cedronella canariensis* Africa / D. H. Linder 2670 / 1926 / "—aromatic"

3714 *Nepeta floccosa* Kashmir / Webster & Nasir 5896 / 1955 / "...foliage aromatic..."

India / W. Koelz 9589 / 1936 / "Plant scented"

3715 *N. glutinosa* India / W. Koelz 9557 / 1936 / "Plants scented"

3716 *N. salviaefolia* Kashmir / Webster & Nasir 5917 / 1955 / "...evil-smelling"

3717 *Nepeta* indet. Kashmir / R. R. Stewart 18833 / 1939 / "...very aromatic"

3718 *Glechoma hederacea* var. *micrantha* U.S.: Ky. / Blumer & Gutermeth 215 / 33 / "'Cat Mint'"

3719 *G. hederacea* China / Lingnan Herb. 254.1249 / 1924 / "Used in Chinese medicine" / "'Tau kwat siu'... 'Go thru bone dissolve'; 'Un ti fung'... 'Garden ground wind'"

3720 *Prunella vulgaris* U.S.: Ala. / L. V. Porter s.n. / 1937 / "'Heal-all'"

China / T. T. Yü 919 / 1932 / "medicine"

China / K. P. To 2854 / 1919 / "medicinal value"

3721 *P. vulgaris* var. *lanceolata* U.S. or Canada / G. Gilbert s.n. / 94 / "'Self-Heal'"

U.S.: Mass. / E. F. Fletcher s.n. / 1909 / "'Heal-All'"

U.S.: Penn. / S. P. Sharples 208 / 1858–64 / "'Heal All'"

U.S.: Ohio / R. J. Webb 13 / 1895 / "'Heal-all'"

U.S.: Wis. / I. A. Lapham s.n. / no year / "'Heal-All'"

3722 *P. vulgaris* var. *lanceolata* forma *iodocalyx* U.S.: N.J. / Halstead 67 / 1891 / "'Heal-all,' 'Self-heal,' 'Blue-curls'"

3723 *Leonotis nepetaefolia* W.I.: Dominica / W. H. & B. T. Hodge 3401 / 1940 / "'Gros tête': tea from leaves used for fever"

W.I.: Dominica / W. H. & B. T. Hodge 2949 / 1940 / "brew of whole plant used for colds & fever" / "'gros tête'"

Bermuda Is. / F. S. Collins 271 / 1913 / "'sea-eggplant' i.e. sea-urchin plant"

3724 *Paraphlomis rugosa* China / F. A. McClure 1942 / 1926 / "'Paat kok fa'... (Eight cornered flower or Anise flower)"

3725 *Leucas zeylanica* China / W. T. Tsang 21582 / 1932 / "...used as medicine"

3726 *Galeopsis tetrahit* Canada: Ontario / M. W. Chepcsiuk s.n. / no year / "'Hemp nettle'"

3727   *G. tetrahit* var. *bifida* U.S.: Me. / Gray 421 / no year / "'Hemp Net-tle'"

U.S.: Ohio / R. J. Webb 66 / 1895 / "'Common Hemp-Nettle'"

3728   *Lamium amplexicaule* U.S.: Mass. / E. F. Fletcher s.n. / no year / "'Dead Nettle'"

U.S.: N.J. / Halstead 68 / 1891 / "'Dead Nettle,' 'Hen-bit'"

3729   *L. maculatum* U.S.: Ohio / R. J. Webb 610 / 1903 / "'Spotted Dead-Nettle'"

3730   *Leonurus* indet. Nicaragua / V. Grant 7285 / 41 / "Herbage pungent to scent"

3731   *L. sibiricus* China / no collector 12586 or 12454 / 1924 / "'Red flowered Artemisia,' 'Pregnant old Artemisia'"

China / W. T. Tsang 21523 / 1932 / "...leaves used as medicine; roots can be cooked with pork for food"

3732   *Stachys palustris* var. *homotricha* U.S.: Ohio / Webb & Rood 1596 / 1924 / "'Woundwort'"

3733   *S. hyssopifolia* U.S.: Md. / C. C. Plitt 854 / 1904 / "'Hyssop-leaved Hedge Nettle'"

3734   *S. tenuifolia* U.S.: Penn. / S. P. Sharples 211 / 1858–64 / "'Hedge Nettle'"

3735   *S. tenuifolia* var. *hispida* U.S.: Mass. / G. Gilbert s.n. / 1893 / "'Hedge Nettle'"

3736   *S. tenuifolia* var. *platyphylla* U.S.: N.J. / O. H. Brown 10716 / 1918 / "'Rough Hedge-Nettle'"

3737   *S. floridana* U.S.: Ga. / W. H. Duncan 11034 / 1950 / "...rhizomes white, fleshy, edible"

3738   *S. grandidentata* Chile / D. Bertero 669 / 1828 / "Vulgò *Yerba saneta* aus S. *Maria*"

Chile / Morrison & Wagenknecht 17144 / 1939 / "...very malodorous"

3739   *S. lanata* Peru / R. D. Metcalf 30264 / 1942 / "herb, with a pleasant odor like lavender"

3740   *Stachys* indet. China / W. T. Tsang 21912 / 1933 / "Flower...ill-smelling"

3741   *Stachys* indet. India: Manipur / F. Kingdon-Ward 17010 / 1948 / "Whole plant scented"

3742   *S. officinalis* Denmark / N. C. Seringe s.n. or 1873 / 1873? / "...employée en médecine comme nervine, tonique etc...."

3743   *Salvia longifolia* U.S.: Calif. / G. G. C——s.n. / 1884 / "'Black Sage'"

3744   *S. reflexa* U.S.: Tex. / V. L. Cory 50322 / 1945 / "'Lanceleaf Sage'"

U.S.: Utah / F. W. Gould 1374 / 1941 / "Strongly scented weed growing abundantly in farmyard"

Mexico / R. M. Stewart 1574 / 1941 / "'Chilladora'"

Mexico / E. Palmer 327 / 1896 / "'Chia'"

Mexico / E. Palmer 446 / 1896 / "'White Chia': seeds form a cooling drink"

Australia: Queensland / C. T. White s.n. / 1930 / "Regarded locally as very poisonous to stock"

3745 *S. azurea* subsp. *pitcheri* U.S.: Tex. / V. L. Cory 50296 / 1945 / "'Pitcher sage'"

3746 *S. polystachya* U.S.: Calif. / G. G. Coming? s.n. / 1884 / "'White Sage'"

3747 *S. microphylla* var. *wislizeni* U.S.: Ariz. / J. C. Blumer 1762 / 1907 / "Peppermint odor"

3748 *S. apiana* U.S.: Calif. / A. & R. Nelson 3283 / 1939 / "'White Sage'"

3749 *S. carduacea* U.S.: Calif. / A. A. Heller 7651 / 1905 / "A plant of economic importance among the Indians, known as Chia"

U.S.: Calif. / W. H. Brewer 1210 / 1860–62 / "'Greater Chea'"

3750 *S. columbariae* U.S.: Ariz. / H. Cutler 4734 / 1941 / "Strong skunk odor"

U.S.: Calif. / C. C. Kingman s.n. / 1911 / "'Chia,' or 'Sage'"

U.S.: Calif. / J. M. Bigelow or Douglas s.n. / 1800's / "'Chea' of the Indians. The seeds are put into water for a mucilaginous drink, which is very pleasant"

U.S.: Calif. / G. Thurber 607 / 1852 / "Called 'Chia' by the Mexicans who collect the seeds which stirred in water make a cooling mucilaginous drink—"

Mexico / C. Lumholtz 37 / 1910 / "'Chia'"

3751 *S. axillaris* Mexico / Y. Mexia 1514 / 1927 / "Aromatic. Used in cooking. 'Oregano'"

3752 *S. occidentalis* El Salvador / P. C. Standley 22071 / 1922 / "'Herba del Cangro'"

El Salvador / P. C. Standley 19412 / 1921–22 / "'Mozote de gallina' ... Decoction taken for dysentery"

El Salvador / P. C. Standley 21932 / 1922 / "'Trencilla negra'" / "'Mozote de pollo'" / "'Gonce de gallina'"

Puerto Rico / P. Sintenis 208 / 1884 / "'Indiera—Fria'"

3753 *S. ballotaeflora* Mexico / G. B. Hinton 16593 / 44 / "Aromatic"

3754 *S. subincisa* Mexico / E. Palmer 305 / 1896 / "'Chia'"

3755 *S. helianthemifolia* Mexico / Y. Mexia 1583 / 1927 / "Strong mint odor"

3756 *S. lavanduloides* Guatemala / V. Grant 678 / 1940 / "tea of herbage for malaria" / "'Verbena'"

3757 *S. potus* Mexico / E. Palmer 422 / 1904 / "Salvia Chia"

3758 *S. hispanica* Mexico / E. Langlassé 568 / 1898 / "Graines rafraichissantes" / "'Chia'"

Mexico / E. Palmer 967 / 1896 / "'Chia blanco'"

Mexico / U.S. Nat. Herb. Seed Coll. 17757 / no year? / "Plant grown from seed purchased by Dr. E. Palmer in Durango, Mexico"

Mexico / J. G. Schaffner 675 / 1876 / "'Chia'"

Mexico / A. Dugès s.n. / 1895 / "Vulgo: Chia. Cultivée en pleine terre"

Mexico / G. B. Hinton 12384 / 38 / "'Chia de Castilla'"

El Salvador / S. Calderón 527 / 1922 / "Cult." / "'Chan'"

El Salvador / P. C. Standley 22435 / 1922 / "'Chan'... Seeds used for frescos and remedy for liver diseases"

Guatemala / H. von Türckheim 8266 / 1901 / "'Chaaú'"

Cuba / C. F. Baker's Economic Plants of the World 21 / 1907 / "Raised from seeds purchased in the open market at Grenada, Nicaragua, where it was for sale in large quantities under the name of Chia. The seeds placed in water, give off a large amount of clear jelly, which is used flavored with fruit juices to form an agreeable drink. But the seeds should be investigated as to other possible uses, since very large crops can be readily produced"

3759 *S. coccinea* Mexico / V. Grant 518 / 1940 / "'Mito'"

Cuba / R. A. Howard 6641 / 1941 / "Cueba de la macha"

3760 *S. palealis* Mexico / G. B. Hinton 14971 / 39 / "Sap has a strong fishy smell"

3761 *S. hyptoides* El Salvador / P. C. Standley 19541 / 1922 / "Used as poultice for tumors, etc." / "'Hierba de reuma'"

3762 *S. lasiocephala* Mexico / R. S. Ferris 5850 / 1925 / "'Chia del monte'"

3763 *S. amarissima* Mexico / A. Dugès s.n. / 1900 / "'Chia del Cerro'"

3764 *S. sphacelifolia* Mexico / H. S. Gentry 7295 / 1945 / "Acrid odor"

3765 *S. compacta* Mexico / Y. Mexia 8841 / 1937 / "'Chia'"

3766 *S. tiliaefolia* Mexico / J. Gregg 542 / 1848–49 / "'Chia del Campo'"

Mexico / R. M. Stewart 2206 / 1941 / "'Chilla'"

3767 *S. albiflora* Mexico / E. Langlassé 430 / 98 / "Aromatique"

3768 *S. longispicata* Mexico / J. G. Ortega 6498 / 1926 / "'Verbena'"

3769 *S. atropaenulata* Mexico / G. B. Hinton 15413 / 39 / "Stem sticky, fragrant"

3770   *S. microphylla* Mexico / G. B. Hinton 11861 / 38 / "'Mirto de casa'"
Mexico / G. B. Hinton 5237 / 33 / "'Mirto'"

3771   *S. microphylla* var. *neurepia* Mexico / J. G. Schaffner 660 / 1876 /
"'Mirto'"

3772   *S. praestans* Mexico / G. B. Hinton 14765 / 39 / "Leaves very fragrant"

3773   *S. regla* Mexico / E. Palmer 726 / 1898 / "Market of San Luis Potosi"

3774   *S. sessei* Mexico / Y. Mexia 8955 / 1937 / "'Sangre de Toro'"

3775   *S. holwayi* Guatemala / A. F. Skutch 223 / 1933 / "'Santa Elena'"

3776   *S. purpurea* Mexico / Y. Mexia 8992 / 1937 / "...mint odor"
Honduras / Williams & Molina R. 14018 / 48 / "'Cucaracho'"

3777   *S. leucantha* Mexico / G. B. Hinton 4886 / 33 / "Medicinal" / "'Salvia Real'"

3778   *S. longistyla* Mexico / Moore, Jr., Hernandez X., & Porras H. 5597 /
1949 / "'Chuparosa'"

3779   *S. nervata* Guatemala / A. F. Skutch 784 / 1933 / "'Santa Elena'"

3780   *S. californica* Mexico / Carter & Kellogg 2955 / 1950 / "'Salvia Real'"

3781   *S. gravida* Mexico / G. B. Hinton 15677 / 40 / "High scented"

3782   *S. gesneraeflora* Mexico / G. B. Hinton 15502 / 40 / "Leaves highly
scented"

3783   *S. sclarea* Mexico? / A. Dugès 27? / 1895 / "Cultivée" / "'Maro'"

3784   *Salvia* indet. Mexico / T. C. & E. M. Frye 2951 / 1940 / "...juice
milky"

3785   *Salvia* indet. Mexico / A. Dugès s.n. / no year / "Odeur très forte—
Saveur assez agréable mais trop aromatique" / "Thé de Pázcuaro'"

3786   *S. serotina* Bermuda Is. / B. L. Robinson 83 / 1912 / "Bitter varnish-
like odor to foliage"
Bahama Is. / A. E. Wight 183 / 1905 / "Odor of mint..."

3787   *S. micrantha* Jamaica / W. T. Stearn 147 / 1956 / "'Chicken Weed'"

3788   *S. uncinata* Dominican Republic / R. A. & E. S. Howard 9155 / 1946 /
"...flowers...frequently visited by hummingbirds"

3789   *S. riparia* Venezuela / H. Pittier 9926 / 1921 / "Emits a pungent odor"

3790   *S. punctata* var. *glabra* Peru / Y. Mexia 04124 / 1935 / "...pungent
leaves"

3791   *S. pichinchensis* Ecuador / Y. Mexia 7700 / 1935 / "Leaves stuck on
forehead for headache" / "'Quinde-sungana mangapaque'"

3792   *S. macrophylla* Ecuador / W. H. Camp E-2214 / 1945 / "'Manga-paqui
azul'—infusion of lvs. taken for kidneys and liver"

3793 *S. palaefolia* Colombia / Killip & Smith 16619 / 1927 / "'Mastranso' (used to kill insects in animals)"

3794 *S. tubiflora* Peru / F. L. Herrera 1502 / 1917 / "'Ñuccchu'"
Peru / Worth & Morrison 15667 / 1938 / "Odor of sheep!"

3795 *S. oppositiflora* Peru / C. Vargas 191 / 1937 / "'Mucchu'"

3796 *S. praeclara* Bolivia / W. J. Eyerdam 25079 / 1939 / "A sweetly scented shrub. Leaves when crushed, smell like lemon grass"

3797 *S. haenkei* Peru / Mr. & Mrs. F. E. Hinkley 30 / 1920 / "medicinal" / "'Matico'"

3798 *S. calocalicina* Colombia / Killip & Smith 19390 / 1927 / "Remedy for snake bites" / "'Dominíco'"

3799 *S. sarmentosa* Peru / C. Vargas 9738 / 1939 / "Fetid odor"

3800 *S. rypara* Argentina / Eyerdam & Beetle 22722 / 1938 / "...strong aromatic odor, rather unpleasant"
Argentina / Eyerdam & Beetle 22574 / 1938 / "Very aromatic and agreeable mint odor"

3801 *S. stachydifolia* Bolivia / J. West 8288 / 1937 / "...foliage...strongly aromatic"

3802 *S. persicifolia* Brazil / Y. Mexia 4330 / 1930 / "Foliage with mint-like odor"

3803 *S. oxyphora* Bolivia / I. Steinbach 7504 / 1926 / "'Boca del sapo'"

3804 *Salvia* indet. Argentina / Meyer 33871 / 1940 / "planta aromática..."

3805 *Salvia* indet. Argentina / Meyer 33869 / 1940 / "'Verbena morada'"

3806 *Salvia* indet. Argentina / Bailetti 108 / 1918 / "'Matic'"

3807 *S. handelii* China / H. Wang 41672 / 1939 / "...root very fragrant, officinal"

3808 *S. miltiorrhiza* China / Cheo & Yen 27 / 1936 / "For medical use..."

3809 *S. scapiformis* China / To & Ts'ang Lingnan Univ. Herb. 12644 / 1924 / "Used for medicine" / "'Field celery'"

3810 *Salvia* indet. China / Tsiang & Wang 16475 / 1939 / "...rt. very fragrant, medicinal"

3811 *Salvia* indet. Afghanistan / E. Bacon 45 / 1939 / "Used as seasoning; also for stomachache"

3812 *S. farinacea* New Caledonia / M. Baumann 16047 / 1952 / "Plante cultivée à Nouméa"

3813 *Monarda punctata* var. *lasiodonta* U.S.: Tex. / A. Traverse 109 / 1956 / "Strong 'horsemint' odor"

3814 *M. austromontana* Mexico / S. S. White 3341 / 1940 / "'Orégano'"

Mexico / H. S. Gentry 2697 / 1936 / "Used for seasoning foods" / "'Orégano'"

Mexico / H. S. Gentry 1753 / 1935 / "Used as seasoning in meats. Medic. A decoction made from herbage for unsettled stomach" / "'Oregano'"

3815 *M. citriodora* Mexico / R. M. Stewart 1602 / 1941 / "'Mariquita'"

3816 *Sphacele tenuiflora* Peru / F. W. Pennell 13450 / 1925 / ". . . herb, sweet-scented"

3817 *Lepechinia nelsonii* Mexico / Y. Mexia 9006 / 1937 / ". . . pungent odor"

Mexico / G. B. Hinton 15399 / 39 / "Smells like pepper"

Mexico / G. B. Hinton 14961 / 39 / "Leaves highly scented"

Mexico / G. B. Hinton 8970 / 36 / "Bad smelling"

3818 *L. floribunda* Bolivia / W. J. Eyerdam 25033 / 1939 / "Strong, rather unpleasant aromatic odor"

3819 *L. meyeni* Peru / R. D. Metcalf 30265 / 1942 / ". . . delightfully pungent odor . . . Indians make a tea of the leaves for stomach pains"

3820 *Ziziphora clinopodioides* Siberia / F. N. Meyer 911 / 1911 / ". . . not touched by cattle and horses"

3821 *Hedeoma pulegioides* U.S.: Penn. / F. R. Fosberg 15977 / 1938 / "Very strongly aromatic . . ."

3822 *H. nanum* U.S.: Tex. / R. McVaugh 7763 / 1947 / "Odor aromatic"

Mexico / E. Palmer 702 / 1896 / "with odor of Thyme"

Mexico / R. M. Stewart 245 / 1940 / "'Mejoranillo'"

Mexico / W. P. Hewitt 185 / 1947 / "tea of plant is taken for stomach-ache" / "sagey aroma" / "'Mejorana del Campo'"

3823 *H. drummondii* U.S.: Texas-Mexico region / Berlandier 2544 / 1834 / "Vulgo *Toronjil*"

U.S.: Ariz. / Gould & Pultz 3166 / 1945 / "herbage strongly scented"

Mexico / C. H. Muller 2995 / 1939 / "'Poleo'"

Mexico / T. C. & E. M. Frye 2425 / 1939 / "plant with mint-like smell"

Mexico / L. C. Ervendberg 366 / 1858 / ". . . smells and tastes and serves as Peppermint"

3824 *H. floribunda* Mexico / H. S. Gentry 1919 / 1935 / "Used as seasoning for meats. Medic. Decoction made of herbage and drunk for indigestion, etc." / "'Orégano,' Mex. 'Mapá,' W."

3825 *H. glabrescens* Mexico / C. H. Muller 3255 / 1939 / "Used locally as a seasoning for meat and frijoles" / "'Orégano'"

Mexico / G. B. Hinton 16523 / 44 / " . . . pungent odor . . . Condiment sold in the markets for three pesos kilo" / " 'Oregano' "

3826 *H. patrinum* Mexico / Stanford, Retherford & Northcraft 116 / 1941 / ". . . strong menthol-like taste to plant"

3827 *H. piperitum* Mexico / G. B. Hinton 4925 / 33 / "medicinal" / " 'Tabaquillo' "
Mexico / G. B. Hinton 2095 / 32 / "In tea for stomach ache, sold for this purpose" / " 'Tabaquillo' "

3828 *Hedeoma* indet. Mexico / I. M. Johnston 9212 / 1941 / " 'Oregano' "

3829 *Hedeoma* indet. Mexico / R. M. Stewart 1591 / 1941 / " 'Oregano' "

3830 *Hedeoma* indet. Mexico / R. M. Stewart 1597 / 1941 / " 'Polio' "

3831 *Hedeoma* indet. Mexico / H. S. Gentry 6284 / 1941 / "Infused for tea as a tonic potion" / " 'Oregano' "

3832 *Hedeoma* indet. Mexico / W. C. Leavenworth 80 / 1940 / "Very aromatic herb"

3833 *Hedeoma* indet. Mexico / G. B. Hinton 16508 / 44 / "Tea; and a good one" / " 'Menta' "

3834 *H. mandoniana* Peru / F. W. Pennell 13434 / 1925 / ". . . odor of pennyroyal"

3835 *Poliomintha incana* U.S.: N.M. / W. A. Archer 7322 / 1938 / ". . . foliage aromatic"

3836 *Satureia rigida* U.S.: Fla. / E. H. Butts s.n. / 1943 / "Plant aromatic"

3837 *S. douglasii* U.S.: Calif. / G. Lemmon 9 / 1872 / " 'Yerba buena' of Spanish"

3838 *S. calamintha* U.S.: Va. / A. Ryland 3366 / 1934 / " 'Basil thyme' "
Bermuda Is. / B. L. Robinson 5 / 1912 / "With strong but agreeable mint-like odor"

3839 *S. arkansana* Canada: Ontario / Pease & Ogden 24825 / 1935 / "Odor of pennyroyal"

3840 *S. ashei* U.S.: Fla. / L. J. Brass 14625 / 1945 / " 'Pennyroyal' "
U.S.: Fla. / L. J. Brass 14503 / 1945 / ". . . very aromatic"

3841 *S. vulgaris* var. *neogaea* U.S.: Penn. / S. P. Sharples 202 / 1858–64 / " 'Wild Basil' "

3842 *S. georgianum* U.S.: Ga. / R. McVaugh 8621 / 1947 / ". . . odor pungently aromatic"

3843 *S. brownei* Guatemala / Türckheim 373 / 1879 / "Stark nach Pfeffernüsse riechend"
Ecuador / J. N. Rose 22627 / 1918 / "Strong odor of pennyroyal"
Argentina / T. Meyer 2138 / 1936 / ". . . muy aromática"
Venezuela / H. Pittier 9262 / 1921 / ". . . whole plant very aromatic"

3844 *S. laevigata* Mexico / G. B. Hinton 8930 / 36 / "Medicinal tea" / "'Te del monte'"

Mexico / H. S. Gentry 7205 / 1945 / "Decocted for tea for colic" / "'Poleo'"

Mexico / H. S. Gentry 5887 / 1940 / "Highly esteemed by natives for tea" / "'Poleo'"

3845 *S. macrostema* Mexico / W. H. Camp 2606 / 1937 / "Used in native alcoholic drinks"

Mexico / G. B. Hinton 2452 / 32 / "Used as tea" / "'Te del Monte'"

Mexico / G. B. Hinton 523 / 32 / "Very strong perfume"

Mexico / E. Palmer 2774 / 1891 / "Purchased in the market of Culican, Mexico" / "'Poleo'—a strong aromatic plant when made into an infusion is used by females & children for ordinary stomach troubles"

3846 *Satureia* indet. Mexico / J. G. Schaffner 74 / 1876 / "'Oregano del monte'"

3847 *S. viminea* Jamaica / G. R. Proctor 20810 / 1960 / "...aromatic..."

3848 *S. brevicalyx* Peru / J. West 3725 / 1935–36 / "'Oregano de los Inkas'"

3849 *S. brownei* subsp. *eubrownei* Colombia / F. W. Pennell 10777 / 1922 / "...odor of *Hedeoma*"

3850 *S. boliviana* Argentina / S. Venturi 6133 / 1928 / "'Muña'"

Bolivia / J. Steinbach 5845 / 1921 / "Es planta aromatica"

3851 *S. darwinii* Argentina / Eyerdam, Beetle & Grondona 23981 / 1938 / "...fragrant"

3852 *S. elliptica* Peru / F. W. Pennell 14734 / 1925 / "...with strong aromatic odor"

3853 *S. gilliesii* Chile / P. Aravena 33360 / 1942 / "...strong odor, attractive to insects"

3854 *S. nubigena* var. *glabrescens* Colombia / F. W. Pennell 6930 / 1922 / "...odor of *Hedeoma*"

3855 *S. obovata* Chile / D. Bertero 291 / 1828 / "Vulgó Oreganillo"

3856 *S. oligantha* Argentina / L. R. Parodi 8005 / 1927 / "Medicinal(?)" / "'Muña-muña'"

3857 *S. parvifolia* Argentina / S. Venturi 4588 / 1926 / "'Muña-muña'"

Argentina / Y. Mexia 04394 / 1936 / "...strongly pungent foliage" / "'Oreganilla'"

3858 *S. tomentosa* Peru / J. West 8092 / 1936 / "Used medicinally by native population, tea of leaves to cure colds" / "'Panisara'"

3859 *Satureia* indet. Argentina / Meyer 33868 / 1940 / "'Muña'"

3860 *Satureia* indet. Argentina / Meyer 33872 / 1940 / "'Muña-Muña'"

3861 *Satureia* indet. Argentina / Meyer & Bianchi 33866 / 1940 / "'Muña-muña'"

3862 *S. simensis* Belgian Congo / D. H. Linder 2117 / 1927 / "Young leaves aromatic, mint-like in odor when crushed"

3863 *S. spicigera* Asia Minor / E. K. Balls 2023 / 34 / "Leaves . . . strongly aromatic"

3864 *Calamintha microstemon* Mexico / E. K. Balls 4055 / 1938 / "Whole growth strongly aromatic . . . Used by the Indians for making a 'Tizane'"

3865 *Calamintha* sp. nov. Nyasaland / L. J. Brass 17200 / 1946 / ". . . pennyroyal-scented"

3866 *Gardoquia elegans* Ecuador / A. Rimbach 14 / no year / "Plant aromatic, peppermint like. Infusion can be taken as a kind of tea"

3867 *Micromeria brownei* Jamaica / W. T. Stearn 233 / 1956 / ". . . strongly aromatic ('mint-scented') when crushed. Pennyroyal"

3868 *Micromeria* indet. Nyasaland / L. J. Brass 17248 / 1946 / ". . . mint-scented"

3869 *Micromeria* indet. Nyasaland / L. J. Brass 16643 / 1946 / ". . . strongly penny royal-scented"

3870 *Micromeria* indet. Morocco / E. K. Balls B.2443 / 1936 / "Leaves . . . strongly scented" / "'Amizmiz'"

3871 *Kurzamra pulchella* Chile / I. M. Johnston 5954 / 1926 / ". . . herbage very redolent"

3872 *Conradina glabra* U.S.: Fla. / J. A. Lowell or Hb. Chapman 3620 / no year / "'Wild Rosemary'"

3873 *Pogogyne serpylloides* U.S.: Calif. / W. H. Brewer 1179 / 1860–62 / "Strong odor of Pennyroyal"

3874 *P. douglasii* subsp. *typica* U.S.: Calif. / L. Constance 2342 / 1938 / "strongly peppermint-scented"

3875 *Ceranthera linearifolia* U.S.: Ala. / C. Mohr 13 / no year / "highly fragrant"

3876 *Ceranthera* rel. U.S.: Fla. / L. J. Brass 14646 / 1945 / ". . . aromatic"

3877 *Ceranthera* rel. U.S.: Fla. / L. J. Brass 15604 / 1945 / ". . . aromatic"

3878 *Monardella villosa* U.S.: Calif. / G. T. Robbins 1245 / 1943 / ". . . strong mint odor"

3879 *M. douglasii* U.S.: Calif. / H. N. Bolander 2499 / 1866 / "Odor very strong"

3880 *M. lanceolata* U.S.: Calif. / C. C. Kingman s.n. / 1910 / "'Polero,' 'Pennyroyal'"

3881 *M. undulata* U.S.: Calif. / C. B. Wolf 2072 / 1934 / ". . . herbage very aromatic"

U.S.: Calif. / W. H. Brewer 421 / 1860–62 / "Strong odor of peppermint"

3882 *M. arizonica* U.S.: Ariz. / Kearney & Peebles 14220 / 1939 / "foliage odor of pennyroyal"

3883 *M. cinerea* U.S.: Calif. / J. A. Ewan 8312 / 1934 / "... pungently menthodorous"

3884 *Monardella* indet. Mexico / Y. Mexia 1804 / 1927 / "Mint odor"

3885 *Pycnanthemum setosum* U.S.: Ga. / J. S. Harper 262 / 1931 / "'Pennyroyal'"

3886 *P. tenuifolium* U.S.: Penn. / S. P. Sharples s.n. or 200 / 1858–64 / "'Virginian Thyme'"

U.S.: Va. / F. R. Fosberg 15429 / 38 / "Herbage with mint-like odor ..."

U.S.: Ky. / H. Bishop & Bot. Class 65 / 1932 / "'Basil,' 'Narrow-leaved Mountain Mint'"

3887 *P. virginianum* U.S.: R.I. / no collector s.n. / 1840 / "'Virginian Thyme'"

3888 *P. muticum* U.S.: Mass. / E. F. Fletcher s.n. / no year / "'Hairy Mountain Mint'" / "'Mountain Mint—Basil'"

U.S.: N.C. / R. L. Wilbur 3908 / 1955 / "Crushed foliage very pungent"

3889 *P. californicum* U.S.: Calif. / J. G. Lemmon 59 / 187– / "'Wild mint'"

3890 *P. incanum* U.S.: Md. / C. C. Plitt 846 / 1904 / "'Hoary Mountain Mint'"

3891 *P. loomisii* U.S.: N.C. / Channell & Rock 28 / 1956 / "... herbage aromatic"

3892 *Thymus serpyllum* China / Roerich Exped. 37.III / 1934 / "Used medicinally"

China / C. Y. Chiao 2595 / 1930 / "Ls. & st. aromatic"

Kashmir / Webster & Nasir 6140 / 1955 / "aromatic"

3893 *T. pallidus* var. *eriodontus* Morocco / E. K. Balls B2449 / 1936 / "Strongly aromatic"

3894 *T. satureoides* Morocco / E. K. Balls B2480 / 1936 / "Strongly aromatic"

3895 *Bystropogon glabrescens* Peru / F. L. Herrera 1500 / 1927 / "'Guñuca,' 'Guñu-muña'"

3896 *B. mollis* Ecuador / W. H. Camp E-4515 / 1945 / "Plants with heavy and somewhat unpleasant odor"

3897 *B. spicata* Peru / F. W. Pennell 14347 / 1925 / "... aromatic odor"

3898   *B. verticillata* Argentina / A. Burkart 7467 / 1935 / "'Peperina'"

3899   *Bystropogon* indet. Argentina / Budin 37716 / 1909 / "'Muña'"

3900   *Bystropogon* indet. Argentina / Meyer & Bianchi 33873 / 1940 / "Planta aromática"

3901   *B. origanifolius* Canary Is. / O. Burchard 87 / 1904 / "'poleo de monte'"

3902   *B. canariensis* Canary Is. / D. H. Linder 2680 / 1926 / "...aromatic odor"

      Canary Is. / O. Burchard 138 / 1904 / "'Poleo de monte'"

3903   *Minthostachys glabrescens* Peru / Stork & Horton 10906 / 1939 / "Strong odor like mint"

3904   *M. setosa* Bolivia / J. Steinbach 9812 / 1929 / "...aromatisch"

3905   *M. spicata* Peru / R. D. Metcalf 30285 / 1942 / "...with a distinct odor"

3906   *Cunila longiflora* Mexico / G. B. Hinton 11707 / 37 / "'Toronjil'"

3907   *C. polyantha* Mexico / G. B. Hinton 14866 / 39 / "Leaves have a strong sweet odor, and are used to make tea"

3908   *C. lythrifolia* Mexico / Y. Mexia 1589 / 1927 / "'Poleo,' Remedial"

      Mexico / G. B. Hinton 4885 / 33 / "Concoction taken for fright"

      Mexico / G. B. Hinton 2448 / 32 / "Sweet smelling"

3909   *C. pycnantha* Mexico / Y. Mexia 8969 / 1937 / "Used as tea" / "'Poleo'"

3910   *Lycopus americanus* U.S.: Mass. / E. F. Fletcher s.n. / no year / "'Cutleaved Water Hoarhound'"

3911   *L. rubellus* U.S.: Tex. / V. L. Cory 49878 / 1945 / "'stalked water-hoarhound'"

3912   *L. virginicus* U.S.: Ala. / L. V. Porter s.n. / 1939 / "'Water Horehound'"

3913   *Mentha arvensis* Mexico / G. B. Hinton 804 / 32 / "Boiled it is good for colds" / "'Polleo'"

      Mexico / G. B. Hinton 4206 / 33 / "Heated with vinegar rubbed on forehead for colds" / "'Poleo'"

      Philippine Is. / Merrill 840 / 1915 / "...introduced" / "It is universally known in the Philippines under its Spanish name, *yerba buena*"

3914   *M. viridis* Mexico / A. Dugès s.n. / 1893 / "'Te de yerbabuena'"

3915   *M. citrata* Santo Domingo / E. L. Ekman H11796 / 1929 / "'Yerba buena'"

3916   *M. nemorosa* Haiti / E. C. & G. M. Leonard 14433 / 1929 / "Used for tea by Haitians"

3917 *M. sylvestris* Afghanistan / E. Bacon 30 / 1939 / "Eaten sprinkled on *mast* (curds)"

3918 *M. grandiflora* Australia: Queensland / C. T. White 11362 / 1940 / ". . . very strong pungent pennyroyal smell when crushed"

3919 *Micheliella anisata* U.S.: Ala. / Bigelow s.n. / no year / "Called Citronella by the French and is used as tea by them, wherefore the Americans call it French tea. It has a pleasant odor"

3920 *Perilla ocymoides* China / F. A. McClure 23 / 1921 / "Whole plant steeped in water to make wash for feet that have 'itch' from wading in dirty water" / "'Chue Mo Soh'"

3921 *P. nankinensis* China / K. Ling 2359 / 26 / "Medicine"
China / W. T. Tsang 23058 / 1933 / ". . . fruit smoky, edible"
China / W. T. Tsang 21764 / 1932 / ". . . fl. . . . edible"
Japan / E. Elliott 7 / 1946 / "Leaves and seeds for culinary use, generally eaten raw or salted" / "Food Use: leaves commonly salted with Mume-plum and eaten. Inflorescence is eaten raw with fish" / "Medicinal Use: It removes phlegm and increases appetite" / "'Shiso'"

3922 *P. nankinensis* var. *laciniata* China / W. T. Tsang 25906 / 1935 / ". . . fr. edible"

3923 *Orthodon chinensis* China / no collector 12913 / 1924 / "'T'in heung lo' (Heavenly incense bowl)"
China / H. Fung 21080 / 1937 / ". . . aromatic leaves" / ". . . purchased in Flowery Bridge market"

3924 *O. grosseserratum* Japan / S. Suzuki 119 / 1937 / "Leaves . . . odorous of Thymol"

3925 *Elsholtzia cristata* China / Steward, Chiao & Cheo 476 / 1931 / ". . . in temple yard"
India / K. C. Sahni 21951 / 1955 / ". . . aromatic shrub"

3926 *E. patrini* Japan / K. Ichikawa 155 / 1924 / "with bad odor"

3927 *E. eriostachya* Nepal / Polunin, Sykes & Williams 161 / 1952 / "Leaves aromatic, mint like"

3928 *Pogostemon championi* China / no collector 12276 / 1924 / "'Pok ho' (Mint or Peppermint)"

3929 *P. glaber* China / F. A. McClure 1876 / 1921 / "Lois use—gather leaves beat & grind out the water and rub on mosquito bites to relieve the pain and itching"

3930 *P. heyneanus* Philippine Is. / C. O. Frake 767 / 1958 / "Leaves applied to wounds" / "'malbaka'" Sub.

3931 *Dysophylla sampsoni* China / Ng. C. Lam 3379 / 1918 / "to poison the flea"

3932   *D. auricularia* China / K. P. To 6250 / 1920 / "Tea made from it used for washing wounds"

3933   *Hyptis asperifolia* Honduras / Williams & Molina R. 13992 / 48 / "'Flor de Lis'"

3934   *H. atrorubens* Panama / Bro. Heriberto 184 / 1921 / "Tumba-Muerto"

3935   *H. albida* Mexico / Y. Mexia 1776 / 1927 / "Greatly loved by bees" / "'Salvia Blanca'"

      Mexico / Leavenworth & Hoogstraal 1598 / 1941 / "Aromatic..."

3936   *H. capitata* Honduras / A. Molina R. 855 / 1948 / "'Boleta'"

      W.I.: Dominica / W. H. & B. T. Hodge 3380 / 1940 / "'petite bombe' (or 'bom'?)"

      Virgin Is. / Eggers s.n. / 1876 / "'Wild hops'"

3937   *H. emoryi* Baja Calif. / H. S. Gentry 4101 / 1938 / "... much visited by bees and flies" / "'Salvia'"

      Mexico / H. S. Gentry 1156 / 1934 / "Flowers put in ear to relieve ear-ache" / "'Salvia,' Mex."

3938   *H. mutabilis* Mexico / Y. Mexia 8993 / 1937 / "mint odor"

      Mexico / Y. Mexia 1296 / 1926 / "'Chia cimarona'"

      Honduras / A. Molina R. 2644 / 1949 / "'Huele hiede'"

      El Salvador / P. C. Standley 19151 / 1922 / "'Chan montés'"

      El Salvador / P. C. Standley 19391 / 1921–22 / "'Orégano montes'"

      Honduras / A. Molina R. 517 / 1947 / "'Cham.'"

      Colombia / F. W. Pennell 5174 / 1922 / "Plant odorous"

      Bolivia / J. Steinbach 6093 / 1924 / "Fuertamente aromatica"

      Argentina / P. Jorgensen 2241 / 17? / "... olor fuerte"

3939   *H. oblongifolia* Honduras / A. Molina R. 2648 / 1949 / "'Alcamfor'"

3940   *H. polystachya* Mexico / Y. Mexia 1396 / 1927 / "Strong mint aroma"

3941   *H. stellulata* Mexico / E. Langlassé 690 / 98 / "Pl. aromatique"

3942   *H. suaveolens* Mexico / G. B. Hinton 4998 / 33 / "medicinal" / "'Chia'"

      Mexico / R. S. Ferris 5392 / 1925 / "plant with strong odor"

      Mexico / H. S. Gentry 1842 / 1935 / "Cultivated by Warihios; seeds put in water sweetened and drunk" / "'Konivari,' W."

      Mexico / J. G. Ortega 6.393 / 1926 / "'Chan.'"

      Mexico / J. G. Ortega 6.139 / 1926 / "'Verbena'"

      Mexico / U.S. Nat. Herb. Seed Coll. 18810 / no year / Sheet F: "Plants grown from seed purchased by Dr. E. Palmer, Colima, Mexico ...Called *Chan negro*" / Sheet E: "... Called *Chan*"

Mexico / C. Conzatti 3659 / 1919 / "... muy aromatica"

El Salvador / P. C. Standley 20844 / 1922 / "'Chichiguaste'"

El Salvador / P. C. Standley 21052 / 1922 / "'Chichinguaste'"

Costa Rica / Tonduz 13655 / 1899 / "'Chian'"

Panama / I. M. Johnston 675 / 1945 / "Rank strong-smelling"

Mexico / E. Palmer 2005 / 1898 / "Procured at a village near Guadalajara. Seed was planted in greenhouse, Washington" / "'Chia'"

Mexico / A. Dugès s.n. / 1897 / "'Chia de Guadalajara'"

Jamaica / Proctor & Mullings 21976 / 1961 / "Pungently aromatic weed"

W.I.: Curaçao / Curran & Haman 197 / 1917 / "Medicinal—stomach"

Venezuela / H. Pittier 10714 / 1922 / "Odorous and aromatic"

Peru / H. E. Stork 11386 / 1939 / "'Alvaca cimarron'"

Bolivia / J. Steinbach 7033 / 1925 / "... fuerte aroma"

Bolivia / J. Steinbach 5574 / 1921 / "Se usa para aromatizar baños" / "'Matricaria'"

Burma / C. E. Parkinson 11287 or 11257 / 32 / "Leaves used as a condiment" / "'Pinsein'"

Burma / C. E. Parkinson 14062 / 32 / "L. strong smelling when crushed—"

Brit. New Guinea / L. J. Brass 6316 / 1936 / "Strong smelling herb"

Philippine Is. / E. D. Merrill 360 / 1902 / "Plant aromatic"

India or Pakistan: Bengal / C. E. Parkinson 3446 / '34 / "Leaves strong smelling. 'Wild mint'"

3943 *H. tephrodes* Baja Calif. / L. Constance 3168 / 1947 / "... strongly aromatic"

3944 *H. tomentosa* Mexico / E. Langlassé 988 / 1899 / "Aromatique"

3945 *H. verticillata* Brit. Honduras / W. A. Schipp 188 / 1929 / "The leaves crushed are used by caribs to drive —— lice out of hen coops and it seems to be very effective when so used"

El Salvador / P. C. Standley 21926 / 1922 / "Used in baths for rheumatism, insect bites & itch" / "'Verbena'"

Honduras / P. C. Standley 53164 / 1927–28 / "'Barrehorno'"

Panama / P. C. Standley 31230 / 1924 / "... formerly sold in commissaries for tea" / "'John Charles' (Jamaican)"

Cuba / R. A. Howard 6316 / 1941 / "Common aromatic plant"

Jamaica / W. Harris 12641 / 1917 / "'Johen Charles'"

3946 *H. schusteri* Haiti / G. R. Proctor 10722 / 1955 / "Aromatic shrub"

3947   *H. odorata* Peru / Y. Mexia 8148 / 1936 / "'Mapa-Rosa'"
Bolivia / I. Steinbach 7257 / 1925 / "'Matiquito'"

3948   *H. sidaefolia* Peru / Y. Mexia 04038 / 1935 / "...mint odor..."

3949   *H. kuntzeana* Bolivia / W. J. Eyerdam 25322 / 1939 / "Aromatic"

3950   *H. sinuata* Bolivia / J. Steinbach 7063 / 1925 / "'Boraja?'"

3951   *H. carpinifolia* Brazil / Y. Mexia 5860 / 1931 / "...strong mint odor. Used for tea for indigestion" / "'Pedresa'"
Brazil / Williams & Assis 6790 / 1945 / "Cook this and bathe in soup, is good for pains in joints"
Brazil / Y. Mexia 5541 / 1931 / "...mint odor..." / "'Cating da mulata'"

3952   *H. perplexa* Brazil / Y. Mexia 5796 / 1931 / "'Brinquín'"

3953   *H. brevipes* Paraguay / W. A. Archer 4789 / 1936 / "Remedy for diarrhea" / "'cabaracaá'"

3954   *H. decurrens* Philippine Is. / A. L. Zwickey 391 / 1938 / "...stomach-ache" / "'Albaka' (Lan.)"

3955   *Asterohyptis mociniana* El Salvador / P. C. Standley 19153 / 1921–22 / "'Verbena montés'"
Mexico / Y. Mexia 8738 / 1937 / "Used in infusion for bruises"

3956   *A. stellulata* Mexico / G. B. Hinton 13334 / 38 / "Smells like pepper"
Mexico / A. Dugès s.n. / 1895 / "'Yerba del ahito'"

3957   *Marsypianthes chamaedrys* Mexico / E. Langlassé 457 / 98 / "Aromatique"
W.I.: St. Lucia / G. R. Proctor 17629 / 1958 / "...aromatic herb"
Peru / Y. Mexia 6491 / 1932 / "mint odor..."
Fr. Guiana / W. E. Broadway 64 / 1921 / "Plant strong-smelling"

3958   *Plectranthus* indet. China / K. P. To 6251 / 1920 / "Soaked in wine and the wine used for sprains" / "'Tuk Hang Tsin Lai'"

3959   *P. coetsa* Burma / F. G. Dickason 7734 / 1937 / "Aromatic, leaves very bitter"

3960   *Plectranthus* indet. Java / van Steenis 12008 / '40 / "...very fragrant"

3961   *Plectranthus* indet. New Britain / A. Floyd 3509 / 54 / "The fragrant leaves are used in grass skirts for the women" / "'La Malalia' (W. Nakanai)"

3962   *P. forsteri* New Hebrides Is. / S. F. Kajewski 258 / 28 / "Used by natives for sores. Sap taken and mixed with salt water" / "'Oulairyang'"?

3963   *P. australis* AUSTRALIA: Hayman I. / C. T. White 10169 / 1934 / "...very strongly scented"

3964   *Coleus amboinicus* W.I.: Grenadines / R. A. Howard 10755 / 1950 / "'Wild thyme' . . . aromatic"

3965   *C. scutellarioides* Solomon Is. / S. F. Kajewski 2191 / 1930 / "The natives say that the plant is indigenous. The sap is applied to sores and cuts" / "'Meme'"

3966   *C. pumilus* Philippine Is. / R. B. Fox 45 / 1948 / "The stem of this weed is used for the shaft of arrows" / "'Biyaó' or 'Bilaó'" Egóngot

3967   *Coleus* indet. Philippine Is. / W. Beyer 2 / 1948 / "Economic Uses: to remove or pull out sharp sticks or other foreign materials in the skin" / "'Tumul'" Ifugao

3968   *Coleus* indet. Philippine Is. / A. L. Zwickey 706 / 1938 / ". . . heated lvs. applied to foot infections" / "'Paiar' (Lan.)"

3969   *C. barbatus* Uganda / Mrs. M. V. Loveridge 429 / 1939 / "Flower . . . with strong, stinging odor"

3970   *Mesona chinensis* China / W. T. Tsang 22965 / 1933 / ". . . flower . . . edible"

3971   *Mesona* indet. China / W. T. Tsang 28568 / 1938 / ". . . fr. edible"

3972   *Ocimum micranthum* U.S.: Fla. / Stern & Chambers 319 / 1958 / ". . . leaves highly aromatic"

     Mexico / G. Martínez-Calderón 341 / 1940 / "Se usa para Bañar a una madre recién salido de su estado" / "'Albaque Simarron'" / "'Amhág'" Chinantec

     El Salvador / P. C. Standley 19463 / 1921 / "Wads of lvs put in ears for earache" / "'Albahaca montés'"

     Santo Domingo / Pater Fuertes 1115 / 1911 / "'Albahaca'"

     W.I.: Grenada / G. R. Proctor 17209 / 1957 / "Aromatic"

     W.I.: Grenada / W. E. Broadway s.n. / 1904 / "'Balm'"

     W.I.: Dominica / W. H. & B. T. Hodge 3302 / 1940 / "Weed—tea used for sick people" / "'fou basin'"

     Puerto Rico / P. Sintenis 300 / 1884 / "'Albahaca cimarrona'"

     Jamaica / W. Harris 11903 / 1915 / "'Wild Basil'"

     Colombia / R. E. Schultes 3496 / 1942 / "Cult."

     Bolivia / J. Steinbach 6110 / 1924 / "Se ocupa para aromatizar los baños" / "'Alva-Aka'"

     Colombia / H. Schiefer 301 / 1945 / "Sold in local market"

3973   *O. basilicum* El Salvador / P. C. Standley 19609 / 1921–22 / "Put in ears to cure deafness" / "strong odor" / "'Albahaca de gallina'"

     W.I.: Dominica / W. H. Hodge 3200 / 1940 / ". . . foliage fragrant; made into tea"

     Colombia / Killip & Smith 14687 / 1926 / "'Vaca monte'"

China / W. T. Tsang 22986 / 1933 / "...flower...edible"

Philippine Is. / C. O. Frake 530 / 1958 / "Agric. ritual—Leaves put on fungous infections" / "'Baningsulug'" Sub.

3974   *O. suave* W.I.: St. Lucia / G. R. Proctor 18236 / 1958 / "Aromatic..." / "'Frond Bazin'"

Nyasaland / L. J. Brass 17062 / 1946 / "...fragrant"

3975   *Ocimum* indet. Paraguay / K. Fiebrig 5055 / 1908–09 / "aromatisch..."

3976   *Ocimum* indet. Bolivia / J. Steinbach 5294 / 1921 / "'Romerillo'"

3977   *O. canum* Burma / C. E. Parkinson 14959 / 32 / "near mosque" / "'Pin Sem'"

Solomon Is. / S. F. Kajewski 2195 / 1930 / "...growing in native gardens and usually cultivated by them" / "As the leaves of this plant have a very pleasing smell, it is put into the native armelts and worn at festival times. It should have good qualities as a herb" / "'Ketorlo'"

Brit. New Guinea / L. J. Brass 3754 / 1933 / "...plant citronella scented..."

Brit. New Guinea / L. J. Brass 3755 / 1933 / "...clove scented"

3978   *O. sanctum* Solomon Is. / L. J. Brass 2815 / 1932 / "Very aromatic. The natives are fond of the perfume and sometimes carry small sprigs of the plant in their hair or stuck through their plaited arm bands"

Philippine Is. / R. B. Fox 99 / 1950 / "rub flowers on head as perfume" / "'Róko-róko'" Tagb.

Philippine Is. / G. E. Edaño 1618 / 1949 / "An infusion of the roots is given to mothers for childbirth" / "'Saling-Kugon Ma,'" or 'Saling-Kugon'" Ma

Philippine Is. / H. C. Conklin 995 / 1957 / "Economic Uses: sābug" / "'Samirig'" or "'Saminig'" Han.

3979   *Ocimum* indet. Solomon Is. / S. F. Kajewski 1825 / 1930 / "This plant is semi-cultivated by the natives on account of its strong pungent smell" / "'Ket-tor-lo'"

3980   *O. gratissimum* Society Is. / Setchell & Parks 152 / 1922 / "Tall mint ... with strong mint odor... officinal among the Tahitians"

New Caledonia / M. Franc 2353 / 1929 / "'Basilie sauvage'"

3981   *Ocimum* indet. New Hebrides Is. / S. F. Kajewski 65 / 1928 / "Native Gardens"

3982   *Ocimum* indet. New Hebrides Is. / S. F. Kajewski 64 / 1928 / "Native garden..."

3983   *O. americanum* S. Rhodesia / H. J. Davison 52 / '24 / "Aromatic odour"

3984   ***O. minimum*** EUROPE / Mrs. J. E. Thayer s.n. / before 1916 / "Culti-vated in pot by an Italian woman"

3985   ***O. polystachyum*** France / J. Gay s.n. / 1815 / "Cult. Paris"

3986   ***Orthosiphon grandiflorus*** Sumatra / W. J. Lütjeharms 5032 / 1936 / "officineel"

3987   *Labiata* indet. Ecuador / J. N. Rose 22628 / 1918 / "'Yerba buena'"

3988   *Labiata* indet. Bolivia / J. Steinbach 5951 / 1921 / "planta muy aro-matica"

3989   *Labiata* indet. Bolivia / W. J. Eyerdam 25208 / 1939 / "Plant very malodorous . . ."

3990   *Labiata* indet. Argentina / S. Venturi 4741 / 1927 / "'Oreganillo'"

3991   *Labiata* indet. China / K. M. Feng 3221 / 1939 / "In cultivation"

3992   *Labiata* indet. China / R. C. Ching 21740 / 1939 / "In cultivation"

3993   *Labiata* indet. China / T. T. Yü 1489 / 1932 / "seeds for extracting oil"

3994   *Labiata* indet. China / F. T. Wang 20689 / 1930 / "on temple court yard ground"

3995   *Labiata* indet. China / R. C. Ching 21371 / 1939 / "In cultivation"

3996   *Labiata* indet. China / C. Y. Chiao 1228 / 1939 / "Herb growing in temple yard"

3997   *Labiata* indet. China / H. T. Tsai 54524 / 1933 / "common, for medical use"

3998   *Labiata* indet. China / W. T. Tsang 25940 / 1935 / ". . . fr. edible"

3999   *Labiata* indet. China / W. T. Tsang 25806 / 1935 / ". . . fr. edible"

4000   *Labiata* indet. China / W. T. Tsang 26068 / 1936 / ". . . ill-smelling"

4001   *Labiata* indet. Afghanistan / E. Bacon 94 / 1939 / "Used for fodder"

4002   *Labiata* indet. Burma / F. G. Dickason 7733 / 37 / "Aromatic; leaves bitter"

4003   *Labiata* indet. Burma / F. G. Dickason 7702 / 1937 / "Leaves aro-matic; bitter"

4004   *Labiata* indet. Java / van Steenis 7111 / '35 / "Leaf fragrant . . ."

4005   *Labiata* indet. New Guinea / L. J. Brass 27232 / 1956 / "aromatic . . ."

4006   *Labiata* indet. New Guinea / G. Angell 4173 / 51 / "Crushed leaves have a distinct aniseed smell"

4007   *Labiata* indet. New Guinea: Papua / L. J. Brass 21876 / 1953 / "Aro-matic . . ."

4008   *Labiata* indet. New Guinea: Papua / L. J. Brass 21747 / 1953 / ". . . aromatic"

4009 *Labiata* indet. Greece / A. S. P. (Pease) 9055 / 1906 / "Ruins of Tiryns"

4010 *Labiata* indet. Italy / A. S. Pease 8697 / 1905 / "Etruscan tombs"

4011 *Labiata* indet. Turkey / E. K. Balls B2402 / 1935 / "Whole plant strongly aromatic"

SOLANACEAE

4012 *Acnistus arborescens* W.I.: Dominica / W. H. & B. T. Hodge 1531 / 1940 / "medicinal:—remedy for fever & colds" / "'Batard sirio' (sirop?)"

4013 *A. cauliflorus* Argentina / F. C. Hoehne 23311 / 1928 / "'Fructa de sabiá'"

4014 *A. australis* Argentina / J. West 6182 / 1936 / "...juicy, pear flavored fruit...edible" / "'Sacha pera'"

4015 *Hebecladus bicolor* Peru / R. D. Metcalf 30270 / 1942 / "...fruit...eaten by natives"

4016 *Physalis leptophylla* Mexico / H. S. Gentry 4771 / 1939 / "Ripe fruits cooked and eaten" / "'Tomatillo'"

4017 *P. angulata* Japan / E. Elliott 71 17 / '46 / "Fruits are edible, leaves well boiled and eaten in times of scarcity. Medicinal use: Decocted fruit cures throat swelling, fever, and is diuretic. Whole plant is antidote for poison and good for cold"

Siam / Wichian 303 / 1946 / "...whole plant used; decoction used for relieving boils in the human rectum" / "'yong chi pong'"

4018 *P. pubescens* Peru / F. Woytkowski 6101 / 1961 / "...fruit...medicinal, contains the alkaloid fisaline" / "'Bolsa mullaca blanca'"

4019 *P. minima* Caroline Is. / C. C. Y. Wong 148 / 1947 / "The ripe fruits are eaten, when eaten in great quantity gives an intoxicating effect" / "'poowa'"

4020 *Saracha jaltomata* Mexico / H. S. Gentry 2528 / 1936 / "Fruit eaten by Warihios and others" / "'Tomatillo'"

4021 *S. procumbens* Mexico / C. Lumholtz s.n. / 1894 / "(fruit eaten)" / "'Turusi'"

4022 *Capsicum frutescens* Philippine Is. / C. O. Frake 672 / 1958 / "Spice, leaves applied to ulcers and headache" / "'Katumbal' (Sub.)"

4023 *Solanum amazonicum* Mexico / H. S. Gentry 1294 / 1935 / "Seeds used to curdle milk for cheese making" / "'Sacamantica' Mex. 'Parowisi poohaira' W."

4024 *S. bicolor* Mexico / H. S. Gentry 5459 / 1940 / "Fruits roasted & applied to sores & aches" / "'Guastomate' 'Berijena'"

Mexico / Ferris & Mexia 5178 / 1925 / "edible berries" / "'beren-jena'"

4025   *S. citrullifolium* Mexico / A. Miranda 105 / 1947 / "es buena para los granos"

4026   *S. douglasii* Mexico / A. Dugès 395 / 1891 / "Vulgo: Yerbamora"

El Salvador / P. C. Standley 22681 / 1922 / "Cooked & eaten along arenal" / "'Hierba mora'"

4027   *S. gracile* Mexico / P. C. Standley 1033 / 1934 / "Medic. for fevers. Fruit edible" / "'Chichikelite,' Mex. 'Manilochi,' W.   Mambia Mayo"

4028   *S. nodiflorum* El Salvador / P. C. Standley 19219 / 1921–22 / "Lvs. cooked & eaten like spinach" / "'Hierba mora'"

Mexico / E. Palmer 959 / 1896 / "...black berries eaten by children"

4029   *S. nodum* Guatemala / J. A. Steyermark 45208 / 1942 / "Remedy for kidney trouble" / "'sac-yol'"

Honduras / P. C. Standley 54361 / 1927–28 / "Leaves have strong offensive odor when crushed"

4030   *S. rostratum* Mexico / A. López 38 / 1948 / "'Boraja' es bueno para el baño"

4031   *S. umbellatum* El Salvador / P. C. Standley 22175 / 1922 / "...leaves with offensive odor" / "'Tapalayote'"

4032   *S. ciliatum* W.I.: St. Lucia / G. R. Proctor 17805 / 1958 / "'Poison Diable'"

4033   *S. torvum* Philippine Is. / C. Frake 398 / 1958 / "Antidote for poisoning—roots boiled and drunk" / "'Gebulnusa' (Sub.)"

Philippine Is. / M. D. Sulit 3109 / 1949 / "Decoction of roots given to women who gave birth to lessen flow of blood" / "'Bokul' Dialect Bukid."

4034   *S. asperrimum* Peru / Y. Mexia 6485 / 1932 / "...seeds used for skin spots; caustic" / "'Ayac mullaca' (bitter fruit)"

4035   *S. coerulescens* Peru / Stork & Horton 9898 / 1938 / "...stems & leaves malodorous, smelling bitter"

4036   *S. lignicaule* Peru / C. Vargas 9730 / 1939 / "'Atoc papa'"

4037   *S. phyllanthum* Peru / Mr. & Mrs. F. E. Hinkley 9 / 1920 / "Used for wounds" / "'Nuccho blanco' (male)"

4038   *S. asperolanatum* Brazil / Y. Mexia 5049 / 1930 / "...fruit and root used in tea for urinary disorders" / "'Jurubeba'"

4039   *S. lycocarpum* Brazil / L. O. Williams 5642 / 1945 / "Tea made from the flower for cough" / "'Fruta de loba'"

4040   *S. spectabile* Brazil / Y. Mexia 4435 / 1930 / "Said to be poisonous" / "'Joao bravo'"

4041   *S. swartzianum* Brazil / Y. Mexia 4210 / 1930 / "Leaves used in infusion as blood purifier" / "'Mercurinho'"

Brazil / Y. Mexia 5792 / 1931 / "Infusion of leaves used for toothache" / "'Borbaso'"

4042   *S. sisymbriifolium* Brazil / Y. Mexia 4375 / 1930 / "...fruit...edible, much like tomato in taste" / "'João bravo'"

4043   *S. coagulans* China / C. I. Lei 1406 / 1934 / "...fruit poisonous"

4044   *S. verbascifolium* China / W. T. Tsang 474 / 1928 / "...fl. white; medicine" / "'Ka In Ip Shue'"

Solomon Is. / S. F. Kajewski 2276 / 30 / "The plant has two uses, the leaves are rubbed on the skin as an antidote to the native superstitious poisons. 2. The leaves are macerated with water, the mouth being rinced out with the liquid, for sores in the mouth" / "'Toi-loi-uokoi'"

New Britain / Waterhouse 428 / '36 / "Leaf used in native medicine" / "'Auminat'"

4045   *S. rechingeri* Solomon Is. / S. F. Kajewski 2388 / 30 / "The bark is macerated and heated, being then applied to sore legs" / "'Nari'"

4046   *S. uporu* Tonga Is. / T. G. Yuncker 15584 / 1953 / "Leaves used to prepare native medicine" / "'Polo tonga'"

4047   *S. tetrathecum* Australia: Queensland / S. T. Blake 15860 / 1945 / "...plant eaten by stock"

4048   *Lycianthes violaefolium* Bolivia / H. H. Rusby 1875 / 1921 / "Fruit edible"

4049   *Cyphomandra betacea* Colombia / O. Haught 1617 / 1935 / "Whole plant rather fetid"

4050   *C. splendens* Peru / J. West 8044 / 1936 / "Fruit used locally and sold on Cuzco Market as 'tomate'...flavor somewhat tomato-like, but quite strongly aromatic....When cooked, seemed less pleasant to taste than when eaten raw" / "'Tomate'"

4051   *Jaborosa crispa* Bolivia / J. West 6472 / 1936 / "Used medicinally" / "'Tekke-tekke'"

4052   *Datura discolor* Mexico / P. C. Standley 1166 / 1934 / "Medic. Warihios smear the leaves with animal fat (Nasua rarica) and apply to afflicted parts"

4053   *Cestrum dasyanthum* Guatemala / J. A. Steyermark 51932 / 1942 / "...leaves with fetid odor" / "'chip-shi'"

4054   *C. lanatum* El Salvador / P. C. Standley 22670 / 1922 / "...fr. black, sweet. Lvs. ill scented, used to kill fleas and insects on chickens. Branches put in hens' nests to keep away vermin" / "'Polo hediondo'"

4055   *C. latifolium* Surinam / Stahel 152 / 1941 / "...used extensively as vegetable (as spinach) and sold on market"

4056   *Duboisia leichhardtii* Australia: Queensland / H. J. Lam 7670 / 1954 /
       "...fr. poisonous"

4057   *Brunfelsia latifolia* Peru / F. Woytkowski 6170 / 1961 / "...medicin-
       al; principally roo˜ for rheumatism" / "'Chiric-sanango'"

4058   *B. maritima* Peru / Y. Mexia 6444 / 1932 / "Root grated, boiled and
       taken for rheumatism; said to be powerful medicinally; dose followed
       by a chill" / "'Chiric-sanango'"

4059   *Solanacea* indet. Peru / Stork & Horton 9491 / 1938 / "Fr....interior
       edible...Monkeys said to eat fruit" / "'Sanongo'"

SCROPHULARIACEAE

4060   *Verbascum arianthum* Afghanistan / E. Bacon 80 / 1939 / "gusi
       xorak. Used for eye trouble: flowers boiled, liquid put on eyes"

4061   *V. speciosum* Rumania / E. Anderson 113 / 34 / "Used as fish poison
       by peasants"

4062   *Leucophyllum frutescens* Mexico / Bro. Abbon 31 / 1911 / "'Ceniza'
       Bueno par reumas?"

4063   *Calceolaria inamoena* Peru / Mr. & Mrs. F. E. Hinkley 24 / 1920 /
       "Local name: Matico del cerro. Used as a liniment"

4064   *Maurandia geniculata* Mexico / P. C. Standley 1283 / 1935 / "Vernac.
       Rastrillo, Mex. Tahewali, W. Medic. leaf rubbed on pimple or 'jillote'"

4065   *Russelia retrorsa* Mexico / J. West 3505 / 1935 / "Used as a febrifuge
       in form of tea" / "'Espinosillo'"

4066   *R. sarmentosa* Mexico / W. P. Hewitt 14 / 1945 / "Common name
       pinito. Economic uses: some say its tea is bitter & good for fevers,
       but it is not 'pinito.' Others say it is 'pinito' and has no medicinal
       properties"

4067   *Pentstemon speciosum* U.S.: Calif. / W. A. Dayton 448 / 1913 /
       "Common name: Beardtongue. Forage value: Limitedly grazed by
       sheep"

4068   *P. tenuifolius* Mexico / L. R. Stanford et al. 205 / 1941 / "heavily
       grazed by goats"

4069   *Sutera lychnidea* S. Africa / R. N. Parker 4383 / 1948 / "Strong clove
       scent at night"

4070   *Capraria biflora* Mexico / Loesener 3679 / 1903 / "los usan los Indi-
       tos para cocer elotes porque huela bien"
       W.I.: Curaçao / Curran & Haman 6 / 1917 / "Tantsji"
       W.I.: Trinidad / W. E. Broadway s.n. / 1932 / "'De the pays' A local
       medicinal plant"
       Virgin Is. / W. C. Fishlock 270 / 1919 / "Used as a tea"
       Santa Domingo / Pater Fuertes 688 / 1911 / "'Te de Santa Maria'
       vernac."

4071   *Scoparia dulcis* El Salvador / S. Calderón 1098 / 1922 / "'Culantrillo'"
Brit. Honduras / P. H. Gentle 4204 / 1942 / "'pine-ridge anise-seed'"
Mexico / J. G. Ortega 6425 / 1926 / "N. vulgar: Chile de pajaro"
Brit. Honduras / W. A. Schipp 827 / 1931 / "Known by the natives as aniseed"
El Salvador / P. C. Standley 19380 / 1921–22 / "'Culantrillo,' 'Culantro montes'... Used for brooms to kill fleas, hen lice, etc."
Jamaica / W. T. Stearn 511 / 1956 / "Dried leaves made into a tea for babies' gripe. 'Sweet broom'"
W.I.: Dominica / W. H. & B. T. Hodge 3325 / 1940 / "tea from leaves used for bath & cooling tea"
Paraguay / W. A. Archer 4924 / 1938 / "Sold by herb dealers in market, Asuncion. 'typychá caratú.' Insecticide. Used in bath to kill vermin"
Bolivia / J. Steinbach 6114 / 1924 / "Obs. Se usa la infusion de la planta contre contusiones i en baños"
Siam / G. den Hoed 130 / 1946 / "Vern. name: no njo"
Sumatra / R. S. Toroes 5605 / 1933 / "doehoet pira-pira"
Borneo / Purseglove & Shah 4424 / 55 / "Vernac. name: Ramput suntora"
Philippine Is. / H. C. Conklin 18664 / 1953 / "Common name: Saguhūy"
Philippine Is. / H. C. Conklin 18603 / 1953 / "Common name: Talisig sa lāwud"

4072   *Picrorhiza scrophulariaefolia* Sikkim / R. E. Cooper 56 / 1913 / "Used extract from root for febrifuge & stomach medicine"

4073   *Campylanthus salsoloides* Canary Is. / O. Burchard 142 / 1934 / "N. indigenum 'Varilla de Alicante'"

4074   *Escobedia linearis* Mexico / Dressler & Jones 249 / 1953 / "'Azucena,' used to kill fleas"

4075   *E. scabrifolia* Bolivia / J. Steinbach 5200 / 1920 / "n.v. Palillo. La raiz contiene una apreciada tintura para emplea en alimentos"

4076   *Melasma arvensis* Philippine Is. / W. Beyer 6849 / 1948 / "...Medicine for child's whitish tongue. Common name: Kala Dial. Ifugao"

4077   *Gerardia greggii* Mexico / L. R. Stanford et al. 139 / 1941 / "Heavily grazed by goats"

4078   *Striga lutea* China / F. Hom & G. E. G. 19367 / 31 / "Cantonese T'ong Shat Ue. Economic uses: Boiled with fish"

4079   *Castilleja pruinosa* U.S.: Calif. / W. A. Dayton 17485 / 1913 / "Common name Indian paintbrush. Forage value: Grazed moderately by goats & sheep"

4080 *C. patriotica* Mexico / P. C. Standley 2731 / 1936 / "Vern.: Yerba del Sapo. Medicinal, 'por mal de orin'"

4081 *Castilleja* indet. Mexico / R. M. Stewart 2795 / 1942 / "'Yerba de la Almorana'"

4082 *C. communis* Peru / F. E. Hinkley 55 / 1920 / "Local name: Violets. Economic uses: medicinal, (por mal interior)"

4083 *Siphonostegia chinensis* China / Y. W. Taam 17 / 1937 / "Chi-jou-ts'ai. Fruit edible"

4084 *Monochasma monantha* China / To & Groff 54 / 1919 / "Common name (Chinese) Pak Long Ki or Kom tsik ts'o. Economic use: Boiled with pork"

BIGNONIACEAE

4085 *Pseudocalymma macrocarpum* Panama / I. M. Johnston 1342 / 1946 / "...leaves & stem with strong garlic odor"

4086 *P. alliaceum* Peru / G. Klug 3096 / 1933 / "Remedy for rheumatism" / "'Ajosacha'"

4087 *Bignonia capreolata* U.S.: La. / W. D. Reese 1775 / 1958 / "Fls. with acrid odor"

4088 *Cremastus sceptrum* Brazil / Williams & Assis 7343 / 1945 / "Used to make a syrup in alcohol, from the root. Good for syphilis. Take syrup with meals" / "'Cigana'"

4089 *Pyrostegia venusta* Brazil / Williams & Assis 6963 / 1945 / "Flower poisonous to cattle" / "'Cipo de Sao Joao'"

4090 *Cydista diversifolia* Cuba / R. A. Howard 4785 / 1941 / "Fls. . . . odorous. A good honey plant" / "'Leñatero'"

4091 *Catalpa duclouxii* China / E. H. Wilson 640 / 07 / "'Chou shu' = stinking tree"

4092 *Deplanchea tetraphylla* New Guinea / P. van Royan 4526 / 1954 / "Flowers very often eaten by lorrykeets" / "'kapul' (Malaysian)"

4093 *Tabebuia pallida* W.I.: St. Kitts / G. R. Cooley 8797 / 1962 / "Leaves used to make tea. Leaves and bark boiled to make a cure for colds"

4094 *Tabebuia* indet. Colombia / A. E. Lawrance 124 / 1932 / "...bark fragrant"

4095 *Tecoma stans* Mexico / I. L. Wiggins 15326 / 1959 / "Flowers reputed good for digestive and gastric troubles" / "'Palo del Arco'"

4096 *Markhamia* indet. Burma / F. G. Dickason 6804 / 1937 / "...fls. and fruits eaten" / "'pithan'"

4097 *Parmentiera edulis* El Salvador / P. C. Standley 21056 / 1922 / "...fr. roasted and eaten, also remedy for colds" / "'Cuajilote'"

4098 *Bignoniacea* indet. Peru / G. Klug 1972 / 1931 / "Tea made from bark" / "'Clavo huasca'"

PEDALIACEAE

4099 *Sesamum indicum* China / S. K. Lau 400 / 1932 / "Used in making wine" / "'Tsio kok'"

MARTYNIACEAE

4100 *Ibicella lutea* Brazil / Reitz 1728 / 46 / "Cultivada—comestivel"

4101 *Proboscidia peruviana* Peru / Stork & Horton 11356 / 1939 / "Tubers eaten with relish by cattle" / "'Yuca de caballo'"
Peru / A. A. Beetle 26193 / 1939 / "... root said to be good to eat and medicinally valuable"

GESNERIACEAE

4102 *Oreocharis* indet. China / W. T. Tsang 23009 / 1933 / "Used as medicine"

4103 *Didimocarpus heterotricha* China / S. K. Lau 1872 / 1933 / "... the leaves are edible" / "Kap Yin Ip"

4104 *Besleria* indet. Colombia / R. E. Schultes 5361 / 1943 / "Use decoction as plaster for snake bite amongst whites .... Indians say it is poison to drink" / "Maripimbara"

4105 *Cyrtandra* indet. Celebes / A. H. G. Alston 16576 / 1954 / "Used to wrap meat to make it tender" / "'Kole' (Tontembuan)"

4106 *C. auriculata* Philippine Is. / R. B. Fox 5040 / 1948 / "This native medicant is used for fungus irritation on the feet. Alipunga in Tagalog, Gadut in Egongot ... fruit roasted ... pounded to ... powder ... rubbed ... on the parts affected"

4107 *C. cumingii* Philippine Is. / R. B. Fox 5052 / 1948 / "The bark of this tree is scraped into water, and the mixture is ingested by a woman who has just given birth. It is believed that this will cause all the blood following child birth to leave the body"

4108 *Cyrtandra* indet. Philippine Is. / C. O. Frake 37988 / 1957 / "Fruit applied to gangrene (?)" / "gantingan mamamoa. Dialect Sub."

4109 *Cyrtandra* indet. Philippine Is. / C. O. Frake 38244 / 1958 / "Bark applied to wounds" / "gantigantig. Dialect Sub."

4110 *Cyrtandra* indet. Philippine Is. / C. O. Frake 38373 / 1958 / "Leaves applied to ulcers" / "glelaping. Dialect Sub."

4111 *Cyrtandra* indet. Philippine Is. / C. O. Frake 38304 / 1958 / "Leaves applied to infectious swellings" / "Selingka? ulangan Sub."

4112 *Cyrtandra* indet. Philippine Is. / P. Añonuevo 13729 / 1950 / "The bark is scraped and heated for a few minutes and is applied on wounds" / "Banaybanay. Dialect Manobo"

4113 *Cyrtandra* indet. Philippine Is. / M. D. Sulit 9974 / 1949 / "Decoction of roots given to women who gave birth, stop bleeding" / "Lumalagsik. Dialect Bukid."

4114 *Columnea* indet. Ecuador / Y. Mexia 8412 / 1936 / "Used medicinally" / "'Huaco'"

4115 *Codonanthe formicarum* Peru / J. M. Schunke 125 / 1935 / "'Trichinisacha'"

4116 *Kohleria longifolia* El Salvador / S. Calderón 174 / 1921 / "'Digital montes'"

RUBIACEAE

4117 *Loretoa peruviana* Peru / G. Klug 2022 / 1931 / "Fruit edible" / "'Meta guais'"

4118 *Calderonia salvadorensis* El Salvador / P. C. Standley 22312 / 1922 / "Bark turns red when cut" / "'Sangre de chucho,' 'Palo colorado'"

4119 *Chimarrhis venezuelensis* Venezuela / J. A. Steyermark 55408 / 1944 / "'quina blanca'"

4120 *Hintonia latiflora* var. *leiantha* Mexico / H. S. Gentry 6790 / 1942 / "...bark much employed for fevers" / "'Copalquin'"

Mexico / G. B. Hinton (Herb.) 4317 / 33 / "Macerated leaves taken internally for malaria. Boiled leaves in a bath for Pinto" / "'Copalche'"

Mexico / Hinton et al. (Hinton Herb.) 6922 / 34 / "generally used and sometimes exported for malaria"

Mexico / G. B. Hinton (Herb.) 3413 / 33 / "Macerated leaves for malaria & for pinto" / "'Copalche'"

4121 *H. latiflora* Mexico / H. S. Gentry 2348 / 1936 / "Bark boiled with salt and drunk before meals for fevers and as a purgative" / "'Copalquin,' Mex."

Mexico / W. P. Hewitt 269 / 1948 / "Bark mashed, soaked in mescal, and drunk to cut malarial fever" / "'Copalquin'"

Mexico / G. B. Hinton et al. (Herb.) 6333 / 34 / "Medicinal for Pinto" / "'Copalche'"

4122 *H. standleyana* Mexico / E. Langlassé 234 / 1898 / "Arbre à quinquina ... Ecorce excellent febrifuge ..." / "'Quina'"

4123 *Pogonopus tubulosus* Argentina / A. G. Sch. 978 / 1936 / "'Cascarilla' ... Corteza febrifuga"

4124 *Hedyotis longipetala* China / Lingnan (To and Ts'ang) 12419 / 1924 / "'Heavenly incense bowl'"

4125 *H. wulsinii* China / Lingnan (To and Ts'ang) 12023 / 1924 / "Used as drug" / "'Cuttle fish liver'"

4126 *Hedyotis* indet. Hainan / W. T. Tsang 94 / 1928 / "...medicine"

4127 *H. diffusa* Singapore / Hsu, Ho-hsiang B / 1961 / "A medicinal plant used for stomach ulcers in Singapore"

4128 *H. vestita* Philippine Is. / A. L. Zwickey 830 / 1938 / "'Tabidan' (Lan.); pounded lvs. applied externally for headache"
Philippine Is. / C. O. Frake 560 / 1958 / "Roots applied for internal pain" / "'Pepenu'ulangan'" Sub.

4129 *Ophiorrhiza peploides* Fiji Is. / Degener & Ordonez 14216 / 1941 / "Fijian name said to refer to urine because natives formerly grew plant along their village walls where populace often urinated at night"

4130 *Cruckshanksia palma* Chile / J. L. Morrison 16999 / 1938 / "...strong odor of vanilla"

4131 *C. chrysantha* Chile / J. King 9 / no year / "'Yerbos-buenas'"

4132 *Rondeletia suffrutescens* Guatemala / J. A. Steyermark 36342 / 1940 / "Flower with odor of sulfur dioxide"

4133 *R. hirsuta* Jamaica / G. R. Proctor 7361 / 1952 / "...bark & leaves aromatic when bruised"

4134 *R. norlindii* Cuba / E. L. Ekman 19181 / 1924 / "...smells strongly of coumarine"

4135 *R. sylvestris* Jamaica / G. R. Proctor 15658 / 1956 / "...cut stems with licorice odour"

4136 *R. stereocarpa* W.I.: Dominica / J. S. Beard 252 / 1944 / "'Quina'"

4137 *Sickingia salvadorensis* El Salvador / Standley & Padilla V. 2787 / 1947 / "Planted along road" / "'Limpia-dientes'"

4138 *S. tinctoria* Brazil / B. A. Krukoff 1544 / 1931 / "...fls....with strong and pleasant odor of vanilla"

4139 *Lindenia rivalis* Costa Rica / Ch. Tonduz 13947 / 1900 / "S'emploie hâchée pour enlever la gâle aux chiens" / "'Lirio'"

4140 *Wendlandia salicifolia* China / Jos. Esquirol 327 / 1904 / "Les feuilles donnent une infusion théiforme assez employée..."

4141 *Wendlandia* indet. Burma / F. G. Dickason 7736 / 1937 / "Used in sacrifice for sickness & for field sacrifice" / "'Bui Hnam'" / "'Thit nee' (Burmese)"

4142 *W. luzoniensis* Philippine Is. / Vidal 1477bis / before 1959 / "'Malasantol'"

4143 *Elaeagia karstenii* Venezuela / J. A. Steyermark 55416 / 1944 / "'Quinón'"

4144 *Cinchona coccinea* S. America / Ruiz et Pavon s.n. / 1804 / "Vulgo Cascarilla"

4145 *C. henleana* Venezuela / J. A. Steyermark 55096 / 1943 / "...bark ...only slightly bitter..."

4146 *Ladenbergia carua* Bolivia / H. H. Rusby G,1 / 1886 / "Quality of bark—worthless" / "Native name—Mula Cascarilla; Carua" / "English equivalent—Mule Bark, False Bark"

4147 *L. muzoniensis* Venezuela / H. N. Whitford 40 / 1917 / "'Quina'"

4148 *L. pittieri* Venezuela / J. A. Steyermark 56695 / 1944 / "'Quina'... This is the bark considered as 'quina' around this area (Palmita, Bolero, Mesa Bolivar), and is sold in the pharmacies and stores with aguardiente to drink"

Venezuela / J. A. Steyermark 55653 / 1944 / "'Campano'...bark bitter turning slightly brown"

4149 *L. magnifolia* Colombia / A. E. Lawrance 740 / 1933 / "When branches are cut a brownish stain flows therefrom which is very sticky"

4150 *L. undata* Venezuela / J. A. Steyermark 56105 / 1944 / "'Quina amarilla'"

4151 *Remijia firmula* Brazil / R. Froes 12541/235 / 1942 / "'Quina flasa'"?

4152 *R. hispida* Venezuela / J. A. Steyermark 57746 / 1944 / "...bark... somewhat bitter..."

4153 *R. purdieana* Colombia / E. Bonitto, Colombian Cinchona Mission, s.n. / 1943 / no economic data

4154 *R. roraimae* Venezuela / J. A. Steyermark 58599 / 1944 / "Bark used for colds and headaches after boiling in water" / "'darona'"

4155 *Bouvardia erecta* Mexico / E. & E. Seler 846 / 1888 / "'Yerva de S. Juan'"

4156 *B. longiflora* Mexico / A. Dugès 365.G / 1897 / "vulg.: Flor de San Juan"

4157 *B. multiflora* Mexico / A. Dugès 405 A. / no year / "nommée ici huele de noche del cerro (del Valenciana)"

4158 *B. scabra* Mexico / R. S. Ferris 5949 / 1925 / "'Huele de Noche'"

4159 *B. ternifolia* Mexico / R. M. Stewart 1896 / 1941 / "'Mirto'"

Mexico / A. Lopez 34 / 1948 / "'garitos' buena para latos"?

Mexico / G. B. Hinton (Herb.) 801 / 32 / "Called 'Trompetillo' also 'Hierba de San Antonio.' Boiled with cinnamon and taken for nine mornings is good for the heart"

4160 *Heterophyllaea pustulata* Argentina / C. Gayer? (Herb. L. R. Parodi) 12874 / 1934 / "...produce la ceguera de los animales" / "'Cegadera'"

4161 *Manettia divaricata* Peru / J. M. Schunke 12 / 1935 / "Indians chew leaves to make their teeth black" / "'Yanamuco'"

4162 **M. cinchonarum** Colombia / J. Cuatrecasas 19201 / 1944 / "'Beju-quillo'"

4164 **Ferdinandusa goudotrana** Venezuela / J. A. Steyermark 60366 / 1944 / "leaves cooked for use in treating colds" / "'wosta-curán-yek'"

4165 **Exostemma caribaeum** Mexico / Y. Mexia 8909 / 1937 / "Bitter bark used in tea for fevers" / "'Copalche'"
Dominican Republic / E. J. Valeur 697 / 1931 / "'Quina'"

4166 **E. sanctae** W.I.: St. Lucia / P. Beard 1166 / 1945 / "Bark bitter, used as febrifuge" / "'Quina'"

4167 **E. peruviana** Peru / Stork & Horton 10094 / 1938 / "The 'cura' says it is one of the best bee plants here; 'good against malaria'" / "'Asar-cito'"

4168 **Coutarea hexandra** El Salvador / P. C. Standley 21664 / 1922 / "Remedy for fevers" / "'Quina'"
El Salvador / P. C. Standley 20892 / 1922 / "'Quinina'"
El Salvador / P. C. Standley 21048 / 1922 / "Bark remedy for malaria" / "'Quina'"
Guatemala / P. C. Standley 75974 / 1940 / "Remedy for malaria. Leaves very bitter" / "'Quina'"
El Salvador / N. C. Fassett 28918 / 1951 / "'Quina' brought in from nearby by the Institute Gardener" / "Instituto Tropical de Investigaciones Científicas, San Salvador"
Brazil / F. C. Hoehne 2056 / 1919 / "'Murto do Matto'"
Brazil / A. Ducke 369 / 41 / "'Quinaquina'"

4169 **C. pterosperma** Mexico / J. Gonzalez Ortega 6.337 / 1926 / "'Copal-quin'"
Mexico / T. (or I.) S. Brandegee s.n. / 1904 / "'Yerba Buena'"
Mexico / Y. Mexia 8765 / 1937 / "Bark made into tea and taken for malaria" / "'Popalche'"
Mexico / H. S. Gentry 1165 / 1934 / "Medic. bark brewed to make a bitter tea for fevers" / "Copalkini, Mex.   Hutetíyo, W."

4170 **Adina cordifolia** INDIA / M. Fratani s.n. / 1922 / "Fruits . . . comestibles"

4171 **Mitragyna stipulosa** Africa / G. F. S. Elliot 5014 / no year / "Leaves of this used for wrapping Kolas"

4172 **Uncaria pteropoda** Malaya / G. anak Umbai for A. H. Millard (K.L. 1479) / 59 / "Phytochemical Survey of Malaya"

4173 **U. insignis** Philippine Is. / C. O. Frake 680 / 1958 / "Leaves chewed, betel applied to ulcers" / "'guilan nutung'" Sub.

4174 *U. perrottetii* Philippine Is. / C. O. Frake 533 / 1958 / "Roots boiled and drunk for hematuria, and during puerperium" / "'Guilan'" Sub.

4175 *U. talbotii* Liberia / D. H. Linder 480 / 1926 / "Expedition of the Harvard Institute of Tropical Biology and Medicine"

4176 **Neonauclea formicaria** Philippine Is. / C. O. Frake 566 / 1958 / "Bark boiled and drunk for puerperium" / "'Glebalud'" Sub.

Philippine Is. / A. L. Zwickey 21 / 1938 / "'Gimbalod' (Lan.); 'Bankal' (Bis.); infusion of bark for malaria or rheumatism"

4177 **Cephalanthus occidentalis** U.S.: Kans. / F. C. Gates 16994 / 1931 / "fatally poisoning cattle in this & previous years"

U.S.: Tex. / Berlandier? 2574 / 1834 / "vulgo *Rosa de S. Juan*"

4178 **Nauclea** indet. Thailand / T. Smitinand 849 / 51 / "Blaze light brown bitter & astringent, 1 cm. thick" / "Local name—Tawn Dong"

4179 **N. officinalis** Cambodia / Pierre 605 / 1872 / "Ecorce fébrifuge"

4180 **N. orientalis** Philippine Is. / C. O. Frake 343 / 1957 / "Young leaves applied to headache" / "'Glebalud Basak'" Sub.

4181 **Mussaenda hainanensis** Hainan / Swa H. V. 2 / 1932 / "Local people use it to cure cancer" / "'Kam Ngan Fa'"

4182 **M. pubescens** China / C. W. Wang 73631 / 1936 / "for medical use"

China / H. H. Chung Herb. 991 / 1923 / "leaf used as a substitute for tea . . ."

China / G. W. Groff 42 / 1919 / "for medical"

4183 **Mussaenda** indet. Siam / native collector (Royal For. Dept.) 4330 / 48 / "Medicinal roots natives use to remedy small pox" / "'Ka-ber'"

4183A **Mussaenda** indet. Siam / T. Smitinand 704 / 51 / "Used for fever remedy" / "'Bao Ma Muad'"

4183B **Mussaenda** indet. Philippine Is. / K. I. Pelzer 77 / 1950 / "Roots boiled—extract drunk by woman who gave birth recently" / "'Tala-watawa' in Bukidnon dialect"

4184 **M. villosa** var. **herveyane** Malaya / G. anak Umbai for A. H. Millard (K.L.) 1611 / 59 / "Phytochemical Survey of Malaya" / "Poisonous fruits & root"

4185 **M. kajewski** Solomon Is. / S. F. Kajewski 2455 / 1931 / "The leaves and flowers are mixed with lime and applied to the natives' heads to kill lice" / "'Mum-mar-ago'"

4186 **M. philippica** Philippine Is. / C. O. Frake 572 / 1958 / "agricultural ritual" / "'Gibing Gibing'" Sub.

Philippine Is. / C. & C. Frake 197 / 1957 / "Bark used as treatment wounds & ulcers" / "'Buyan'" (or 'Bulagan'?) Subanum

Philippine Is. / R. Reed 1 / 1960 / "tobacco substitute" / "'Mabiun Tama'" Batangar

Philippine Is. / L. E. Ebalo 254 / 1939 / "juice of bark is used as cure for headache" / "'Agbay'" Mang.

4187 *M. raiatensis* Fiji Is. / L. Reay 9 / 1941 / "Fijians grate the stem of the plant and mix it with water for people suffering from chest complaints. Has a bitter flavour" / "'Vobo'"

Fiji Is. / A. C. Smith 6644 / 1947 / ". . . root used as a remedy for asthma" / "'Vombo'"

Fiji Is. / Degener & Ordonez 14185 / 1941 / "Scrape bark (bitter) making tea with it. This is drunk for kidney disease" / "'Bofo'"

Fiji Is. / A. C. Smith 25 / 1933 / "Bark used as a medicine for fever" / "'Mbovu'"

4188 *Schizomussaenda dehiscens* China / C. W. Wang 74949 / 1936 / "medicine"

4189 *Gonzalagunia spicata* W.I.: Martinique / Stehlé 7134 / 1946 / "'Herbe fou fou,' 'Herbe à l'eue'"?

4190 *Coccocypselum lanceolatum* Venezuela / J. A. Steyermark 56376 / 1944 / "Fruit used in treatment of pimples (mezquinos)" / "'Pipo de Dulebra'"?

4191 *Pleiocarpidia enneandra* Malaya / K. M. Kochummen (Kepong) 78960 / 59 / "Phytochemical Survey of Malaya" / "Young shoots edible" / "'Melada'"

4192 *Mycetia javanica* Philippine Is. / P. Añonuevo 8 / 1950 / "The roots boiled for a few minutes and given as drink for a person suffering from colds" / "'Oamagtingon'" Manobo

4193 *Urophyllum trifurcum* Malaya / Gadoh (K.L.) 1458 / 59 / "Phytochemical Survey of Malaya"

4194 *Urophyllum* indet. Philippine Is. / C. O. Frake 718 / 1958 / "Bark applied to ulcers" / "'pilit'" Sub.

4195 *Sabicea vogelii* Sierra Leone / G. F. S. Elliot 4175 / no year / "Used for gonorrhoea by Natives . . ."

4196 *Sommera subcordata* Mexico / Y. Mexia 537 / 1926 / "Fr. said . . . to be edible" / "'Huamuchil'"

4197 *Retinophyllum truncatum* Colombia / Schultes & Cabrera 12464 / 1951 / "Fruits red, edible" / "Taiwano = bov-feé"

4198 *Tarenna mollis* Malaya / G. anak Umbai for A. H. Millard (K.L.) 1510 / 59 / "Phytochemical Survey of Malaya" / "Poisonous" / "'Kayu Berteh' (Temuan)"

4199 *T. stenantha* Philippine Is. / C. O. Frake 738 / 1958 / "Leaves applied for external pain" / "'Gesulu'" Sub.

4200 *T. sambucina* Tonga Is. / T. G. Yuncker 15238 / 1953 / "...bark used in medicines" / "'Manonu'" Tongan

Fiji Is. / O. Degener 15350 / 1944 / "For sore stomach, scrape bark, put in water, strain & drink. Add salt to improve flavor" / "'Vaka-ru(m)bi ni davui'" Failevu

Fiji Is. / O. Degener 15401 / 1941 / "Squash bark in water and drink as tonic after sickness" / "'Warumni davui'" Ra

4201 *Randia armata* El Salvador / P. C. Standley 21423 / 1922 / "fr.... eaten" / "'Crucito,' 'Torolillo'"

4202 *R. cinerea* Mexico / G. B. Hinton (Herb.) 10239 / 37 / "Leaves medicinal" / "'Chaqua'"

4202A *R. echinocarpa* Mexico / H. S. Gentry 2272 / 1936 / "Natives relish fruit" / "'Papache,' Mex."

4203 *R. lundelliana* Brit. Honduras / P. H. Gentle 4083 / 1942 / "'Wild Lime'"

4204 *R. laevigata* Mexico / H. S. Gentry 2379 / 1936 / "Fruit ripens in November and eaten by natives" / "'Sapuchi,' Mex."

4205 *R. rosei* Mexico / H. S. Gentry 1466 / 1935 / "'Papache Boracho'"

4206 *R. thurberi* Mexico / Y. Mexia 154 / 1925 / "Fruit...yellow & edible when ripe" / "'Papache'"

4207 *R. watsoni* Mexico / H. S. Gentry 1499 / 1935 / "...fruit largely ex-crescent. Natives eat the black slimey mass eagerly, bitter sweet" / "'Papache,' Mex. 'Hosakola,' W."

4208 *R. mitis* Venezuela / H. Pittier 12373 / 1927 / "'Sajadito'"

4209 *R. formosa* Venezuela / Curran & Haman 1312 / 1917 / "'Cafecillo'"

4210 *R. cochinchinensis* Cochinchinae / L. Pierre 154 / 1865 / "Ecorce employée contre la fièvre dite des bois" / "Nomen anamiticum: ta vó"

4211 *Randia* indet. India / J. Fernandes 1620 / 1950 / "Fruit used for stupefying fish"

Indo-China / W. T. Tsang 28975 / before 1940 / "...fr. edible"

4212 *R. scortechinii* Malaya / G. anak Umbai for A. H. Millard (K.L.) 1376 / 59 / "Phytochemical Survey of Malaya"

4213 *R. coffaeoides* Fiji Is. / A. C. Smith 42 / 1933 / "Leaf used as medicine for fever" / "'Vono ni mbengga'"

Fiji Is. / A. C. Smith 1922 / 1934 / "Tea from leaves used for sore throats" / "'Mothe ni vai'"

4214 *Gardenia jasminoides* China / T. N. Hsiung 489 / 1929 / "fruits can be used to make yellow dyestuff and used as drug also"

China / W. T. Tsang 25485 / 1935 / "...fr. yellow, edible"

China / W. T. Tsang 21226 / 1932 / "...fruit yellow, edible; used as medicine"

China / S. K. Lau 4064 / 1934 / "can be made as medicine and dye-stuff"

China / F. A. McClure 52 / 1921 / "dye & medicine made of seeds" / "'Weng Chi'" Cantonese

4215 *G. stenophylla* China / W. T. Tsang 24115 / 1934 / "...fr. yellow, edible"

4216 *Gardenia* indet. China / C. W. Wang 76325 / 1936 / "fruit green, edible"

India / J. Fernandes 1413 / 1950 / "Milk of the plant is applied to cuts;... 'Nagarkuda' (Marathi)"

4217 *G. curranii* Philippine Is. / M. D. Sulit 4549 / 1952 / "Pounded fruits used for poisoning fish in rivers and in salt sea" / "'Bayag-usa'" Tag.

4218 *G. storckii* Fiji Is. / O. Degener 15064 / 1941 / "Fijians chew buds as chewing gum" / "Fijians wash 2 inch length of root, cut into little pieces, put into cocoanut cloth and steep in cold water. Lung or constipation patient drinks this as cure. It is drastic like castor oil" / "'Bolavatu'" Serua

4219 *Gardenia* indet. New Hebrides Is. / S. F. Kajewski 362 / 1928 / "Leaves put near fire rubbed and put on sore while hot" / "'Neace-ya-vat'"

4220 *Casasia clusiaefolia* Bahama Is. / O. Degener 18776 / 1946 / "'7 year apple'; overripe fruit edible"

4221 *Posoqueria latifolia* Brit. Honduras / W. A. Schipp 530 / 1930 / "...clusters of yellow fruits edible"

4222 *Sherbournia calycina* Africa / G. F. S. Elliot 4110 / no year / "...used by natives for coughs, etc."

4223 *Hypobathrum fasciculatum* Philippine Is. / M. Santos 1 / 1949 / "Fruit edible"

4224 *Tricalysia* Burma / F. G. Dickason 8422 / 1939 / "wild coffee plant" / "'Hkurt'"

4225 *T. sessilis* Philippine Is. / L. E. Ebalo 557 / 1940 / "'cafe-cafe'"

4226 *Amaioua corymbosa* Brit. Honduras / P. H. Gentle 3140A / 1940 / "'wild coffee'"

Brit. Honduras / P. H. Gentle 2649 / 1938 / "'bastard coffee'"

4227 *Duroia eriopila* Brit. Guiana / A. C. Smith 2792 / 1937 / "...fruit edible"

4228 *D. hirsuta* Colombia / R. E. Schultes 3320 / 1942 / "Bark, when tied on arm, forms blisters" / "'Solimán'"

4229 *Alibertia edulis* Brit. Honduras / P. H. Gentle 3564 / 1941 / "fruits edible" / "'wild guava,' 'sul-sul'"

4231 *A. elliptica* Brazil / Y. Mexia 5822 / 1931 / "... fruit edible" / "'Marmelado de cachorro'"

4232 *A. steinbachii* Bolivia / J. Steinbach 5219? /1920 / "La fruta ... sirven para comer, son dulces"

4233 *Heinsia benguelensis* Portuguese W. Africa / A. G. Curtis 144 / 1923 / "Children suck honey from flowers. Witch doctors dig roots"

Angola / M. A. Pocock 372 / 1925 / "Fls. (or Frs. or lvs.?) chewed in mouth and pulp applied to cuts & sores (M'Bunde)" / "Bushmen smoke leaves" / "Native—Mulangu"

4234 *Hamelia axillaris* Panama / Cooper & Slater 187 / before 1928 / "'Guayaba negro'"

4235 *H. patens* Yucatan / J. Bequaert 6 / 1929 / "Used for sores of skin"

Cuba / Hodge, Howard & Godfrey 4038 / 1940 / "Leaves bitter"

4236 *H. xorulliensis* Mexico / Y. Mexia 542 / 1926 / "A decoction of leaves used for washing wounds" / "'Campancillo'"

4237 *Rhabdostigma schlechteri* S. Africa: Transvaal / J. Borle (Nat. Herb.) 64JB? / 19 / "Fruit edible"

4238 *Airosperma trichotomum* Fiji Is. / F. R. Fosberg 18285 / 1937 / "Used medicinally as a cathartic" / "'sila sila'"

4239 *Vangueria infausta* S. Africa: Transvaal / E. H. Wilson s.n. / 1922 / "fruit edible"

4240 *V. edulis* Kenya / E. H. Wilson 73 / 1921 / "fruit edible"

4241 *Canthium dicoccum* Hainan / S. K. Lau 1857 / 1933 / "fruit, edible"

4242 *C. parvifolium* Hainan / S. K. Lau 107 / 1932 / "Fruit not edible"

4243 *C. monstrosum* Philippine Is. / C. O. Frake 806 / 1958 / "Roots boiled and drunk for rigid abdomen" / "'delupung' in Sub. dialect"

4244 *Pygmaeothamnus chamaedendron* S. Africa: Transvaal / A. P. Goosens 1057 / 1932 / Attached to this collection is a translation of a letter from the Director of Veterinary Services, Dept. of Agric., Pretoria, to the Chief, Division of Plant Industry, Pretoria: there had been a report that many animals on the farm on which the above collection was made had succumbed to plant poisoning. 'Gousiekte' was suspected; the 'Gousiekte' plant, devoured in large quantities by sheep and cattle over a long period, had been indicated possibly as a toxic species. Two bags of the plant had been collected and fed to two sheep; by the end of one month the sheep had a high fever, and one of them suffered diarrhea for some days. The final result is not indicated in the letter.

4245 *Ancylanthos fulgidus* Angola / Mrs. L. S. Tucker 42 / 1924 / "Medicine for chest trouble"

4246 *Guettarda viburnoides* Brazil / Y. Mexia 5583 / 1931 / "... fruit yellow when ripe, edible ..." / "'Angico'"

4247 *G. steyermarkii* Venezuela / J. A. Steyermark 56601 / 1944 / "'salvio quino'"

4248 *Antirrhoea coriacea* W.I.: Dominica / D. Taylor s.n. / 1942 / "Bark boiled and decoction used for washing sores" / "'Akukwa'"

4249 *Timonius longitubus* Solomon Is. / S. F. Kajewski 2541 / 1931 / "If there are pains in the stomach the bark is macerated and rubbed on the affected spot" / "'Tehor'"

4250 *Timonius* indet. Philippine Is. / C. O. Frake 850 / 1958 / "Roots boiled and drunk for pasmu?—indigestion" / "'Giangiang'" Sub.

4251 *Chiococca alba* Mexico / H. S. Gentry 6608 / 1941 / "Juice of fruit decocted & employed as a purgative"
W.I.: Martinique / H. & M. Stehlé 6052 / 1943 / "Liane des sorciers"

4252 *Chione venosa* W.I.: Grenada / W. E. Broadway s.n. / ca. 1897 / "Its alleged aphrodisiacal properties contained in the roots have been studied in London and the U.S. of America" / "'Violette,' 'Bois bandé'"

4253 *Pavetta graciliflora* Malaya / G. anak Umbai for A. H. Millard (K.L.) 1455 / 59 / "Phytochemical Survey of Malaya" / "Stupefying"

4254 *P. indica* Malaya / K. M. Kochummen (Kepong) 71994 / 58 / "Phytochemical Survey of Malaya" / "'Jarum-Jarum'"

4255 *Ixora coccinea* India / J. Fernandes 1093 / 1950 / "Fruit ... when ripe eaten" / "'Pitkuli' (Marathi)"

4256 *I. stricta* Indochina / M. Poilane 1174 / 1920 / "Les feuilles et les fleurs sont employées pour faire tisane combat les maux de ventre. Les fruits à maturités sont comestibles"

4257 *Ixora* indet. Malay Peninsula / Gadoh (K.L.) 1281 / 59 / "Phytochemical Survey of Malaya" / "'Nayaram Bukit' (Temaran)"

4257A *Ixora* indet. N. Borneo / Kwanting (N. Borneo For. Dept.) A485 / 1948 / "It is used for medicine as headache" / "'Piriok-Piriok' (Brune Kimanis)" / "'Tagandapon' (Dusun Kimanis)"

4258 *I. macrophylla* Philippine Is. / C. O. Frake 497 / 1958 / "Medicine for rigid abdomen" / "'Malpa'" Sub.
Philippine Is. / C. O. Frake s.n. / 1958 / "Bark boiled and drunk during puerperium" / "'Gulian'" Sub.
Philippine Is. / C. O. Frake s.n. / 1958 / "Leaves applied for headache" / "'pintad'" Sub.

4259 *Ixora* indet. Philippine Is. / R. B. Fox 36 / 1950 / "medicine—see Sungkód—It—Biyangúnan (Babay)"

4259A *Ixora* indet. Philippine Is. / R. B. Fox 18 / 1950 / "medicine" / "Sungkód—It—Biyangúnan" Tagb.

4259B **_Ixora_** indet. Philippine Is. / C. O. Frake 488 / 1958 / "Bark boiled & drunk for puerperium" / "'paginugun'" Sub.

4260 **_I. amplicaulis_** Fiji Is. / O. Degener 15016 / 1941 / "In Serua for elephantiasis people drink tea made from the bark" / "'Laubu'" Savatu

4261 **_I. triantha_** Caroline Is. / C. C. Y. Wong 511 / 1948 / "The stem is used in the medicine for the swelling of veins (_fely nigof_) the bark is scraped and squeezed, then mixed with the water of a young coconut before the mixture is drunk" / "'Gethemuc'"

4262 **_Strumpfia maritima_** Bahama Is. / O. Degener 19024 / 1946 / "used as smudge to repel mosquitoes" / "'Grankini'"

4263 **_Psychotria heterochroa_** Dominican Republic / E. J. Valeur 115 / 1929 / "'Palito de Culebra'"

4264 **_P. barbiflora_** Brazil / Williams & Assis 6867 / 1945 / "'Herva de Rato.' Poison"

4265 **_P. capitata_** Brazil / L. O. Williams 5261 (Assis & Moreira) 1945 / "'Herva de rato' Poison. Kills cattle if eaten and water taken"

4266 **_P. pinularis_** Colombia / Killip & Smith 14245 / 1926 / "snake antidote"

4267 **_P. elmeri_** Brit. N. Borneo / Evangelista & Arsat (B.N.B. For. Dept.) 995 / 1929 / "Fruit can be eaten when ripe" / "'Bintuka'"

4268 **_P. viridiflora_** Indonesia: Bangka / Kostermans & Anta 339 / 1949 / "Fr. white. Infusion as medicament (bathing) for women" / "'pätjong'"

4269 **_Psychotria_** indet. Indonesia: Bangka / Kostermans & Anta 1088 / 1949 / "Fls. white. Infusion is used as a bath for women after childbirth" / "'kaju badja'"

4270 **_P. olivacea_** Solomon Is. / S. F. Kajewski 2520 / 1931 / "The leaves are macerated and boiled in hot water, then being applied to sore legs" / "'Gongor'"

Solomon Is. / S. F. Kajewski 1884 / 1930 / "Leaves are eaten with betel nut to relieve pain in the stomach"

Solomon Is. / S. F. Kajewski 2194 / 30 / "The sap of this vine is used to cure a sickness of the penis, probably a kind of gonorrhoea. The natives swear that a plant Cutope causes this sickness, disregarding the thought of bacterial infection" / "'Torge-galla'"

4271 **_P. schmielei_** Solomon Is. / S. F. Kajewski 2386 / 1930 / "The bark of this tree is scraped and boiled with water and then drunk for paines in the stomach" / "'Pourre-manguta'"

4272 **_Psychotria_** indet. Solomon Is. / S. F. Kajewski 2697 / 1932 / "The sap of this shrub is mixed with water and given to people with sore mouths"

4273   *P. cuernosensis* Philippine Is. / C. O. Frake 559 / 1958 / "leaves applied to burns" / "'Telengisu'" Sub.

4274   *P. luzoniensis* Philippine Is. / C. O. Frake 637 / 1958 / "pith applied to wound" / "'Nugu'"? Sub.

Philippine Is. / C. O. Frake 534 / 1958 / "roots scraped and applied to wounds" / "'Ngayan'" Sub.

4275   *P. manillensis* Philippine Is. / Sulit & Conklin 5153 / 1953 / "Decoction of roots medicine excessive menstruation" / "'Tagpô Arabábá'" Mang. (Hanunoo)

4276   *P. merrittii* Philippine Is. / C. O. Frake 495 / 1958 / "Bark applied to wounds" / "'Ginsim'" Sub.

4277   *P. membranifolia* Philippine Is. / C. Frake 490 / 1958 / "Leaves applied for spleenomegaly" / "'Gatayatay'" Sub.

4278   *Psychotria* indet. Philippine Is. / C. O. Frake 622 / 1958 / "Bark applied for internal pain" / "'Penubulen Jayn'" Sub.

4278A *Psychotria* indet. Philippine Is. / C. O. Frake 538 / 1958 / "Medicine for skin eruptions" / "'Tangkuluran'" Sub.

4278B *Psychotria* indet. Philippine Is. / C. O. Frake 727 / 1958 / "Leaves applied to ulcers" / "'Tubalan-ulangan'" Sub.

4279   *P. macrocalyx* Fiji Is. / O. Degener 15393 / 1944 / "Squash leaf and squeeze juice (without added water) into eye as eye medicine" / "'Langaingai'" Ra.

4280   *P. tephrosantha* Fiji Is. / F. R. Fosberg 18280 / 1937 / "Leave decoction used for sickness, 'nimbatha,' of bowels or stomach" / "'nai vaka reva nimbatha'"

4281   *P. archboldiana* Fiji Is. / O. Degener 15316 / 1941 / "Fijians crush leaf, add water and drink liquid as remedy for T.B." / "'Vunga'" Serua

4282   *Calycosia petiolata* Fiji Is. / F. R. Fosberg 18217 / 1937 / "Bark used medicinally for toothache" / "'makamakandora'"

4283   *Palicourea elongata* Brazil / B. A. Krukoff 1348 / 1931 / ". . . crushed flowers are used for poisoning rats"

4284   *P. gardneriana* Brazil / Williams & Assis 6872 / 1945 / "Poison"

Brazil / Williams & Assis 6806 / 1945 / "Herva de rato de grande. Poison"

4285   *P. guianensis* Brazil / Ducke 1458 / 1943 / "'Herva de rato grande'"

4286   *P. marcgravii* Brazil / A. Gehrt (Herb.?) 28708 / 1930 / "'Herva de rato'"

4287   *P. radiana* Brazil / A. Gehrt (Herb.?) 25255 / 1930 / "'Herva de rato'"

4288　*P. rigida* Bolivia / J. Steinbach 5046 / 1920 / "La raiz es un poderoso remedio contra impurezas de la sangre" / "'La Reina'"

Brazil / L. O. Williams 5464 / 1945 / "Gives a tea good for the heart" / "'Congonha bate caixa'"

4289　*Palicourea* indet. Brazil / Ducke 129 / 1936 / "'herva de rato'"

4289A　*Palicourea* indet. Brazil / Froes 12696/61 / 1942 / "'Flor de Rato'"

4290　*Rudgea subsessilis* Brazil / Williams & Assis 6627 / 1945 / "Fruit made of rat poison" / "'Herva de rato'"

4291　*R. viburnoides* Brazil / Y. Mexia 5580 / 1931 / "A tea made from leaves used as blood-purifier" / "'São Bernadhino'"

4292　*Chasalia curviflora* China / A. To (Herb) 3113 / 1918 / "leaves used rubbed on bruises, for broken legs (?)" / "'Pok Kwat Isó'"

4293　*Geophila herbacea* Tonga Is. / T. G. Yuncker 15317 / 1953 / "Leaves used in medicine" / "'Tono'"

Fiji Is. / A. C. Smith 6356 / 1947 / "leaves used as a cough remedy" / "'Totondro'"

Fiji Is. / O. Degener 15237 / 1941 / "For earache Fijians squash juice of leaf into ear" / "'Totondro'" Serua

Fiji Is. / O. Degener 15381 / 1941 / "To dye hair black, Fijians squash leaf, put in bottle with coconut oil, and put on hair" / "'totonroni-vikau'" Tailevu

4294　*Cephaelis tinctoria* Venezuela / J. A. Steyermark 55256 / 1944 / "Used to give a color to Chimo, a tobacco juice made here" / "'Cafe-cito'"

4295　*Lasianthus oblongus* Malaya / G. anak Umbai for A. H. Millard (K.L.) 1500 / 59 / "Phytochemical Survey of Malaya" / "'stupefying'"

4296　*Lasianthus* indet. Malaya / G. anak Umbai for A. H. Millard (K.L.) 1589 / 59 / "Phytochemical Survey of Malaya"

4297　*L. morus* Philippine Is. / Sulit & Conklin 5119 / 1953 / "The Man-guans feed the fruits to chickens to avoid getting away and become wild" / "'Uli-kayo'" Mang. (Hanunoo)

4298　*Amaracarpus solomonensis* Solomon Is. / S. F. Kajewski 2394 / 30 / "When an infant has a white tongue (constipation), the bark is macerated and given to the child to drink" / "'Pourre'"

4299　*Hydnophytum stewartii* Solomon Is. / S. F. Kajewski 2389 / 30 / "The leaves are applied without any preparation to sore legs" / "'Copa'"

Solomon Is. / L. J. Brass 2548 / 1932 / "The fleshy stock inhabited by great numbers of small brown ants"

4300　*H. formicarium* Philippine Is. / Edaño & Gutiérrez 240 / 1957 / "medicinal leaves decoct for closing (or cleaning?) boils"

4301 *Paederia chinensis* Japan / E. Elliott 85 / "46 / "Medicinal Use: Fruit bruised and the juice well rubbed into the portion having cold injury" / "'Hekusokazura'"

4302 *P. foetida* Philippine Is. / H. C. Conklin 391 / 1953 / "Econ. uses: toothache, either vine or 'kaybasád'; also makes teeth black" / "'Kantút'"

4303 *Coprosma persicaefolia* Fiji Is. / O. Degener 15353 / 1941 / "If Fijians have 'sore body,' they squash leaves in coconut oil, stain and rub themselves with the oil" / "'Timo' in Tailevu"

4304 *Coussarea hydrangeifolia* Bolivia / J. Steinbach 6685 / 1924 / "Se fabrica de hojas y flores un té parecido al de la yerba paraguaya y de propiedades tónicos" / "'Matico'" or "'Yerba de Soria'"

4305 *C. penetantha* Colombia / J. Triana 1714 / before 1893 / "Vulgo, Flor de Muerto"

4306 *C. pilosiflora?* Ecuador / R. C. Gill 8 / 1942 / "Medicinal qualities" / "'Pepa Venemosa'"

4307 *Faramea occidentalis* Brit. Honduras / P. H. Gentle 3713 / 1941 / "'wild coffee'"

El Salvador / P. C. Standley 20107 / 1922 / "'Cafecillo'"

El Salvador / S. Calderón 1387 / 1922 / "'Cafecillo'"

Mexico / Y. Mexia 1174 / 1926 / "'Cafecillo'"

Cuba / C. Wright 241 / 1856–57 / "'wild coffee'"

Cuba / E. P. Killip 44176 / 1954 / "Used for stomach trouble" / "'Palo caja'"

Venezuela / H. Pittier 11840 / 1925 / "Fruta de paloma; guariche, cafecillo"

4308 *F. celata* Ecuador / J. A. Steyermark 53773 / 1943 / "'café de montaña'"

4309 *F. longifolia* S. America: Upper Amazon & tributaries / J. W. H. Traill s.n. or 408 / before 1876 / "'Café rana'"

4310 *F. martiana* Brazil / A. Gehrt 5609 / 1921 / "'Casco da Vacca'"

4311 *F. maynensis* Colombia / A. E. Lawrance 626 / 1933 / "Birds seem to shun this fruit"

4312 *F. orinocensis* Venezuela / L. Williams 13321 / 1940 / "'Cafecillo'"

4313 *Coelospermum reticulatum* Australia / C. T. White 10569 / 1936 / "Local name Medicine Bush, said to be very poisonous"

4314 *Morinda citrifolia* W.I.: St. Thomas / Eggers 754 / 1882 / "vulg. Painkiller"

W.I.: Dominica / W. H. & B. T. Hodge 1647 / 1940 / "Leaves used as poultice" / "Known as 'pain-killer'"

W.I.: Dominica / W. H. & B. T. Hodge 2954 / 1940 / "...leaves used to wrap around rheumatic joints" / "'Feuille froide'"

Tonga Is. / T. G. Yuncker 15367 / 1953 / "leaves used for medicine..." / "Tongan name: Nonu"

Tonga Is. / T. G. Yuncker 15038 / 1953 / "various parts used in medicines, the leaves as a remedy for boils" / "Tongan name: Nonu"

Africa / G. F. S. Elliot (Sierra Leone Boundary Commission) 5278 / no year / "Leaf forms a very good purge" / "'Bungbo'" or "'Bumbo'"

4315 *M. bracteata* Philippine Is. / A. L. Zwickey 786 / 1938 / "lvs. in treatment of cuts and sores" / "'Bunga' (Lan.)"

4316 *M. longiflora* Africa / G. F. S. Elliot (Sierra Leone Boundary Commission) 4186 / no year / "'Ogidogbo' a well-known native medicine for fever, recognised as really very good by all"

4317 *Diodia hyssopifolia* Brazil / B. A. Krukoff 1070 / 1931 / "Medicinal plant (similar use as sarsaparilla)" / "'Setesangrias'"

4318 *Borreria latifolia* Brit. Honduras / W. A. Schipp 656 / 1930 / "...natives use the plant as a remedy for snake bites"

Mexico / G. Martínez-Calderón 166 / 1940 / "Se toma cuando siente dolor en los riñones" / "'Rinonina'" Castillian; "'Riñog'" Chinanteco

4319 *B. ocimoides* Brit. Honduras / W. A. Schipp 719 / 1931 / "...used by Waiki Indians for snake bite remedy"

4320 *B. verticillata* Mexico / E. Palmer 729 / 1898 / "'Sanguinaria blanco,' used as a purge and in kidney diseases"

4321 *B. laevis* W.I.: Dominica / W. H. & B. T. Hodge 3372 / 1940 / "Used for tea for colds" / "'l'herbe acouette'"

4322 *B. articularis* Philippine Is. / G. E. Edaño 1684 / 1949 / "The stem and leaves is used as a medicine for pain in the breast. The part used is ground into fine particles, soaked in water and the liquid given as drink" / "'Magmomong'" Ma

4323 *Galium circaezans* U.S.: Mass. / E. F. Fletcher s.n. / 1913 / "'Wild Liquorice'"

4324 *G. bungei* Philippine Is. / W. Beyer (Herb.) 8845 / 1948 / "Heated leaves & stems placed between teeth to treat toothache" / "'Banata'"? Ifugao

4325 *G. mollugo* EUROPE / N. C. Seringe (Herb. Helveticum) 2337 / no year / "Fleurs pectorales, antispasmodiques, conseillées contre l'epilepsie..."

4326 *G. odoratum* EUROPE / N. C. Seringe (Herb. Helveticum) 2325 / no year / "...Herbe et fleurs séches diurétiques, aperitives"

4327 *G. verum* EUROPE / N. C. Seringe (Herb. Helveticum) 2330 / no year / "Fleurs pectorales, diurétiques..."

4328 *Rubia cordifolia* China / W. T. Tsang 21766 / 1932 / "Root as medicine to heal wound" / "'Hung Sz Min'"

Hainan / W. T. Tsang 79 / 1928 / "... medicines" / "'Hung Sz Sin T'ang'"

4329 *R. tinctorum* EUROPE / N. C. Seringe (Herb. Helveticum) s.n. / no year / "... Racine apéritive, diurétique, astringente, etc...."

4330 *Relbunium hypocarpium* Jamaica / G. E. Nichols 147 / 1903 / "'Cinchona'"

4331 *Rubiacea* indet. Mexico / E. Langlassé 221 / 1898 / "Le fruit, de forme rond, selon les indigènes l'employe contre les maux de tête"

4332 *Rubiacea* indet. Brazil / R. Froes 11818 / 1940 / "Flowers mixed with food are used to kill rats, dogs, etc." / "'Flor de Rato'"

4333 *Rubiacea* indet. Siam / P. Charoenmayu 494 / 49 / "Medicinal roots"

4334 *Rubiacea* indet. Siam / native collector S.24 Royal For. Dept. 4505 / '48 / "Medicinal roots used in child birth" / "'Sarn Khao Tawk'"

4335 *Rubiacea* indet. S. New Britain / K. J. White N.G.F. 10058 / 58 / "Bark scraping used to heal wounds. Native name Igey"

4336 *Rubiacea* indet. Angola / L. S. Tucker 43 / 1924 / "Roots used for medicine for chest disease" / "'Apambaolongombe'"

CAPRIFOLIACEAE

4337 *Sambucus simpsonii* Cuba / F. Marie-Victorin 21602 / 1943 / "Planté pour fins médicinales autour des bohios"

W.I.: Dominica / D. Taylor 5 / 1941 / "Suyeau (Sureau): used for chills, colds etc.—as a 'tea'"

4338 *S. mexicana* Mexico / A. Lopez 40 / 1948 / "'Yerba para la tos'" / "'Sauco'"

Colombia / W. A. Archer 718 / 1930 / "Blossom used in medicinal drink" / "'Sauco blanco'"

4339 *S. oreopola* Guatemala / J. A. Steyermark 49904 / 1942 / "Flowers boiled in water and the infusion drunk is considered to relieve catarrh and colds" / "'salco'"

4340 *S. australis* Argentina / Eyerdam, Beetle & Grondona 23360 / 1938 / "Fruit commonly used for making wine" / "'Seco'"

4341 *S. mexicana* var. *bipinnata* Colombia / Killip & Smith 16820 / 1927 / "Used for catarrh and grippe" / "'Sauco'"

Peru / Y. Mexia 8132 / 1936 / "Leaves used in infusion for cough" / "'Sauco'"

4342 *S. peruviana* Bolivia / I. Steinbach 6584 / 1924 / "Se usa la infusion para sacar sudar" / "Sauco"

4343 *S. sieboldiana* China / W. T. Tsang 20973 / 1932 / "... medicine"

4344   *S. williamsii* China / J. Hers 663 / 1921 / "'su kin shu' much used by physicians for curing sprains etc."

4345   *S. racemosa* China / J. Hers 212 / 1917 / "local name: shan siu kiu (or) niu la shih 'purgative for cattle'"

4346   *S. wightiana* Kashmir / R. R. Stewart 5814 / 1920 / "fruit ... natives call it poisonous. We ate it cooked"

4347   *Viburnum dilatatum* Japan / E. Elliott 26 / '46 / "Children eat the fruit.... Young leaves may be eaten cooked with rice" / "Gamazumi"

4348   *V. fordiae* China / W. T. Tsang 23945 / 1934 / "... fr. edible" / "Fo Chai Tsz Shue"

4349   *V. mongolicum* China / J. Hers 2728 / 1923 / "... bark sometimes used as 'amadou'" / "nuan mu tiao"

4350   *V. foetidum* Burma / J. F. Rock 1969 / 1922 / "tree used for fevers"

4351   *Lonicera affinis* China / W. T. Tsang 21324 / 1932 / "... fruit, black, edible" / "Sai Ip Kam Ngan Fa"

4352   *L. reticulata* China / W. T. Tsang 21202 / 1932 / "... flower, yellow, fragrant, edible" / "Kam Ngan Fa"

VALERIANACEAE

4353   *Valeriana urticaefolia* Mexico / Y. Mexia 8979 / 1937 / "Browsed by deer"

4354   *V. puichella* Costa Rica / Holm & Iltis 572 / 1949 / "... rootstock aromatic"

4355   *V. microphylla* Ecuador / W. H. Camp 4004 / 1945 / "'Valerian.' Has the reputation of being the best of the group medicinally"

4356   *Patrinia villosa* China / W. T. Tsang 24196 / 1934 / "... fr. edible" / "'Fu Chai'"

China / W. T. Tsang 21538 / 1932 / "... leaves edible" / "'Fu Chai Ma'"

China / W. T. Tsang 22727 / 1933 / "... flower white fragrant, it is eaten to remove poison from the system" / "'Fu Tsai Tsoi'"

CUCURBITACEAE

4357   *Apodanthera roseana* Mexico / T. S. Brandegee s.n. / 1904 / "'Yerba Buena'"

4358   *Melothria scabra* El Salvador / P. C. Standley 22613 / 1922 / "'Sandia de culebra'"

4359   *M. mucronata* China / S. K. Lau 435 / 1932 / "Fr. edible" / "'Thing T'o tsz'"

4360   *Melothria* indet. Philippine Is. / L. E. Ebalo 248 / 1939 / "... leaves bitter, used as a cure for black eye" / "'pamiat' Mang.

4361   *Anguria magdalenae* Colombia / Killip & Smith 14260 / 1926 / "'Contra cempie'"

4362   *A. umbrosa* Venezuela / H. Pittier 7931 / 1918 / "N.v. Pasana. Rhizome said to be poisonous"

4363   *Ibervillea tripartita* U.S.: Tex. / E. Marsh 342 / 37 / "'Snake Vine'"

4364   *I. lindheimeri* Mexico / C. V. Hartman 811 / 1891 / "Fruits said to be very poisonous"

4365   *Momordica charantia* Cuba / C. F. Baker 99 / 1907 / "The Chinese are said to cook the flesh of the fruit, and also to eat this and the red arils in salads. It is known in Cuba as Cundeamor"

4366   *M. charantia* var. *abbreviata* Mexico / W. G. Wright 1206 / 1889 / "...fruit gathered & sold for food for cage birds"

4367   *M. acuminata* N. Borneo / Goklin BNB For. Dept. 2805 / 1933 / "Native medicine" / "'Pangara jalil' (Dusun)"

4368   *M. cochinchinensis* Philippine Is. / C. O. Frake 849 / 1958 / "Young leaves applied for internal pain" / "'Tabulu' Sub."

4369   *Luffa cylindrica* Japan / E. Elliott 8 / 46 / "...melon...leaves... eaten. Sap oozing from the cut surface of the stem is gathered and used to cure cough and phlegm" / "'Hechimo'"

4370   *Cucumis anguria* Mexico / P. C. Standley 2357 / 1936 / "Warihios decoct the roots for stomach troubles and eat the tender immature fruits" / "'Melon de Coyote'"

4371   *Bryonopsis affinis* Solomon Is. / L. J. Brass 2706 / 1932 / "Sap said by the natives to cause blindness"
Solomon Is. / S. F. Kajewski 2519 / 31 / "...when a man has diarrhoea the leaves and stems are boiled in water and drunk to cure it"

4372   *Gymnopetalum panicaudii* China / H. Fung 20457 / 1932 / "...edible" / "'Shan Sai Kwa'"

4373   *Lagenaria leucantha* China / Cheo & Yen 278 / 1936 / "...fruit edible"

4374   *Hodgsonia heteroclita* China / T. T. Yü 20518 / 1938 / "Edible"

4375   *Calicophysum brevipes* Venezuela / H. Pittier 7381 / 1917 / "...fr.... contains an acrid, poisonous milk"

4376   *Schizocarpum filiforme* Mexico / Y. Mexia 918 / 1926 / "Vine. 'Estropajillo.' Bitter tasting. Eaten by cattle"

4377   *S. palmeri* Mexico / P. C. Standley 1032 / 1934 / "'Zandia de Culebra,' Mex. 'Si-no-ja-ru,' W."

4378   *Cayaponia racemosa* El Salvador / P. C. Standley 19724 / 1922 / "Said to be poisonous, especially for cattle" / "'Sandia de culebra,' 'Hierba coral,' 'Camara,' 'Taranta'"

4379   *C. citrullifolia* Paraguay / W. A. Archer 4935 / 1937 / "Roots used as blood purifier" / "'tayuyá'"

4380   *C. diversifolia* Argentina / W. A. Archer 4620 / 1936 / "Frt. edible" / "'maracuya'"

4381   *Elaterium aliatum* El Salvador / P. C. Standley 19177 / 1921–22 / "Young shoots cooked . . . & eaten" / "'Cuchinito'"

4382   *E. gracile* El Salvador / P. C. Standley 21441 / 1922 / "Reputed poisonous to cattle" / "'Taranta'"

4383   *Cyclanthera tonduzii* Guatemala / Williams & Molina 9221 / 1947 / "Cultivated for fruit" / "'Caiba'"

4384   *Cucurbitacea* indet. Mexico / M. Stewart 1546 / 1941 / "'Comida de Vibora'"

CAMPANULACEAE

4385   *Codonopsis lancifolia* China / Y. W. Taam 40 / 1937 / "Fruit edible" / "'Tiao-lo kue'"

4386   *Pentaphragma spicatum* China / W. T. Tsang 26862 / 1936 / ". . . fr. edible"

4387   *Sphenoclea zeylanica* U.S.: Tex. / V. L. Cory 50767 / 1945 / ". . . a favorite forage plant of cattle"

4388   *Centropogon cornutus* W.I.: Tobago / D. Abbott 4 / 1959 / "'Deer Meat'"

4389   *C. calycinus* Ecuador / Y. Mexia 7701 / 1935 / "Used as a love potion" / "'Imachutegangue'"

4390   *Siphocampylus corymbiferus* Brazil / Y. Mexia 5709 / 1931 / "Said to be poisonous to stock" / "'Herva de rato'"

4391   *Lobelia fenestralis* Mexico / E. Palmer 713 / 1898 / "Sold . . . for heart diseases"

4392   *L. laxiflora* var. *mollis* El Salvador / P. C. Standley 19268 / 1921–22 / "Decoction used as vomitive, also for internal inflammation and ulcers" / "'Diente de chucho,' 'Lengua de chucho'"

4393   *L. decurrens* Peru / F. Woytkowski 6085 / 1960 / ". . . fruit green; highly toxic; no animals eat it"

4394   *L. zeylanica* China / W. T. Tsang 21576 / 1932 / ". . . used to make medicine" / "'Fai Yeung Tso'"

4395   *L. philippinensis* Celebes / Eyma 389 / 1937 / ". . . milkjuice white, should be cause of blindness"

GOODENIACEAE

4396   *Scaevola plumieri* W.I.: Barbuda / J. S. Beard 367 / 1944 / ". . . fruit said to be eaten by gulls" / "'Gull feed'"

4397   *S. frutescens* Solomon Is. / S. F. Kajewski 2241 / 30 / "The leaves are crushed and given with coconut to the native dogs, to make their scent very keen for possums" / "'Porga'"

Caroline Is. / C. C. Y. Wong 162 / 1947 / "The fruit is used for *säfein sät*, sea medicine, by pounding the fruit which is placed in a coconut cloth and squeezing the juice on coconut meat in a bowl, then eating the coconut meat. The white flower is used as medicine for *meseta*, conjunctivitis (Elbert says the berries are used for inflamed eyes)" / "'remes'"

4398   *S. floribunda* Fiji Is. / L. Reay 14 / 1941 / "Used as stomach medicine and as hair dye" / "'Turulevu'"

COMPOSITAE

4399   *Struchium vaillantii* Brit. Guiana / A. S. Hitchcock 16621 / 1919 / "Called ant-bush"

4400   *Pacourina edulis* Dominican Republic / R. A. & E. S. Howard 9852 / 1946 / "...plant strong scented"

4401   *Centratherum muticum* Brazil / Y. Mexia 5388 / 1930 / "...foliage has a mint odor"

4402   *Erlangea glabra* Tanganyika Terr. / Mrs. M. V. Loveridge 691 / 1939 / "Flower ... with strong and unpleasant odor"

4403   *Blanchetia heterotricha* Brazil / D. Bento Pickel 3121 / 1932 / "Cultivated ..."

4404   *Vernonia novaeboracensis* U.S.: Mass.(?) / G. Gilbert s.n. / 1894 / "(Iron-weed)"

U.S.: N.J. / J. A. Kelsey s.n. / 1892 / "Iron-weed"

4405   *V. glauca* forma *longiaristata* U.S.: Va. / M. Whitesel 3272 / 34 / "Iron weed"

4406   *V. altissima* U.S.: Ohio / R. J. Webb 396 / 1899 / "Tall Ironweed"

4407   *V. baldwinii* var. *interior* U.S.: Tex. / V. L. Cory 50796 / 1945 / "'ironweed'"

4408   *V. deppeana* El Salvador / P. C. Standley 22770 / 1922 / "Remedy for pain in stomach" / "'Suquináy'"

4409   *V. leiocarpa* Guatemala / P. C. Standley 61234 / 1938 / "'Barrete'"

Honduras / J. B. Edwards 168 / 1932 / "'Acerillo'"

El Salvador / P. C. Standley 20119 / 1922 / "Remedy for asthma" / "'Palo de asma'"

4410   *V. liatroides* Mexico / E. Seler 307 / 1887 / "'Hamalacatlocotli'"

4411   *V. palmeri* Mexico / J. Gonzales Ortega 5.618 / 1925 / "'Tacotillo'"

4412   *V. patens* Mexico / C. & E. Seler 1979 / 1896 / "'Cordon de vieja'"

El Salvador / P. C. Standley 21227 / 1922 / "'Pié de zope'"

El Salvador / P. C. Standley 20534 / 1922 / "'Suquinayo'"

El Salvador / S. Calderón 158 / 1921 / "'Palo blanco'"

Costa Rica / Dodge & Thomas 6430 / 1930 / "'Tuete'"

Costa Rica / P. H. Allen 5909 / 1951 / "Crushed leaves reported used to stop nose bleed, being crushed and thrust up the nostril" / "'Tuete'"

4413 *V. salvinae* Guatemala / P. C. Standley 68303 / 1939 / "'Araña'"

4414 *V. sinclairi* Mexico / Y. Mexia 755 / 1926 / "'Vara de Cuervo'"

4415 *V. tortuosa* Mexico / R. Cárdenas 513 / 1910 / "'Rica'"

Honduras / P. C. Standley 56060 / 1928 / "Fls. . . . with strong vanilla odor"

El Salvador / S. A. Padilla 283 / 1922 / "'Aroma'"

Costa Rica / Dodge, Hanckel & Thomas 6460 / 1930 / "Teute (native name)"

4416 *V. triflosculosa* El Salvador / P. C. Standley 21810 / 1922 / "'Sanquillo, Suquináy prieto'"

El Salvador / P. C. Standley 22667 / 1922 / "'Rájate bien'"

El Salvador / S. Calderón 346 / 1922 / "'Rájate bien'"

4417 *V. baccharoides* Bolivia / Rusby & White 110 / 1921 / "Strong odor of vanilla"

Bolivia / Rusby & White 606 / 1921 / "Strong odor of vanilla" / "'Lejía'"

4418 *V. hoveaefolia* Brazil / Y. Mexia 5551 / 1931 / "Used in infusion for stomach cramps" / "'Linahaso do campo'"

4419 *V. scorpioides* Brazil / Y. Mexia 4939 / 1930 / "Used as remedy for mange" / "'Herva de Coelho'"

4420 *V. squamulosa* Argentina / Venturi 326 / 1919 / "'Santa Rosa'"

4421 *V. cinerea* Solomon Is. / S. F. Kajewski 2187 / 1930 / "The leaves of this plant are macerated and applied to picaninnies heads for pains in that region" / "'Ilipah'"

4422 *V. cuneata* Solomon Is. / S. F. Kajewski 2500 / 1931 / "When the natives have fever, the roots are macerated and the sap drank to cure it" / "'Kaka-por'"

4423 *Vernonia* indet. Philippine Is. / C. O. Frake 596 / 1958 / "ritual" / "'silikmata-ulangan'"

4424 *Vernonia* indet. Philippine Is. / W. Beyer 15 / 1948 / "Heated mashed leaves is rubbed against tung, to treat whitish tung of children" / "'Kahing'"

4425   *Vernonia* indet. Philippine Is. / M. D. Sulit 4331 / 1951 / "Juice and leaves said to be medicinal on newly cut wound" / "'Mabuas'" Bis

4426   *V. polyura* Nyasaland / L. J. Brass 16374 / 1946 / "...leaves...bitter, used in native medicine" / "'Fusa' (Chinyanja)"

4427   *Oliganthes discolor* Colombia / O. Haught 6259 / 1948 / "Leaves... eaten by cattle where in reach"

4428   *Lychnophora brunioides* Brazil / V. A. Moreira 5739 / 1945 / "Mix with alcohol for a 'tonica'"?

4429   *Elephantopus hypomalacus* El Salvador / P. C. Standley 23582 / 1922 / "'Hierba de pincel'"

4430   *E. mollis* Mexico / Y. Mexia 1783 / 1927 / "Hairs on stem irritating"
Guatemala / P. C. Standley 75866 / 1940 / "'Contrayerba'"
Philippine Is. / C. O. Frake 604 / 1958 / "Leaves applied to wounds" / "'bingkenutung'"? Sub.

4431   *E. spicatus* El Salvador / P. C. Standley 19997 / 1922 / "Remedy for dysentery" / "'Oreja de coyote'"
El Salvador / P. C. Standley 20453 / 1922 / "'Amor seco'"
El Salvador / S. Calderón 128 / 1921 / "'Oreja de chucho'"
El Salvador / P. C. Standley 21990 / 1922 / "'Escoba'"
Colombia / H. Schiefer 306 / 1945 / "Used for diarrhea" / "bought in market. Cali, Valle del Cauca"

4432   *E. scaber* China / W. T. Tsang 23076 / 1933 / "...used as medicine"
China / W. T. Tsang 21591 / 1932 / "...can be used as medicine"

4433   *Rolandra fruticosa* W.I.: Dominica / W. H. & B. T. Hodge 3278 / 1940 / "lvs. used as tea for sores..." / "'tête negresse'"

4434   *Piqueria trinervia* Mexico / E. Palmer 596 / 1898 / "Market of San Luis Potosi"

4435   *Phania curtissii* Cuba: Isla de Pinos / E. L. Ekman 11888 / 1920 / "...whole plant aromatic"

4436   *Ophryosporus bipinnatifidus* Peru / W. J. Eyerdam 25155 / 1939 / "Aromatic fragrance"

4437   *O. charua* Argentina / Eyerdam & Beetle 22194 / 1938 / "Has strong but not unpleasant odor"

4438   *O. scabrellus* Mexico / H. S. Gentry 1984 / 1935 / "Odorous, visited by beetles and Hymenoptera"

4439   *O. paradoxus* Chile / Worth & Morrison 16704 / 1938 / "Odoriferous"

4440   *O. piquerioides* Peru / Y. Mexia 8125 / 1936 / "...pleasantly pungent..."

4441 *Alomia microcarpa* Costa Rica / P. C. Standley 32704 / 1924 / "'Santa Lucía'"

4442 *A. fastigiata* Brazil / Hoehne & Gehrt. / 17504 / 1926 / "'Mata-pasto'"

4443 *Ageratum conyzoides* Guatemala / V. Grant 597 / 1940 / "herbage has minty odor"

El Salvador / P. C. Standley 19220 / 1921–22 / "Remedy for gonorrhoea" / "'Mejorana'"

China / G. W. Groff 16 / 1919 / "Used to feed fish" / "'An ii fa' Ha' Tchûn Tsô'"

Siam / Wichian 307 / 1946 / ". . . leaves foetid smelling"

Philippine Is. / E. D. Merrill 35 / 1902 / ". . . with somewhat the odor of Hierochloa"

4444 *A. corymbosum* Guatemala / P. C. Standley 61206 / 1938 / "'Mejorana'"

4445 *A. oerstedii* Costa Rica / E. W. D. Holway 324 / 1915 / "'Santa Lucia'"

4446 *Stevia trifida* Mexico / H. S. Gentry 7316 / 1945 / "Strong fragrant odor . . ."

Mexico / H. S. Gentry 7211 / 1945 / "Pleasantly aromatic"

Mexico / E. Langlassé 33 / 1898 / "Racine et fleurs en infusion contre la dysenterie" / "'Manzanilla del agua'"

4447 *S. aschenborniana* Mexico / G. B. Hinton 13380 / 1938 / "Plant fragrant"

4448 *S. polycephala* Guatemala / A. F. Skutch 170 / 1933 / "Foliage has faint odor of *Calycanthus carolinianus*"

4449 *S. salicifolia* Mexico / E. K. Balls 4120 / 1938 / "Leaves emitting strong-scented gum when crushed"

4450 *S. salicifolia* var. *exaristata* Mexico / A. Dugès 475 / 1893 / "'Yerba de la mula'"

4451 *S. lucida* Mexico / A. Dugès 274 / 1893 / "'Yerba del aire'"

Venezuela / A. Jahn 546 / 1921 / "'Chirca'"

Colombia / Killip & Smith 17171 / 1927 / "(Used for inflammation)"

4452 *S. berlandieri* Mexico / W. C. Leavenworth 149 / 1940 / "Aromatic . . ."

4453 *S. glandulosa* Mexico / Y. Mexia 1476-a / 1927 / "Used medicinally in fevers"

4454 *S. pilosa* Mexico / C. G. Pringle 3478 / 1890 / "'Flor de María'"

4455 *S. rhombifolia* Guatemala / J. A. Steyermark 51753 / 1942 / "'flor de plata'" / "'tuán'"

Mexico / Leavenworth & Hoogstraal 1670 / 1941 / "Fragrant shrub . . ."

Guatemala / V. Grant 632 / 1940 / "herbage gives aromatic odor" / "Tea for 'dismenorrea'" / "'Peracon'"

4456 *Stevia* indet. Mexico / T. Morley 653 / 1946 / "...herbage with pungent odor"

4457 *S. elongata* var. *caracasana* Venezuela / A. Jahn 1098 / 1922 / "'Molinillo'"

4458 *Stevia* indet. Venezuela / W. Gehriger 6 / 1930 / "'Chirca'"

4459 *S. puberula* Peru / Macbride & Featherstone 1288 / 1922 / "A tea substitute and stomach medicine" / "Foliage fragrant" / "'Limalima'"

4460 *Stevia* indet. Argentina / Periano 58429 / 1936 / "'albahaca del campo'"

4461 *Trichocoronis rivularis* Mexico or Central America: Monterey / Dr. Gregg 50 / 1847 / "Used for mouthwash"

4462 *Trichogonia villosa* Brazil / Y. Mexia 5707 / 1931 / "...crushed leaves are rubbed on tick bites" / "'Inxota'"

4463 *Eupatorium purpureum* U.S.: Va. / Fernald & Long 8488 / 1938 / "Drying plant with odor of Vanilla"

4464 *E. compositifolium* U.S.: S.C. / H. W. Ravenel? s.n. / no year / "dog fennel"

4465 *E. capillifolium* U.S.: Tex. / V. L. Cory 49847 / 1945 / "'hogweed'"
Cuba / C. Wright 2810 / 1860–64 / "Emits bruised the odor of Fennel"
Bahama Is. / R. A. & E. S. Howard 10218 / 1948 / "...foliage aromatic"

4466 *E. leucolepis* U.S.: Tex. / V. L. Cory 49774 / 1945 / "'justice weed'"

4467 *E. aromaticum* U.S.: Va. / A. Ryland 3797 / 1934 / "'Smaller White-Snakeroot'"
U.S.: Va. / J. B. Lewis 2977 / 1933 / "'White Snakeroot'"
U.S.: N.C. / P. O. Schallert s.n. / 1921 / "'White Sanicle'" / "'White Snake-root'"
U.S.: N.C. / P. O. Schallert s.n. / 1921 / "'Smaller White Snake-root'"

4468 *E. verbenaefolium* U.S.: Mich. / A. K. & G. C. Harrison s.n. / 1908 / "'White Snakeroot'"

4469 *E. rugosum* U.S.: Va. / M. Canada 2981 / 1932 / "'White Snakeroot'"
U.S.: Ohio / E. L. Moseley s.n. / 1909 / "This plant causes trembles and milksickness!"
U.S.: Ala. / L. V. Porter s.n. / 1938 / "This plant is said to poison cattle" / "'White Snakeroot'"

U.S.: Minn. / E. A. Mearns s.n. / 1891 / "Collected for the Army Medical Museum" / " 'White Snake-root' "

4470   *E. rugosum* var. *angustatum* U.S.: Tex. / G. B. Ryan s.n. / 1931 / "So-called 'Poison Weed' from . . . Edwards Plateau. Reported as very destructive to sheep"

4471   *E. sessilifolium* U.S.: Penn. / J. A. Shafer 1492 / 1901 / " 'Upland Boneset' "

4472   *E. perfoliatum* f. *truncatum* U.S.: Va. / Ellyson & Puette 3706 / 1934 / " 'Vervain Throughwort' "

4473   *E. wrightii* U.S.: Tex. / E. Mash 358 / 37 / " 'Snake Weed' "

Mexico / A. López 18 / 1947 / " 'Oregano' "

4474   *E. incarnatum* U.S.: Ark. / H. C. Benke 4808 / 1928 / " 'Pink Thoroughwort' "

4475   *E. azureum* Mexico or Central America: Monterey / Dr. Gregg 52 / 1847 / "(Used for astringent poultice)"

4476   *E. glabratum* Mexico / C. Ehrenberg 400.-b / no year / " 'Yerba del Golpe' "

4477   *E. leucocephalum* Guatemala / P. C. Standley 68433 / 1939 / " 'No-chebuena' "

4478   *E. mairetianum* Mexico / E. Langlassé 50 / 1898 / " 'Amargosillo' "

4479   *E. monanthum* Mexico / H. S. Gentry 5651 / 1940 / ". . . odorous shrub"

4480   *E. odoratum* Honduras / P. C. Standley 55020 / 1928 / " 'Rey del todo, Crucito' "

Puerto Rico / Shevslohn? 69 / 1914 / ". . . herbage ill scented. . ."

W.I.: Trinidad / W. E. Broadway s.n. / 1932 / "A medicinal plant" / " 'Christmas bush' "

W.I.: Grenada / R. A. Howard 10685 / 1950 / ". . . foliage aromatic"

W.I.: Grenada / R. Day 24 / 1908 / " 'Christmas Bush' "

4481   *E. oresbioides* Guatemala / J. A. Steyermark 34705 / 1940 / " 'flor de celeste' "

4482   *E. pazcuarense* Mexico / E. Langlassé 40 / 1898 / " 'Amargosillo' "

4483   *E. phoenicolepis* El Salvador / P. C. Standley 21550 / 1922 / " . . . lvs. with strong odor"

4484   *E. prunellaefolium* Guatemala / J. A. Steyermark 50332 / 1942 / " 'heliotropa de monte' "

4485   *E. pycnocephalum* Guatemala / S. F. Blake 7576 / 1919 / "Leaves used for a sudorific tea" / " 'té' "

Costa Rica / Dodge, Hanckel & Thomas 6453 / 1930 / "Santa Lucía native name"

El Salvador / S. Calderón 156 / 1921 / "'Mejorana'"

El Salvador / P. C. Standley 22176 / 1922 / "Used for eye affections" / "'Mejorana'"

El Salvador / P. C. Standley 22518 / 1922 / "remedy for erysipelas" / "'Mejorana'"

4486   *E. quadrangulare* Mexico / Y. Mexia 1280 / 1926 / "Hollow stem . . . inhabited by ants"

El Salvador / P. C. Standley 19203 / 1921–22 / "Eaten by cattle" / "'Chimaliote hueco'"

4487   *E. tuerckheimii* Guatemala / J. A. Steyermark 42561 / 1942 / "Leaves . . . smelling of helianthoid-like odor . . ."

4488   *Eupatorium* indet. Nicaragua / W. R. Maxon 7661 / 1923 / "'Conejito'"

4489   *E. bahamense* Bahama Is. / O. Degener 19120 / 1946 / "Boil leaves, add salt and drink for bellyache; 'cat tongue'"

4490   *E. brachychaetum* Cuba / Ekman 16247 / 1923 / "Smells strongly of coumarine when dried"

4491   *E. dalea* Jamaica / W. Harris 6667 / 96 / "'Cigar Bush'"

4492   *E. portoricense* Puerto Rico / Shevslohn? 118 / 1914 / ". . . foliage ill scented"

4493   *E. pseudo-dalea* Cuba / E. L. Ekman 16537 / 1923 / ". . . smells of coumarine when drying"

4494   *E. triste* Jamaica / F. Shreve s.n. / 1906 / "Cinchona (introduced)"

4495   *E. sciatraphes* Venezuela / H. Pittier 7882 / 1918 / "Medicinal, pectoral" / "'Niquibao'"

4496   *E. acuminatum* Colombia / Kalbreyer 856 / 1878 / ". . . much used to scent tobacco" / "'Apio del Monte'"

4497   *E. inulaefolium* forma *suaveolens* Colombia / Killip & Smith 16717 / 1927 / "(used to cure sick animals)" / "'Almorduz'"

Bolivia / J. Steinbach 7050 / 1925 / "'Chilcas de los Barbechos'"?

Brazil / Y. Mexia 4569 / 1930 / "'Santa Cruz'"

Uruguay / Rosengurtt B1251 / 1937 / ". . . aromático"

4498   *E. paezense* Colombia / J. Triana 3 or 1171 / 1853 / "'Yerba de los Uribes'" / "'Sanalo-todo'"? / "'venturosa morada'"

4499   *E. vitalbae* Colombia / O. Haught 2728 / 1939 / "Plant has strong but not unpleasant odor"

Peru / Y. Mexia 8106 / 1936 / "'Camote Huano'"

4500   *E. diplodictyon* Peru / J. F. Macbride 4296 / 1923 / "Herbage pungent"

4501　*E. pentlandianum* Peru / F. L. Herrera 1401 / 1927 / "'tayac-chchillca'"

　　　Peru / F. L. Herrera 1478 / 1924 / "'Kkiuña'"

　　　Peru / J. West 3857 / 1935 / "'Jasmin'"

　　　Peru / F. L. Herrera 771 / no year / "'tarac-chama'"

4502　*E. persicifolium* Peru / Macbride & Featherstone 1287 / 1922 / "'Ancuchuta'"

　　　Peru / F. L. Herrera 1122 / 1926 / "'chamanuai'"

4503　*E. rhodotephrum* Peru / Macbride & Featherstone 1508 / 1922 / "'Macha macha'"

4504　*E. sternbergianum* Peru / Stork & Vargas 9328 / 1938 / "Not browsed by cattle or even goats"

　　　Peru / Killip & Smith 22150 / 1929 / "Used in flavoring in cooking" / "'Marmaquillo'"

　　　Peru / F. L. Herrera 1474 / 1927 / "'Manca-ppaqui'"

4505　*E. crenulatum* Bolivia / J. Steinbach 5330 / 1921 / "'Chilca'"

4506　*E. hecatanthum* Bolivia / J. Steinbach 6763 / 1924 / "Muy aromático"

　　　Argentina / T. Meyer 296 / 1931 / "Los tallos y las hojas emanan al ser restregados un olor fuerte y poco agradable"

4507　*E. laevigatum* Bolivia / J. Steinbach 2731 / 1916 / "'chilca'"

　　　Bolivia / J. Steinbach 7064 / 1925 / "'Chilca umbrelliforme'"

　　　Brazil / Edwall 16280 / 905 / "'Cambará de meia legua'"

4508　*E. simillimum* Bolivia / J. Steinbach 6647 / 1924 / "Del cocimiento de las hojas dice que se saca un linda color azul-indigo" / "'Anilero'"

4509　*E. subscandens* Bolivia / J. Steinbach 7086 / 1925 / "'Chilca blanda'"?

4510　*E. intermedium* Brazil / Y. Mexia 4432 / 1930 / "Slight pungent odor. Leaves are cooked for greens" / "'Capisol'"

4511　*E. itatiayense* Brazil / Y. Mexia 4209 / 1930 / "Visited by bees. 'Cruz branca'"

　　　Brazil / Zikán 12 / 1921 / "'Cambara'"

4512　*E. ivaefolium* Brazil / Y. Mexia 5651 / 1931 / "'Asápéche do miudo'"

4513　*E. macrophyllum* Brazil / A. Chase 8433 / 1925 / "...heavy odor"

4514　*E. squalidum* Brazil / Y. Mexia 5526 / 1931 / "'Asapéche'"

4515　*E. buniifolium* Uruguay / C. Osten 4469 / 1906 / "'chilca'"

　　　Argentina / A. T. Hunziker 6660 / 1946 / "'romerillo crespo'"?

　　　Argentina / A. Burkart 1896 / 1928 / "...follaje perfumado..."

　　　Argentina / W. Lossen 265 / no year / "'Romerito'"

　　　Argentina / F. Kurtz 9280 / 1897 / "'Chilquilla'"

4516   *E. clematideum* Argentina / Eyerdam & Beetle 22601 / 1938 /
"...lvs. aromatic"

4517   *E. crenulatum* var. *tucumanense* Argentina / A. G. Schulz 184 / 1931
/ "Planta muy olorosa"
Argentina / S. Venturi 1802 / 1922 / "'N̄atchará'"

4518   *Eupatorium* indet. Argentina / E. L. Ekman 1126 / 1908 / "'Martires chico'"

4519   *E. salvia* Chile / D. Bertero 988 / 1829 / "'Salvia macho'"
Chile / J. West 3944 / 1935 / "...aromatic foliage"

4520   *E. japonicum* Japan / E. Elliott 10 / 1946 / "Young leaves are boiled
and eaten well seasoned"

4521   *E. stoechadosmum* Japan / E. Elliott 66 / 1946 / "Medicinal use:
root decoction is antidote to poisoning; also regulates menstruation" /
"'Fujibokamo'"

4522   *Critonia dalea* Jamaica / W. Harris 8235 / 1901 / "'Cigar Bush'"

4523   *Eupatoriastrum nelsonii* var. *cardiophyllum* El Salvador / S. Calderón
1679 / 1923 / "The plant is highly esteemed as a depurative ... its
root ... pungent and aromatic.... The leaves are slightly bitter ..." /
"'Sunsunpate, Raíz barbona'"

4524   *Mikania scandens* U.S.: Mass. / E. F. Fletcher s.n. / before 1923 /
"'Climbing Hemp-weed'"
U.S.: Va. / M. Ryland 3171 / 1933 / "'Climbing Hemp-weed'"
U.S.: Tex. / V. L. Cory 49723 / 1945 / "'Climbing boneset'"

4525   *M. cordifolia* Mexico / C. & E. Seler 279 / 1888 / "'toxichec cimarron'"?
Mexico / E. Kerber 336 / 1883 / "'Barba de chivo'"
Peru / Y. Mexia 8042 / 1936 / "'Huaco verde'"

4526   *M. houstoniana* Mexico / C. & E. Seler 5475 (396) / 1911 / "Ruinas
de Palenque, Chiapas"

4527   *M. micrantha* Mexico / E. Langlassé 840 / 1899 / "Pl. répandant une
mauvaise odeur ..."
Peru / J. M. Schunke 54 / no year / "'Camotillo'"
Fiji Is. / O. Degener 15084 / 1941 / "Leaves squashed and put on
sores as poultice. Also acts as styptic. Squashed leaves put on insect
bites" / "'Bosuthu'" Serua

4528   *M. biformis* Brazil / F. C. Hoehne 24268 / 1929 / "'Guaco'"

4529   *M. glomerata* Brazil / F. C. Hoehne 3454 / 1919 / "'Guaco'"

4530   *M. lindbergii* Brazil / Y. Mexia 5793 / 1931 / "'Cipó de Sao Juão'"

4531   *M. officinalis* Brazil / A. Löfgren 16394 / 88 / "'Caraçáo de Jesus'"?

4532 *M. sessilifolia* var. *regnellii* Brazil / Y. Mexia 5719 / 1931 / "Flower made into bitter tea for stomach trouble" / "'Quássa'"

4533 *M. smilacina* Brazil / Williams & Assis 6928 / 1945 / "Good for rheumatism, to bathe with, and to drink" / "'Balsamo'"

4534 *Mikania* indet. Brazil / D. Nogueira 263 / 1917 / "'Coração de Jesus'"

4535 *Trilisa paniculata* U.S.: Fla. / L. J. Brass 15703 / 1945 / "...aromatic..."

4536 *Brickellia greenei* U.S.: Calif. / D. D. Keck 4882 / 1938 / "With strong sweetish odor"

4537 *B. oblongifolia* var. *linifolia* U.S.: Utah / A. Carter 1542 / 1940 / "...herbage aromatic"

4538 *Brickellia* indet. U.S.: Colo. / J. Ewan 14578 / 1942 / "...herbage fragrant"

4539 *B. botterii* Mexico / E. K. Balls 4290 / 1938 / "Aromatic"

4540 *B. laciniata* Mexico / F. Shreve 9334 / 1939 / "'barra dulce'"

4541 *B. lanata* Mexico / Y. Mexia 876 / 1926 / "'Sanguinaria.' A medicine for the blood is made from this plant. A drink, 'Tepache,' is made from the root"

4542 *B. veronicaefolia* Mexico / G. B. Hinton 16886 / 49 / "Pungent piney odor"

4543 *Barroetia laxiflora* Mexico / H. S. Gentry 1040 / 1934 / "'San Jual del Monte'"

4544 *Kuhnia eupatorioides* var. *corymbulosa* U.S.: Tex. / V. L. Cory 49732 / 1945 / "'false boneset'"

4545 *Liatris acidota* U.S.: Tex. / V. L. Cory 50073 / 1945 / "'button snakeroot'"

4546 *L. pycnostachya* U.S.: Minn. / E. A. Mearns 142 / 1888 / "'Button snakeroot'; 'Blazing-Star'"
U.S.: Tex. / V. L. Cory 50045 / 1945 / "'button snakeroot'"

4547 *L. graminifolia* var. *graminifolia* U.S.: N.C. / P. O. Schallert s.n. / 1921 / "'Gar Feather'; 'Devil's Bit'; 'Backache-root'"

4548 *L. graminifolia* var. *dubia* U.S.: Va. / E. Venzey 4870 / 35 / "'Button Snakeroot'"

4549 *L. gracilis* U.S.: Fla. / G. Gilbert s.n. / 1911 / "'Button Snakeroot'— 'Blazing Star'"

4550 *L. punctata* var. *punctata* U.S.: Tex. / V. L. Cory 50061 / 1945 / "'button snakeroot'"

4551 *L. punctata* var. *coloradensis* U.S.: Colo. / C. N. S. Horner s.n. / no year / "'Slender Button Snakeroot'"

4552 *Grindelia microcephala* var. *microcephala* U.S.: Tex. / V. L. Cory 51121 / 1945 / "'gum plant'"

4553 *G. squarrosa* var. *nuda* U.S.: Tex. / V. L. Cory 50179 / 1945 / "'gum plant'"

4554 *G. oölepis* U.S.: Tex. / V. L. Cory 51421 / 1945 / "'gum plant'"

4555 *G. oxylepis* var. *oxylepis* Mexico? / Dr. Gregg 69 / 1847 / "Resinous plant"

4556 *Gutierrezia sarothrae* U.S.: Utah / Andrews & Noble s.n. / 1936 / "'Match-brush' or 'Snakeweed'"
U.S.: Utah / E. H. Graham 9980 / 1935 / "'Horse Feed'"

4557 *G. gilliesii* Argentina / Eyerdam, Beetle & Grondona 23490 / 1938 / "Apparently not much grazed by sheep"

4558 *Amphipappus fremontii* U.S.: Calif. / C. B. Wolf 6678 / 1935 / "Herbage pleasantly sweetly scented"

4559 *Heterotheca subaxillaris* U.S.: Fla. / L. J. Brass 15380 / 1945 / "...somewhat odorous annual..."

4560 *H. inuloides* Mexico / Y. Mexia 2650 / 1929 / "'Arnica'"

4561 *H. inuloides* var. *rosei* Mexico / W. C. Leavenworth 262 / 1940 / "'Arnica'"

4562 *H. leptoglossa* Mexico / H. S. Gentry 1974 / 1935 / "Slender herb a meter high with an odor like camphor. Medic. A decoction made from flowers for colds and grip" / "'Gordolobo'" Mex.

4563 *Chiliotrichiopsis keideli* Argentina / S. Venturi 4639 / 926 / "'Trompo'"

4564 *Chrysopsis* indet. (*Heterotheca subaxillaris*?) U.S.: N.C. / W. B. Fox 3953 / 1950 / "...aromatic..."

4565 *Solidago curtisii* U.S.: Fla. / L. V. Porter s.n. / 1938 / "Leaves sweet scented"

4566 *S. velutina* Mexico / Stanford, Retherford & Northcraft 239 / 1941 / "...heavily grazed by goats"

4567 *S. microglossa* Argentina / T. Meyer 77 / no year / "...nociva para el ganado"

4568 *S. japonica* Japan / E. Elliott 65 / 1946 / "Food Use: Leaves boiled and eaten with sauce" / "'Akinokirinsō'"

4569 *Haplopappus fremontii* U.S.: Colo. / W. A. Kellerman 4286 / 1905 / "kidney cure"

4570 *H. bloomeri* U.S.: Calif. / W. A. Dayton 461 / 1913 / "Forage value: Unpalatable"

4571 *H. eastwoodae* U.S.: Calif. / S. F. Blake 10129 / 1927 / "Weakly pine-scented"

4572 *Haplopappus* indet. U.S.: Fla. Keys / Brizicky & Stern 342 / 1956 / "...leaves aromatic, viscid..."

4573 *H. spinulosus* Mexico / E. Palmer 62 / 1898 / "Sold in market as Yerba de la vivora, as a blood purifier"

Mexico / Stanford, Retherford & Northcraft 196 / 1941 / "...heavily grazed by goats"

Mexico / Dr. Gregg 6 / 1841 / "Yerba de la vibora in decoction used for pains in stomach—also in disorders of uterus detention of cotamenia, etc."

4574 *H. pulchellus* Chile / J. West 3933 / 1935 / "...aromatic foliage"

4575 *H. remyanus* Chile / R. Wagenknecht 4340 / 1941 / "Su olor peculiar se nota desde lejos" / "Es visitado por mariposas de la familia Satyridae y muchos dípteros"

4576 *Haplopappus* indet. Chile / R. Wagenknecht 18586 / 1940 / "'bailahuén'"

4577 *Chrysothamnus viscidifolius* U.S.: Wyo. / A. Nelson 618 / 1894 / "Poison Spider"

4578 *C. albidus* U.S.: Nev. / H. M. Hall 12194 / 1925 / "Odor, rank; depressed resin-pits evident..."

4579 *C. pulchellus* ssp. *baileyi* U.S.: Tex. / V. L. Cory 50392 / 1945 / "'rabbitbrush'"

4580 *C. nauseosus* ssp. *graveolens* U.S.: Calif. / W. A. Dayton 658 / 1913 / "'rabbit-brush'"

4581 *Pteronia incana* S. Africa / R. Marloth 12107 / 1925 / "...very aromatic, good stock food" / "'Bees' karroo'"

4582 *P. onobromoides* S. Africa / R. Marloth 12398 / 1925 / "Used by the Hottentots mixed with grease for the skin"

4583 *Cyathocline lyrata* Burma / F. G. Dickason 1022 / '32 / "Fl. lav., aromatic"

4584 *Egletes viscosa* Bolivia / J. Steinbach 6601 / 1924 / "...aromatica"

4585 *Calotis hispidula* Australia / C. T. White 9532 / 1933 / "'Bindy-eye'; one of the worst pests in the district"

4586 *Townsendia hookeri* Canada: Saskatchewan / W. Spreadborough 5027 / 1894 / "Medicine Hut"

4587 *Aster spectabilis* U.S.: Mass. / Fernald & Long 17512 / 1918 / "Fresh plant with strong tar-like odor"

4588 *A. puniceus* var. *oligocephalus* N.W. Newfoundland / Fernald & Long 29133 / 1925 / "'Pit-nagen'—smoked by Indians"

4589 *A. tanacetifolia* U.S.: Tex. / V. L. Cory 50372 / 1945 / "'tansy aster'"

4590 *A. exilis* Mexico / C. G. Pringle 13271 / 1904 / "Lecheria"

4591 *A. incisa* Japan / E. Elliott 144 / 1946 / "Food Use: Young leaves are eaten cooked" / "Medicinal Use: Stem and leaf are diuretic, and cure insect bite" / "'Yomena'"

4592 *A. scaber* Japan / E. Elliott 104 / 1946 / "Food Use: Young leaves are reported to be edible" / "'Shirayamagiku'"

4593   *A. tataricus* Japan / E. Elliott 93 / 1946 / "Medicinal Use: dried decocted root cures cough, removes phlegm and is said to stop bleeding after birth of child" / "'Shion'"

4594   *Erigeron annuus* U.S.: Mass. / E. F. Fletcher s.n. / before 1923 / "'Daisy Fleabane'" / "'Sweet Scabious'"

4595   *E. alamosanus* Mexico / H. S. Gentry 1351 / 1935 / "'Durasnillo' W."

4596   *E. karvinskianus* Mexico / C. Conzatti 2474 / 1909 / "Empleado en infusion como Thé"

4597   *E. bonariensis* Puerto Rico / P. Sintenis 382 / 1884 / "'Alegrillo'"

4598   *E. canadensis* Japan / E. Elliott 138 / 1946 / "Food Use: Leaf boiled, chopped, and dried, can be cooked with rice" / "'Hime-mukashiyomo-gi'"

4599   *Diplostephium meyenii* Peru / Mr. & Mrs. F. E. Hinkley 72 / 1920 / "Local name 'Manzanella'"

4600   *Diplostephium* indet. Venezuela / W. Gehriger 473 / 1930 / "... hojas aromáticas"

4601   *Gundlachia corymbosa* Bahama Is. / G. R. Proctor 8919 / 1954 / "'Jamaica trash'"

         W.I.: Barbuda / J. S. Beard 369 / 1954 / "'Yam bush'"

         Bahama Is. / C. B. Lewis s.n. / 1954 / "Resinous shrub" / "'Jamaica trash'"

4602   *Olearia argophylla* Tasmania / L. Rodway H398 / 1929 / "'Native Musk'"

         Australia: New S. Wales / no collector s.n. / 1898 / "'Musk'"

4603   *O. dentata* Australia: New S. Wales / W. A. W. de Beuzeville 44(10) / 1916 / "'Parrot Flower'"

4604   *O. racemosa* Australia / Smith & Everist 932 / 1940 / "Locally known as 'Turkey Bush'"

4605   *Conyza filaginoides* Mexico / A. Dugès s.n. / 1893 / "'Gordolobo'"

4606   *C. lyrata* El Salvador / P. C. Standley 20349 / 1922 / "... offensive odor. Decoction of lvs. remedy for malaria" / "'Talilla'"

         Panama: San José Is. / I. M. Johnston 1332 / 1946 / "Strong scented, clammy herb ..."

4607   *C. triplinervis* Brazil / Y. Mexia 4134 / 1929 / "'Santa Anna'"

4608   *Heterothalamus brunioides* Argentina / E. Seler s.n. / 1910 / "'Romerillo'"

4609   *Baccharis sarothroides* U.S.: Calif. / C. B. Wolf 4345 / 1932 / "Bees attracted in large numbers"

         Mexico / S. S. White 4867 / 1941 / "'Romerillo'"

4610   *B. glutinosa* Mexico / R. S. Ferris 5928 / 1925 / "'jarella'"

Mexico / R. S. Ferris 5474 / 1925 / "'jarrillas'"
Mexico / Y. Mexia 526 / 1926 / "'Jarilla'"
Mexico / W. P. Hewitt 369 / 1948 / "Leaves aromatic" / "'Jarrilla'"
El Salvador / P. C. Standley 19605 / 1921–22 / "'Chilca'"
El Salvador / P. C. Standley 21434 / 1922 / "'Sauce'"
Peru / Stork & Horton 10807 / 1939 / "...lvs. protected with resin"

4611   *B. thesioides* Mexico / H. S. Gentry 1232 / 1935 / "Warihios report medicinal properties"

4612   *B. pilosa* Mexico / W. P. Hewitt 361 / 1948 / "'Aceitilla'"

4613   *B. alamosana* Mexico / H. S. Gentry 4795 / 1939 / "Infused as a potion for ailments" / "'Yerba del Palma'"

4614   *B. conferta* Mexico / E. K. Balls 4052 / 1938 / "Strongly aromatic"

4615   *B. rhexioides* El Salvador / L. V. Velasco 8855 / 1905 / "'Guarda-Barranco'"

4616   *B. trinervis* El Salvador / P. C. Standley 22566 / 1922 / "'Tapa-barranca'"
Guatemala / J. A. Steyermark 49377 / 1942 / "Cook leaves and place in bath for a woman 10 days after she has had a child; give her 3 baths daily. This is considered as a health measure" / "'Santo Domingo'"
Argentina / T. Meyer 2089 / 1936 / "'Chilca'"

4617   *B. halimifolia* Cuba / R. A. Howard 6229 / 1941 / "Foliage resinous"
Australia: Queensland / C. T. White 12636 / 1945 / "Locally regarded as a very serious pest" / "'Groundsel'"

4618   *B. myrsinites* Dominican Republic / E. J. Valeur 1012 / 1933 / "'Cura Maguey'"
Dominican Republic / R. A. & E. S. Howard 9129 / 1946 / "...leaves ...resinous above"
Dominican Republic / M. Fuertes s.n. / 1912 / "'Palo de Toro'"
Haiti / W. M. Mann s.n. / no year / "Ant-inhabited plant"

4619   *B. scoparius* Jamaica / W. R. Maxon 10091 / 1926 / "'Bitter broom'"

4620   *B. microphylla* Venezuela / A. Jahn 516 / 1921 / "'Sanalotodo'"

4621   *B. polyantha* Ecuador / O. Haught 3144 / 1942 / "...strongly resinous odor"

4622   *B. alnifolia* Peru / Mr. & Mrs. F. E. Hinkley 71 / 1920 / "medicinal" / "'Chilca'"

4623   *B. fevillei* Peru / Macbride & Featherstone 638 / 1922 / "'Chilco'"

4624   *B. floribunda* Peru / Macbride & Featherstone 1631 / 1922 / "Leaves bound on cuts, bruises, etc." / "'Chilca'"

Peru / J. West 3735 / 1935 / "'Ullccochilca'"

Peru / Stork & Horton 10387 / 1939 / "many flower heads infected by gall insect"

4625 *B. patiensis* Bolivia / J. Steinbach 6995 / 1925 / "'Chilca'"

4626 *B. scandens* Bolivia / Mandon 19 / 1859 / "'Chilca'"

4627 *B. marginalis* Chile / E. E. Gigoux s.n. / 1885 / "'Chilca-Dadin'"

Chile / Bertero 830 / 1835 / "'Chilquilla del Rio'"

Chile / I. M. Johnston 4854 / 1925 / "'Dadin'"

Argentina / Y. Mexia 04362 / 1936 / "'Chulco'"

4628 *B. paniculata* Chile / R. Wagenknecht 4489 / before 1943 / "'romero común'"

4629 *B. rhomboidalis* Chile / G. Looser 2472 / 1932 / "'romero gaucho'"

4630 *B. rosmarinifolia* Chile / Bertero 79, 834 & 1413 / 1835 / "'Romero,' 'Romerillo,' 'Romero de la tierra'"

Chile / I. M. Johnston 4914 / 1925 / "'Romero'"

4631 *B. artemisioides* Argentina / W. Lossen 196 / no year / "'Pichara blanca'"

4632 *B. articulata* Argentina / T. Meyer 184 / 1931 / "Se lo emplea en medicina para la dispepsia y enfermedades del estómago" / "'carqueja'"

4633 *B. cordifolia* Argentina / S. Venturi 1163 / 931 / "'Nio'"

Argentina / T. Meyer 186 / 1931 / "Es nocivo para el ganado" / "'mio-mio,' 'romerillo'"

4634 *B. lanceolata* Argentina / L. R. Parodi 8792 / 1928 / "'Chirca'"

4635 *B. myrtilloides* Argentina / L. R. Parodi 8067 / 1927 / "'Pajaro bobo'"

4636 *B. ulicina* Argentina / A. L. Cabrera 4355 / 1938 / "'Yerba de la oveja'"

Argentina / J. B. Correa 37040 / 41 / "'Pichana amarga'"

4637 *Baccharis* indet. Argentina / L. R. Parodi 7799 / 1927 / "'Chilca dulce'"

4638 *B. calvescens* Brazil / Y. Mexia 4921 / 1930 / "'Alegrin'"

4639 *B. dracunculifolia* Brazil / Williams & Assis 6739 / 1945 / "'Alecrim'"

Brazil / Y. Mexia 4169 / 1929 / "'Alecrin do matto'"

4640 *B. genistelloides* Brazil / Williams & Assis 6283 / 1945 / "Make a tea of it for intestines" / "'Carqueja'"

Brazil / Y. Mexia 4843 / 1930 / "Tea of leaves used as stomach tonic" / "'Carqueja'"

4641 *B. helichrysoides* Brazil / Y. Mexia 4212 / 1930 / "Stems and foliage used in infusion to wash wounds"

4642 *B. platypoda* Brazil / Y. Mexia 5794 / 1931 / "'Alécrín'"

4643 *B. trinervis* var. *rhexioides* Brazil / Y. Mexia 4603 / 1930 / "'Cascadin'"

4644 *Parastrephia lepidophylla* Peru / J. West 7138 / 1936 / "...resinous, highly inflammable" / "'Tola'"

4645 *Blumea balsamifera* China / G. W. Groff 4417 / 1919 / "Medicinal value" / "'Tai fung ngai'"
Hainan / W. T. Tsang 108 / 1928 / "...medicinal"
Philippine Is. / Mendoza & Convocar 112 / 1949 / "Febrifuge"
Philippine Is. / A. L. Zwickey 18 / 1938 / "...infusion of leaves for malaria, and after a long period without food, also as bath after child-birth" / "'Handelebon' (Bis.); 'Salimbawangan' (Lan.)"

4646 *B. riparia* var. *megacephala* China / W. T. Tsang 23049 / 1933 / "...used as medicine"

4647 *B. lacera* India / N. L. Bor 16090 / 43 / "An aromatic composite in open country"
Philippine Is. / E. D. Merrill 672 / 1903 / "...Very aromatic"

4648 *B. mollis* Indo-China / R. W. Squires 766 / 1932 / "...strong mint" / "'Growe'"
India: Punjab / R. R. Stewart 13854 / 1934 / "turpentine smell"

4649 *Blumea* indet. Philippine Is. / I. P. Paniza 103 / 1948 / "Medicinal"

4650 *Blumea* indet. Philippine Is. / Mendoza & Convocar 1 / 1949 / "Leaves for febrifuge" / "'Bagasbaz'" Bis.

4651 *Laggera alata* Nyasaland / L. J. Brass 16892 / 1946 / "...aromatic"

4652 *Pluchea borealis* U.S.: Colo. / Engelmann & Sargent s.n. / 1880 / "'Arrow wood'"

4653 *P. camphorata* U.S.: Va. / Fernald & Long 5116 / 1935 / "...plant with strong catty odor"
U.S.: Ga. / A. Cronquist 4183 / 1946 / "...mildly but not unpleasingly odorous"?

4654 *P. purpurascens* U.S.: Fla. / A. Traverse 620 / 58 / "Whole plant has fetid odor"
Dominican Republic / R. A. & E. S. Howard 9641 / 1946 / "Aromatic herb..."
Puerto Rico / P. Sintenis 1041 / 1885 / "'Palo-seco'"
W.I.: Guadeloupe / Stehlé & Quentin 305-W. / 1936 / "'Tabac diable'"

4655 *P. purpurascens* var. *succulenta* U.S.: Mass.? / F. H. Peabody s.n. / 1892 / "'Saltmarsh Fleabane'"

U.S.: Md. / C. C. Plitt 689 / 1903 / "'Salt-marsh Fleabane'"

U.S.: Tenn. / H. F. L. Rock 934 / 1957 / "Plants foetid-smelling"

4656 *P. rosea* U.S.: Fla. / A. Traverse 556 / 1958 / "Plant has strong resinous odor"

4657 *P. odorata* El Salvador / P. C. Standley 22530 / 1922 / "Remedy for stomach affections" / "'Siguapate'"

W.I.: St. Lucia / R. A. Howard 11625 / 1950 / "... used medicinally for colds"

W.I.: Trinidad / W. E. Broadway 51 / 1933 / "(a well known, local, medicinal plant)" / "'Gueritoute'"

4658 *P. salicifolia* Mexico / L. C. Ervendberg 343 / 1858 / "... smells very strong"

4659 *P. suaveolens* Bolivia / J. Steinbach 6762 / 1924 / "Aromatica" / "'Fresadilla grande'"

4660 *Tessaria absinthioides* Argentina / T. Meyer 3510 / 1941 / "'bobo'"

4661 *T. dodoneaefolia* Argentina / T. Meyer 4170 / 1942 / "'Chilca'"

Argentina / L. R. Parodi 8489 / 1928 / "'Chirca'"

4662 *T. integrifolia* Bolivia / J. Steinbach 5585 / 1927 / "El ganado apetece la hoja, especialmente el caballar" / "'Parajobobo'"

Peru / H. H. Rusby 1715 / 1885 / "Ex. Herb. Parke, Davis & Co."

Peru / A. Weberbauer 7224 / 1917 / "'pájaro bobo'"

4663 *Tessaria* indet. Argentina / T. Meyer 33065 / 1940 / "'Bobo'"

4664 *Tessaria* indet. Bolivia / H. H. Rusby 791 / 1921 / "'Parajobobo'"

4665 *Epaltes brasiliensis* W.I.: Martinique / H. & M. Stehlé 6146 / 1942 / "Medicinale" / "'Ti carré,' 'Herbe à fer'"

4666 *E. australis* Hainan / W. T. Tsang 64 / 1928 / "... medicinal drug plant"

4667 *Pterocaulon hassleri* Bolivia / J. Steinbach 5329 / 1921 / "'Frezadilla blanca'"

4668 *P. alopecuroideum* Bolivia / J. Steinbach 6988 / 1925 / "En la noche aparecen ser algo luminosas estas plantas, dando a la donde avundan un aspecto caracteristico" / "'Fresadilla blanca'"

4669 *P. lorentzii* Bolivia / J. Steinbach 6686 / 1924 / "'Frezadilla'"

4670 *P. purpurascens* Bolivia / J. Steinbach 5167 / 1920 / "'Frezadilla negra'"

4671 *Antennaria* indet. Burma / F. G. Dickason 7720 / 1937 / "Plant has pungent odor" / "Sts. used to stuff pillows" / "'Tek Rawl'"

4672 *Achyrocline satureoides* Bolivia / J. Steinbach 6104 / 1924 / "La infusion de esta planta favorece la digestion; es de agradable gusto" / "'Viravira'"

4673 *Gnaphalium obtusifolium* U.S.: Tex. / V. L. Cory 53914 / 1947 / "'cudweed'"

4674 *G. obtusifolium* var. *helleri* U.S.: Ga. / A. Cronquist 4750 / 1947 / "Plants pleasantly aromatic"

4675 *G. chilense* U.S.: Calif. / S. F. Blake 9983 / 1927 / "Aromatic, but not sweet-scented"

4676 *G. leucocephalum* U.S.: Ariz. / Harrison & Kearney 6215 / 29 / "—odor of 'lemon verbena'"

4677 *G. microcephalum* U.S.: Calif. / H. M. Hall 9966 / 1915 / "Herbage balsamic-viscid and pleasingly fragrant"

4678 *G. texanum* U.S.: Tex. / E. Marsh 347 / 37 / "'Rabbit Tobacco'"

4679 *G. uliginosum* U.S.: Mass. / E. F. Fletcher s.n. / before 1924 / "'Low Cudweed'"

U.S.: Penn. / S. P. Sharples 149 / 1858–64 / "'Marsh Cudweed'"

U.S.: Ohio / A. N. Rood 348 / 1898 / "'Low Cudweed'"

4680 *G. purpureum* U.S.: Va. / Angus & Heywood 4472 / 35 / "'Purplish Cudweed'"

U.S.: Ohio / R. J. Webb 453 / 1901 / "'Purplish Cudweed'"

4681 *G. attenuatum* Mexico / J. G. Ortega 4440 / 1921 / "'Gordolobo'"

El Salvador / P. C. Standley 19260 / 1921–22 / "Cooked leaves used as poultice for swellings" / "'Papelillo'"

4682 *G. bourgovii* Mexico / H. S. Gentry 2737 / 1936 / "Herbage decocted for stomach ailments" / "'Mansanilla del Rio,' Mex."

4683 *G. chartaceum* Mexico / Y. Mexia 8989 / 1937 / "'Gordo-Lobo'"

4684 *G. decurrens* Mexico / G. B. Hinton 803 / 32 / "Good for burns" / "'Gordo Lobo'"

4685 *G. leptophyllum* Mexico / H. S. Gentry 2656 / 1936 / "Medic., either as decoction or infusion (tea) for children and for 'empache'"

4686 *G. linearifolium* Mexico / H. S. Gentry 5329 / 1939 / "'Gordolobo'"

4687 *G. oxyphyllum* Baja Calif. / H. S. Gentry 4319 / 1939 / "'Gordolobo'"

4688 *G.* cf. *purpurascens* Mexico / Krueger & Gillespie 11 / 40 / "Known as Anis and used by natives for tea"

4689 *G. rhodanthum* Mexico / C. & E. Seler 2277 / 1896 / "'Sanrióm'"

4690 *G. roseum* Mexico / G. B. Hinton 2427 / 32 / "Leaf used to heal wounds" / "'Gordo lobo'"

Mexico / Y. Mexia 8957 / 1937 / "'Gordo-Lobo'"

4691 *G. schraderi* Mexico / Y. Mexia 8843 / 1937 / "'Gordo-Lobo'"

4692 *G. semiamplexicaule* Mexico / A. Dugès 450 / 1893 / "Emollient" / "'Gordolobo'"

4693 *G. sprengelii* Mexico / A. Dugès s.n. or 2 / 1893 / "autre gordolobo"

4694 *G. viscosum* Guatemala / J. A. Steyermark 43039 / 1942 / "...leaves with odor of licorice ..."

4695 *Gnaphalium* indet. Mexico / M. Martínez s.n. / 1940 / "'Gordolobo'"

4696 *G. araucanum* Argentina / A. L. Cabrera 7321 / 1941 / "Muy aromático. Ola a limón"

4697 *G. cheiranthifolium* Peru / J. F. Macbride 4944 / 1923 / "Pleasant odor"

4698 *Gnaphalium* indet. Bolivia / W. J. Eyerdam 24852 / 1939 / "Aromatic odor"

4699 *Gnaphalium* indet. Peru / Worth & Morrison 15722 / 1938 / "Intensely aromatic..."

4700 *Gnaphalium* indet. Peru / F. W. Pennell 13294 / 1925 / "...pungent odor"

4701 *Gnaphalium* indet. Peru / F. W. Pennell 13472 / 1925 / "...with fragrant odor of *G. polycephalum*"

4702 *Gnaphalium* indet. Argentina / T. Meyer 3517 / 1941 / "Planta aromatica"

4703 *Gnaphalium* indet. Bolivia / J. Steinbach 5350 / 1921 / "Es planta tónica y estomacal. Se usa con mucha frecuencia" / "'Vira-vira'"

4704 *Gnaphalium* indet. Chile / F. W. Pennell 12935 / 1925 / "...odor of *G. polycephalum*"

4705 *G. multiceps* China / W. T. Tsang 20769 / 1932 / "...edible"

4706 *G. luteo-album* Africa: Polela Natal / R. A. Dyer 3286 / 35 / "Thought to be harmful to sheep in the early stages of growth if eaten"

4707 *Cassinia subtropica* Australia: Queensland / C. T. White 12716 / 34 / "...leaves strongly scented when crushed"

4708 *Petalacte coronata* S.W. Africa / MacOwan 116 / 1882 / "'False Bay'"

4709 *Phaenocoma prolifera* S.W. Africa / MacOwan 200 / 1883 / "'False Bay'"

4710 *Helichrysum cassinioides* Australia: Queensland / C. T. White 9501 / 1933 / "...said to be eaten freely by horses"

4711 *H. diosmifolium* Australia: Queensland / C. T. White 6073 / 1929 / "Lvs. very strongly scented"

4712 *H. ericoides* S. Africa / M. A. Pocock S.28 / 1926 / "Decoction from leaves used for colds & stomach trouble" / "'Klip rhenoster'"

4713 *H. baccharoides* Tasmania / C. T. White 8258 / 1932 / "'Kerosene Bush'"

4714 *Inula cappa* China / W. T. Tsang 21522 / 1932 / "...roots used as medicine"

4715 **Anvillea radiata** Morocco / E. K. Balls B2637 / 1936 / "Strongly aromatic"

4716 **Odontospermum maritimum** Algeria / Lerat s.n. / 1870 / "Nos soldats mangent les feuilles en guise d'epinards"

4717 **Lagascea helianthifolia** var. *suaveolens* Mexico / H. S. Gentry 1426 / 1935 / "Two species of hummingbirds frequent visitors to this plant"

4718 **L. decipiens** Mexico / H. S. Gentry 1258 / 1935 / "Attractive to butterflies. Medic. herbage cooked in water to make wash for insect stings and rattlesnake bites" / "'Confituria Grande,' Mex. 'Tusuli,' W."

4719 **Clibadium appressipilum** Panama / P. H. Allen 857 / 1938 / "...leaves crushed and used to poison fish"

4720 **C. erosum** W.I.: Martinique / H. & M. Stehlé 5806 / 1945 / "'Enivrage'"

4721 **C. sylvestre** W.I.: Dominica / W. H. Hodge 3195 / 1940 / "...the leaves and all parts pounded—the result being used as a fish poison in pools of streams" / "'enivrage'"
Brit. Guiana / A. C. Smith 2182 / 1937 / "Leaves and infl. used as fish poison...Cultivated..." / "'Conami'"
Brazil / B. A. Krukoff 5212 / 1933 / "(Used as 'fish poison')"
Peru / J. M. Schunke 310 / 1935 / "Used as a fish poison. At Indian rustic habitation" / "'Huaca'"
Brit. Guiana / J. S. de la Cruz 1228 / 1921 / "Used for poisoning fish"
Brit. Guiana / A. S. Hitchcock 17607 / 1920 / "...fish poison, the leaves used"

4722 **C. microcephalum?** Ecuador / Y. Mexia 8467 / 1936 / "'Salvia'"

4723 **C. strigillosum** Peru / Killip & Smith 26834 / 1929 / "Leaves and fruit used as fish poison" / "'Barbasco'"

4724 **C. surinamense** Surinam / W. A. Archer 2647 / 1934 / "'Kanami-ran' (Carib) (means bastard conami)"
Brazil / R. Froes 1738 / 1932 / "...the leaves used as a fish poison" / "'Cunamby'"

4725 **Clibadium** indet. Brazil / Ducke 1323 / 1943 / "Ichthyotoxica"

4726 **Clibadium** indet. Brit. Guiana / A. S. Hitchcock 17658 / 1920 / "Fish poison"

4727 **Ichthyothere peruviana** Peru / Y. Mexia 6506a / 1932 / "Whole plant macerated and used as fish-poison" / "Cultivated" / "'Huaca'"
Peru / Y. Mexia 6254 / 1931 / "Used as fish poison leaves and flowers crushed" / "'Huáca'"

4728 **I. terminalis** Brazil / Williams & Assis 5893 / 1945 / "Kills cattle"

4730 *Polymnia eurylepis* Venezuela / W. Gehriger 10 / 1930 / "'Anime'"

4731 *P. glabrata* Peru / F. L. Herrera 1480 / 1927 / "Medicinal"

Peru / Macbride & Featherstone 1554 / 1922 / "Pungent . . . Used for colds" / "'Poque'"

Peru / R. D. Metcalf 30279 / 1942 / ". . . odor of tar"

4732 *P. pyramidalis* Colombia / Killip & Smith 18043 / 1927 / "'Anime'"

4733 *Polymnia* indet. Venezuela / H. Pittier 12942 / 1928 / "'Aníme negro'"

4734 *Guardiola tulocarpus* Mexico / Y. Mexia 1183 / 1926 / "Used for fevers, leaves" / "'Tabardillo'"

4735 *Melampodium cinereum* Mexico / Stanford, Retherford & Northcraft 186 / 1941 / ". . . heavily grazed by goats"

4736 *M. perfoliatum* Mexico / E. Kerber 103 / 1882 / "'Porlocotillo'"

4737 *Melampodium* indet. Mexico / H. S. Gentry 7005 / 1944 / "'Manzanilla del Coyote'"

4738 *Melampodium* indet. Mexico / R. M. Stewart 414 / 1941 / "'Manzanilla'"

4739 *M. camphoratum* Brit. Guiana / A. C. Smith 2175 / 1937 / "Plant used for tea"

4740 *Acanthospermum australe* Paraguay / W. A. Archer 4687 / 1936 / "Used for foot sores" / "Sold by herb dealer in market, Villarica" / "'tapequé'"

4741 *A. hispidum* Peru / H. E. Stork 11423 / 1939 / "Burros eat it" / "'Verdelago'"

Brazil / Y. Mexia 4776 / 1930 / ". . . leaves eaten by stock" / "'Rebenta carneiro'"

4742 *Silphium laciniatum* U.S.: Minn. / E. A. Mearns s.n. / 1891 / "'Rosinweed' / 'Compass-Plant'"

4743 *Berlandiera lyrata* Mexico / H. S. Gentry 2504 / 1936 / "Roots medicinal either as a decoction or infusion for stomach troubles. Carried & sold to the drug stores below" / "'Coronilla,' Mex."

4744 *Lindheimera texana* U.S.: Tex. / R. McVaugh 7735 / 1947 / ". . . odor strong of Helianthus annuus"

4745 *Parthenium hysterophorus* U.S.: Tex. / V. L. Cory 50899 / 1945 / "'Santa Maria feverfew'"

Mexico / V. Grant 504 / 1940 / "Natives make bitter tea to stimulate appetite" / "foliage has faint musty odor" / "'Amargoso'"

Yucatan / J. Bequaert 83 / 1929 / "'altaniza'"

Mexico / J. G. Ortega 5.859 / 1925 / "'Hierba del burro'"

W.I.: Dominica / W. H. Hodge 3189 / 1940 / ". . . tea from foliage used for fever; planted about native hut: Bataka"

Argentina / Eyerdam & Beetle 22621 / 1938 / "...lvs. aromatic when crushed"

Bolivia / J. Steinbach 6344 / 1924 / "Es popular on la medicina casera, especialmente en la compostura de untos" / "'Chupuruhúme'"?

4746 *P. incanum* U.S.: N.M. / E. O. Wooton 47 / 1907 / "In Northern Mexico rubber is extracted from it, though not so extensively as from the Guayule, *Parthenium argentatum* A. Gray. It is sometimes used as a medicine for stomach troubles. Commonly known here as Variola"

Mexico / W. P. Hewitt 244 / 1947 / "spicey, sagey fragrance"

Mexico / A. López 81 / 1948 / "'Mariola' buena para el estomago"

Mexico / G. B. Hinton 16574 / 44 / "Used to adulterate bales of guayule. Tea for stomach-ache" / "'Mariola'"

4747 *P. integrifolium* U.S.: Ohio / F. E. Ford 1435 / 1922 / "'Fever-few'"

4748 *Parthenice mollis* Mexico / H. S. Gentry 6102 / 1941 / "Herbage with rancid odor"

4749 *Iva annua* U.S.: Tex. / A. Traverse 238 / 1956 / "resinous odor of plant..."

4750 *I. frutescens* U.S.: Fla. / C. H. Baker 550 / 1918 / "Leaves thickish, strong-smelling & bitter..."

4751 *I. axillaris* var. *robustior* U.S.: Calif. / Alexander & Kellogg 3813 / 1944 / "Aromatic odor..."

U.S.: Calif. / Phillips & Sargent s.n. / 1878 / "'Bad Weed'"

4752 *I. acerosa* U.S.: Utah / E. H. Graham 9973 / 1935 / "Reputed poisonous to stock in winter..."

U.S.: N.M. / C. Kluckhohn s.n. / 1938 / "...known to Indians"

4753 *Hymenoclea monogyra* Mexico / A. Miranda 123 / 1948 / "'Romerio'"

4754 *Ambrosia trifida* U.S.: Colo. / R. W. Woodward s.n. / 1881–83 / "This plant is called 'Wild Hemp' about Bessemer.... It is not *Cannabis*"

4755 *A. artemisiaefolia* var. *elatior* U.S.: N.J. / Halstead 30 / 1891 / "'Ragweed,' 'Hog-weed,' 'Roman Wormwood'"

4756 *A. artemisiaefolia* forma *villosa* U.S.: Mass. / E. F. Fletcher s.n. / Herb. E. F. Fletcher 1845–1923 / "'Roman Wormwood'"

4757 *A. psilostachya* U.S.: Tex. / A. Traverse 221 / 1956 / "...foliage has strong aromatic odor"

4758 *A. cumanensis* El Salvador / P. C. Standley 20587 / 1922 / "...offensive odor" / "'Altamisa'"

Mexico / E. Langlassé 207 / 1898 / "Aromatique"

4759 *A. hispida* Bahama Is. / R. A. & E. S. Howard 10011 / 1948 / "'Bay vine'"

Bahama Is. / C. B. Lewis s.n. 1954 / "'Bay tansy'"

Bahama Is. / O. Degener 18827 / 1946 / "...tea drunk as cathartic" / "'Bay Geranium'"

4760 *A. paniculata* Jamaica / W. Harris 7259 / 98 / "'Wild Tansy'"

W.I.: Dominica / W. H. Hodge 3190 / 1940 / "Growing around native hut; tea from leaves used for fever ..."

4761 *A. velutina* Dominican Republic / R. A. & E. S. Howard 9577 / 1946 / "...plant aromatic"

4762 *Franseria tenuifolia* U.S.: Tex. / Lindheimer 471 / 1850 / "Very bitter"

U.S.: Tex. / L. H. Shinners 7306 / 1945 / "Crushed leaves aromatic"

4763 *F. ambrosioides* U.S.: Ariz. / E. Palmer 120 / 1867 / "strong rancid smell"

Mexico / H. S. Gentry 1336 / 1935 / "Medic. Roots cooked in water for 'female trouble'" / "'Chichura,' Mex."

4764 *F. confertiflora* Mexico / H. S. Gentry 8539 / 1948 / "With rank herbage around trees"

4765 *F. cordifolia* Mexico / H. S. Gentry 5403 / 1940 / "...with acrid juice"

4766 *F. artemisioides* Ecuador / A. Rimbach 29 / no year / "...covered with glandular aromatic hairs"

Ecuador / Dr. Rose 13 / 22 / "'Altamisa,' 'malco'"

Peru / J. West 3832 / 1935 / "Used medicinally (carminative)" / "'Marcco'"

Colombia / Killip & Smith 19676 / 1927 / "'Artimisa'"

4767 *F. meyeniana* Peru / Eyerdam & Beetle 22132 / 1938 / "*Artemisia*-like odor"

Peru / Mr. & Mrs. F. E. Hinkley 26 / 1920 / "Local name 'Romero del cerro'"

Peru / Mr. & Mrs. F. E. Hinkley 79 / 1920 / "'Ajenjo'"

4768 *Xanthium strumarium* U.S.: Penn. / S. P. Sharples 139 / 1858–64 / "'Clot Weed'"

4769 *X. italicum?* Mexico / J. Gregg 457 / 1848–49 / "'Cardilla'"

4770 *X. macrocarpum* Mexico / V. L. Portillo 33 / 1908 / "'Cochinilla'"

4771 *Xanthium* indet. Mexico / R. M. Stewart 2173 / 1941 / "'Cadillo'"

4772 *X. spinosum* Argentina / T. Meyer 33076 / 1940 / "'chusca'"

4773 *Xanthium* indet. T. Meyer 4173 / 1942 / "'abrojo macho'"

4774 *Podanthus mitiqui* Chile / G. Montero 83 / 1921 / "'Mitrin' or 'Mitriu'"

Chile / G. Montero O.83A / 1926 / "'mitriu . mitique'"

Chile / P. Aravena 33357 / 1942 / "... whole plant strongly odorous"

Chile / D. Bertero 470 / 1828 / "'Mitria'"

4775 *Zinnia elegans* Mexico / Y. Mexia 8744 / 1937 / "'San Miguel'"

4776 *Z. multiflora* Mexico / A. López 16 / 1947 / "sirve para el mal de los ojos" / "'Herba del Indio'"

Mexico / C. & E. Seler 34 / 1888 / "'malacatillo'"

Argentina / R. Paz 36 / 1941 / "'Chinita'"

4777 *Rumfordia floribunda* Mexico / E. Langlassé 83 / 1898 / "'Palo gogo'"

4778 *R. penninervis* Guatemala / J. A. Steyermark 35843 / 1940 / "Leaves boiled and used as tea beverage" / "'te'"

4779 *Siegesbeckia jorullensis* Mexico / G. B. Hinton 15694 / 1940 / "Stem and leaves sticky, fragrant"

4780 *Siegesbeckia* indet. Argentina / L. R. Parodi 11047 / 1933 / "'grasilla'"

4781 *S. orientalis* Philippine Is. / A. L. Zwickey 238 / 1938 / "Used to stop bleeding"

4782 *Jaegeria hirta* Brazil / Y. Mexia 4784 / 1930 / "'Picão'"

4783 *Eclipta alba* China / G. W. Groff 4055 / 1919 / "Leaves boiled with congel (or congu?) good for indigestion of small boys" / "'Pak fa pang ki kuk'"

4784 *Montanoa tomentosa* Mexico / C. & E. Seler 1428 / 1895 / "'yerva de la parida'"

4785 *M. subtruncata* Mexico / E. Langlassé 621 / 1898 / "'Flor de San Francisco'"

4786 *M. pauciflora* Guatemala / P. C. Standley 58306 / 1938 / "'Flor de Concepción'"

4787 *M. affinis* Mexico / Y. Mexia 8842 / 1937 / "'Bosque Amargo'"

4788 *M. hibiscifolia* Guatemala / P. C. Standley 61292 / 1938 / "'Cajete'"

4789 *Montanoa* indet. Venezuela / H. Pittier 12749 / 1928 / "'Aníme blanco; la madera es suave ...'"

4790 *M. pyramidata* Brazil / Y. Mexia 4739 / 1930 / "Bee plant ... 'Flor de Maio'"

4791 *Isocarpha oppositifolia* El Salvador / V. Grant 726 / 1940 / "herbage has faint lemon-ish odor"

Venezuela / A. Jahn 657 / 1921 / "'Botón de cáncer'"

4792 *I. divaricata* Venezuela / H. M. Curran s.n. / 1917 / "Used as remedy for malarial fever" / "'quinina de pobre'"

4793 *Ratibida mexicana* Mexico / H. S. Gentry 2712 / 1936 / "Roots decocted for bites and other ailments, as a wash and as a potation" / "'Howinowa,' W."

4794 *Wulffia baccata* Peru / Stork & Horton 9460 / 1938 / "...rank herb"

4795 *Wulffia* indet. Bolivia / J. Steinbach 5071 / 1921 / "visitadas por Papidios del grupo Aristolochia"

4796 *Iostephane heterophylla* Mexico / H. S. Gentry 6349 / 1941 / "... roots tuberous and decocted for medicine; makes women fertile" / "'Escosionero'"

Mexico / H. S. Gentry 2547 / 1936 / "Roots valued for medicinal properties; dug, carried, and sold to druggists below..." / "'Escosionero'"

Mexico / J. G. Schaffner 253 / 1876 / "'Raiz del manzo'"

4797 *Zaluzania triloba* Mexico / E. Palmer 757 / 1898 / "Purchased in market" / "City of Zacatecas"

4799 *Balsamorhiza terebinthacoa* U.S.: Ore. / Spalding s.n. / no year / "...root...taste & adhesiveness..."

U.S.: Ore. / Spalding 43 / no year / "...Smell of Root like turpentine in taste and adhesiveness..."

U.S.: Ore. (now Idaho) / Spalding s.n. or 42 / no year / "Bark of root gives a pitch like pine in taste and smell. Root peeled and baked for food"

4800 *Borrichia frutescens* U.S.: Fla. / A. Traverse 715 / 1958 / "Resinous odor"

4801 *B. arborescens* W.I.: Barbuda / A. C. Smith 10445 / 1956 / "...infusion of leaves drunk as a cure for poisoning from eating certain kinds of fish locally..."

4802 *Wedelia trilobata* Brit. Honduras / W. A. Schipp 689 / 1931 / "Creoles used the plant as a tonic when boiled"

Guatemala / C. C. Deam 39 / 1905 / "Locally called Button-flower and used for stomach and kidney trouble"

4803 *Wedelia* indet. Brazil / B. A. Krukoff 1266 / 1931 / "'Camará de flecha'"

4804 *Wedelia* indet. Philippine Is. / A. L. Zwickey 774 / 1938 / "...tea of lvs. for stomachache" / "'Tikaua'"

4805 *W. biflora* New Hebrides Is. / S. F. Kajewski 196 / 1928 / "Used by natives by cooking fish with leaves to give it a flavour"

4806 *W. strigulosa* Fiji Is. / A. C. Smith 5 / 1933 / "Used for medicine" / "'Lawati'"

Fiji Is. / O. Degener 15113 / 1941 / "Leaves edible when cooked" / "'Wulewule' (Serua)"

4807 *Wyethia mollis* U.S.: Calif. / L. Constance 2451 / 1938 / "...foliage resinous-scented"

4808 *W. angustifolia* U.S.: Calif. / S. F. Blake 10050 / 1927 / "Root with much the smell of Tetragonotheca helianthoides"

4809 *Tithonia rotundifolia* El Salvador / P. C. Standley 22568 / 1922 / "'Acate'"

Yucatan / J. Bequaert 82 / 1929 / "'Arnica' (Castillano)"

El Salvador / P. C. Standley 19714 / 1922 / "Decoction of lvs. used for fevers" / "'Árcabo,' 'Barabaja'"

El Salvador / S. Calderón 135 / 1921 / "'Acate,' 'Chilicacate'"

4810 *Gymnolomia scaberrima* El Salvador / P. C. Standley 19272 / 1921–22 / "Lvs. used in baths for fevers and colds" / "'Pulagaste'"

4811 *Viguiera porteri* U.S.: Ga. / J. D. Smith s.n. / 1883 / "Juices somewhat resinous"

4812 *V. cordifolia* Mexico / H. S. Gentry 8432 / 1948 / "'Romerillo'"

4813 *V. decurrens* Mexico / C. V. Hartman 562 / 1891–92 / "Used for poisoning fish" / "'Na-ka-ró-ri' Sar"

4814 *V. dentata* var. *lancifolia* Mexico / H. S. Gentry 2937 / 1936 / "Musky sunflower odor"

4815 *V. linearis* Mexico / H. S. Gentry 8424 / 1948 / "'Romerillo'"

4816 *V. montana* Mexico / H. S. Gentry 1288 / 1935 / "Medic. Women put leaves on stomach to cause menses flow, 'por sale la sangre.' E.B." / "'Ariosa,' Mex. 'Wachomo,' W."

4817 *Viguiera* indet. Mexico / R. M. Stewart 1010 / 1941 / "'Romerillo'"

4818 *Viguiera* indet. Mexico / R. M. Stewart 1874 / 1941 / "'Romerillo'"

4819 *Viguiera* indet. Mexico / R. M. Stewart 1559 / 1941 / "'Romerillo'"

4820 *Viguiera* indet. Mexico / R. M. Stewart 1272 / 1941 / "'Romerillo'"

4821 *Viguiera* indet. Mexico / R. M. Stewart 1286 / 1941 / "'Romerillo'"

4822 *V. revoluta* Chile / P. Aravena 33383 / 1942 / "'Maravilla'"

4823 *V. weberbaueri* Peru / Worth & Morrison 15723 / 1938 / "Plants heavily grazed by stock"

4824 *Syncretocarpus sericeus* Peru / Goodspeed & Metcalf 30222 / 1942 / "...herb...with a very pleasant sweet odor"

4825 *Helianthus debilis* U.S.: Fla. / J. D. Smith s.n. / 1882 / "(Juice resinous)"

4826 *H. grosseserratus* subsp. *maximus* U.S.: Mo. / Palmer & Steyermark 41225 / 1933 / "(Precocious blooming, probably due to insect injury)"

4827 *H. californicus* U.S.: Calif. / H. N. Bolander s.n. / 186– / "It has a pretty large tuberous root of a strong terebinthine smell—"

U.S.: Calif. / Alexander & Kellogg 2609 / 1941 / "...tuber has strong sage smell"

4828 *H. tuberosus* China / F. A. McClure 20575 / 1937 / "Tubers dried and pickled as a relish"

4829 *Perymenium strigillosum* El Salvador / P. C. Standley 19261 /
1921–22 / "Juice remedy for eye diseases" / "'Tatascame,' 'Palo de
tisate'"

4830 *Melanthera aspera* El Salvador / P. C. Standley 21288 / 1922 /
"'Orozuz'"

4831 *M. nivea* Ecuador / Y. Mexia 8466 / 1936 / "Bruised leaves used on
wounds to stop bleeding" / "'Yuyuesabalo'"

4832 *Flourensia cernua* U.S.: Tex. / Thurber 115 / no year / "'Smells like
Hops'"
U.S.: Tex. / Marsh 199 / 37 / "'Tar-bush'"
Mexico / A. López 21 / 1947 / "es medicinal para la estomago" /
"'Ojasen'"

4833 *F. resinosa* Mexico / H. E. Moore, Jr. 1265 / 1946 / "Viscid shrub . . .
Edible exudate chewed like chicle"

4834 *Flourensia* indet. Mexico / I. M. Johnston 8698 / 1941 / ". . . leaves
glutinous, with strong unpleasant odor"

4835 *Flourensia* indet. Mexico / I. M. Johnston 8788 / 1941 / "Odor of
herbage very strong and somewhat disagreeable"

4836 *Flourensia* indet. Mexico / R. M. Stewart 1694 / 1941 / "'Ojasen'"

4837 *F. angustifolia* Peru / Macbride & Featherstone 1018 / 1922 /
"Sweet-scented"

4838 *F. macrophylla* Peru / J. F. Macbride 2929 / 1923 / "Foliage has odor
suggesting *Ceanothus velutinus*"

4839 *F. thurifera?* Chile / R. Wagenknecht 18471 / 1939 / "'Incienso'"
Chile / E. E. Gigoux s.n. / 1886 / "'Maravilla del Campo'"

4840 *Spilanthes americana* Mexico / C. G. Pringle 3643 / 1890 / "'Flor de
Maria'"

4841 *S. ocymifolia* El Salvador / L. V. Velasco 8851 / 1905 / "'Duerme-
lengua'"
Mexico / Y. Mexia 1237 / 1926 / "Eaten by stock"

4842 *S. urens* W.I.: Martinique / H. & M. Stehlé 6158 / 45 / "'Creosote,'
'Bouton d'Or'"
Mexico / Eyerdam & Beetle 8671 / 1938 / ". . . juice in stem has acrid
taste"
Colombia / Killip & Smith 14322 / 1926 / "'Adormidera'"

4843 *Salmea scandens* Guatemala / J. A. Steyermark 33598 / 1940 /
"'flora culantro de montaña'"
El Salvador / P. C. Standley 22384 / 1922 / "'Duerme-boca'"

4844 *Notoptera guameri* Yucatan / C. L. Lundell 1007 / 1931 / ". . . vanilla
scented flowers"

4845   *Encelia californica* U.S.: Calif. / W. H. Brewer 187 / 1860–62 / "'Bad smell'"

4846   *Simsia annectens* Mexico / Y. Mexia 1832 / 1927 / "Stems slightly stinging hairs"

4847   *S. calva* Mexico / R. M. Stewart 1706 / 1941 / "'Cuajadora'"

4848   *Pionocarpus madrensis* Mexico / H. S. Gentry 2707 / 1936 / "Roots employed for rheumatism and other diseases" / "'Kachana,' Mex."

4849   *Zexmenia brevifolia* Mexico / Stanford, Retherford & Northcraft 237 / 1941 / "...heavily grazed by goats"

4850   *Z. frutescens* El Salvador / P. C. Standley 19189 / 1921–22 / "Wood gives white ashes used by spinning women to keep fingers smooth" / "'Tisate'"

      Brit. Honduras / H. H. Bartlett 11424 / 1931 / "...strong turpentine odor"

4851   *Z. podocephala* Mexico / H. S. Gentry 2468 / 1936 / "The tuberous roots are highly regarded for their medicinal properties; decocted for stomach ailments" / "'Pioniya,' Mex."

4852   *Oyedaea verbesinioides* Panama / P. H. Allen 2866 / 1941 / "'pererina'"

4853   *Verbesina chihuahuensis* Mexico / Stanford, Retherford & Northcraft 228 / 1941 / "...heavily grazed by goats"

4854   *V. crocata* Mexico / G. B. Hinton 5707 / 1934 / "'Belladona'"

      Mexico / Y. Mexia 8949 / 1937 / "Leaves bruised to make cool drink" / "'Capitaneja'"

      Mexico / E. Langlassé 727 / 1899 / "Fournit tisane fébrifuge" / "'Capitonya' (Médicinal)"

4855   *V. greenmanii* Mexico / G. B. Hinton 13507 / 1938 / "'La Capitaneja'"

      Mexico / Y. Mexia 1353 / 1927 / "Remedy" / "'Tacote amarillo'"

      Mexico / E. Langlassé 605 / 1898 / "'Cardillo amarillo'"

4856   *V. myriocephala* Honduras / P. C. Standley 55039 / 1928 / "'Tabaquillo,' 'Tabaco de monte'"

4857   *V. pinnatifida* Mexico / G. B. Hinton 2384 / 1932 / "Root boiled in water cures tooth ache" / "'Capita negra'"

4858   *V. punctata* E. Salvador / P. C. Standley 19155 / 1921–22 / "'Chimaliote blanco,' 'Tabaquillo'"

4859   *V. serrata* Mexico / Y. Mexia 1291 / 1926 / "'Capitanejo'"

4860   *V. australis* Argentina / W. Lossen 29 / no year / "'Santa Maria'"

4861   *V. simulans* Venezuela / W. Gehriger 23 / 1930 / "...resinoso..." / "'Resinoso'"

4862   *Calyptocarpus vialis* Mexico / V. Grant 507 / 1940 / "infusion for diarrhoea" / "'Cochineta'"

4863   *Coreopsis verticillata* U.S.: N.C. / Channell & Rock 39 / 1956 / "With anise-like odor when fresh"

4864   *C. fasciculata?* Peru / Mr. & Mrs. F. E. Hinkley 51 / 1920 / "'Manzanilla del cerro'"

4865   *Dahlia coccinea* Mexico / H. S. Gentry 6302 / 1941 / "Tubers eaten raw by natives" / "'Jicama'"
Mexico / G. B. Hinton 9439 / 1936 / "root edible" / "'Cherihuesca'"

4866   *Dahlia* indet. Mexico / C. Lumholtz 1047 / 1891 / Separate label says: "Bávis. Tar. name see 5 Sept. 94. Root medicinal. Good for empacho del estomago. Tar. and Mexic . . ."

4867   *Hidalgoa ternata* Mexico / G. Martínez-Calderón 91 / 1940 / "Se usa cuando tiene viento en los ojos"

4868   *Glossogyne tenuifolia* Hainan / C. I. Lei 1038 / 1934 / "Used for medicine"

4869   *Bidens ostruthioides* Mexico / G. B. Hinton 2737 / 32 / "'Te de Chivo'"

4870   *B. pilosa* Mexico / C. & E. Seler 733 / 1888 / "'Té de milpa'"
W.I.: Dominica / W. H. Hodge 3285 / 1940 / ". . . leaves squeezed and juice used for eye-irritation-cure"
China / G. W. Groff 4117 / 1919 / "Medical use"
Hainan / C. I. Lei 1385 / 1934 / "Used for medicine"

4871   *B. pilosa* var. *bimucronata* f. *odorata* Mexico / A. Dugès 471 / 1891 / "'Aceitilla'"
Mexico / C. V. Hartman 564 / 1892 / "As young eaten with pinole and salt, boiled" / "'Se-pāē,' Tar."

4872   *B. urbanii* Mexico / Y. Mexia 1305 / 1926 / "Slightly ill smelling"

4873   *Bidens* indet. Mexico / Y. Mexia 8748 / 1937 / ". . . grazed by cattle. 'Estrellita'"

4874   *Bidens* indet. Mexico / C. Lumholtz 1044 / 1894 / "Eaten cooked before flowering" / "'Se-pé' Tar."

4875   *Bidens* indet. Mexico / C. Lumholtz 1040 / 1894 / "Eaten raw, mostly with salt" / "'Xú-ve' Tar."

4876   *B. cynapiifolia* W.I.: Curaçao / Curran & Haman 194 / 1917 / "Medicinal—stomach" / "'Jeerba die pataaka'"

4877   *B. subalternans* Argentina / T. Meyer 294 / 1931 / "'amor seco'"

4878   *B. triplinervis* Peru / Eyerdam & Beetle 22136 / 1938 / "lvs. pungent odor"

4879   *Bidens* indet. Argentina / S. Venturi 6108 / 928 / "'Lactilla'"

4880   *Bidens* indet. Bolivia / J. Steinbach 6270 / 1924 / "'Sanaana'"

4881   *B. frondosa* Japan / E. Elliott 34 / 1946 / "FOOD USE: Young leaves and stems are eaten cooked" / "MEDICINAL USE: Stem and leaf good for cold, fever, and dysentery" / "'Sendangusa'"

4882   *B. pilosa* var. *minor* Philippine Is. / A. L. Zwickey 1026 / 1941 / "... crushed lvs. used to poultice boils" / "'Nagumamo'"

4883   *B. mathewsii* Pitcairn Is. / H. St.John 15003 / 1934 / "... foliage with sweet odor of Alyxia olivaeformis"

       Pitcairn Is. / Fosberg & Clark 11276 / 1934 / "Leaves have parsnip odor"

4883A *B. urticifolia* Honduras / E. C. Becker 4 / 1949 / "'Amargoso'"

4884   *Cosmos parviflorus* Mexico / C. Lumholtz 563 / 1892 / "Eaten raw when young" / "'Ku-kū-ve' Tar."

4885   *C. pringlei* Mexico / H. S. Gentry 2688 / 1936 / "The tuberous roots are much sought for their medicinal properties" / "'Bavisa,' Mex."

4886   *C. sulphureus* Mexico / J. G. Ortega 4450 / 1921 / "'San Miguel'"

       El Salvador / P. C. Standley 19235 / 1921–22 / "'Flor de muerto,' 'Botón de oro'"

       Brazil / Y. Mexia 4548 / 1930 / "'Cravo de Morto'"

4887   *C. caudatus* Colombia / Killip & Smith 15443 / 1926 / "'Mapolá'"

4888   *Calea megacephala* Mexico / E. Kerber 101 / 1882 / "'Zaratillo blanco'"

       Mexico / W. E. Safford s.n. / 1918 / "Drug market of the city of Mexico" / "A Mexican drug plant ..." / "'Atanasia amarga'" / "'Prodigiosa'"

4889   *C. integrifolia* El Salvador / P. C. Standley 19252 / 1921–22 / "'Santo Domingo'"

4891   *C. urticifolia* var. *axillaris* Mexico / G. B. Hinton 2948 / 1932 / "Concoction taken for stomach ache—Sold for this—Bitter" / "'Prodigiosa'"

       Mexico / G. B. Hinton 3038 / 1932 / "Alcoholic solution of leaves used for stomachache & intoxicant" / "'Prodigiosa'"

       Mexico / E. K. Balls 5490 / 1938 / "'Bandarilla,' 'Jalapa'"

4892   *Galinsoga parviflora* Dominican Republic / R. A. & E. S. Howard 8602 / 1946 / "'Romerilla'"

4893   *Madia glomerata* U.S.: Colo. / D. D. Keck 878 / 1930 / "Peppermint odor ..."

       U.S.: Calif. / D. D. Keck 1715 / 1932 / "... strong odor of Hemizonia bicolor"

4894 *Hemizonia halliana* U.S.: Calif. / Keck & Stockwell 3252 / 1935 / "... herbage strongly, pungently but pleasantly odorous"

4895 *H. fasciculata* U.S.: Calif. / S. F. Blake 843 / 1927 / "Resinous odor" U.S.: Calif. / M. Rodman s.n. / 1897 / "'Tar Weed'"

4896 *H. lobbii* U.S.: Calif. / D. D. Keck 2560 / 1933 / "In grassy field where trampled and chewed by stock"

4897 *H. pungens* subsp. *laevis* U.S.: Calif. / Parish Bros.? 528 / no year / "'Tar Weed'"

4898 *H. congesta* subsp. *lutescens* U.S.: Calif. / S. F. Blake 10348 / 1927 / "Rank odor"

4899 *H. heermannii* U.S.: Calif. / J. A. Ewan 8045 / 1933 / "Honey-scented herbage"

4900 *Calycadenia multiglandulosa* U.S.: Calif. / S. F. Blake 10312 / 1927 / "Rank-scented"

4901 *Layia heterotricha* U.S.: Calif. / Keck & Clausen 3070 / 1935 / "... odor pronounced"

4902 *Baileya multiradiata* Mexico / Dr. Gregg 33 / 1847 / "Used as a poultice for pains, etc." / "'Talampacate' (or Rosa Amarilla)" Mexico / W. P. Hewitt 194 / 1947 / "'Mala Mujer'"

4903 *Perityle leptoglossa* Mexico / Wiggins & Rollins 96 / 1941 / "...malodorous"

4904 *Laphamia lindheimeri* U.S.: Tex. / R. McVaugh 8282 / 1947 / "... plant with faint lemon odor"

4905 *L. dissecta* Mexico / W. P. Hewitt 178 / 1947 / "Plant is aromatic (sagey)..."

4906 *Oxypappus scaber* Mexico / Y. Mexia 1350 / 1927 / "'Manzanilla del Campo'"

4907 *Flaveria bidentis* Peru / Stork, Horton & Vargas 10578 / 1939 / "Rather rank smelling"

4908 *Villanova pratensis* El Salvador / P. C. Standley 19223 / 1921–22 / "... with odor like mice"

4909 *V. titicacensis* Colombia / Killip & Smith 17403 / 1927 / "'Matricaria'"

4910 *Amblyopappus pusellus* Chile / G. Looser 2902 / 1933 / "planta aromatica ..."

4911 *Bahia ambrosioides* Chile / D. Bertero 839 / 1829 / "'Mançanilla Simarona'" Chile / I. M. Johnston 4913 / 1925 / "'yerba raton'"

4912 *Schkuhria anthemoidea* var. *guatemalense* El Salvador / S. Calderón 962 / 1922 / "'Escoba amarga.' Used to kill fleas"

4913   *S. pinnata* var. *pinnata* Bolivia / C. W. Hein 562? / 1943 / "The uses of this plant are to destroy fleas and to fight all sorts of diseases, including malaria. It is taken as an infusion, stems and seed being poured into hot water; this same liquid serves to wet the floor of the rooms that are to be disinfected" / "'Piquipichana' (Flea-broom)"

Argentina / W. Lossen 56 / no year / "'Matapulga'"

Peru / Macbride & Featherstone 1294 / "Medicinal. 'Concha lagua'"

4914   *S. pinnata* var. *abrotanoides* Peru / Stork & Horton 10793 / 1939 / "Used to exterminate lice, also medicinally for kidneys and liver"

4915   *S. pinnata* var. *octoaristata* Bolivia / J. Steinbach 6937 / 1925 / "Remedio" / "'Canchalágua'"

Peru / Macbride & Featherstone 275 / 1922 / "Used to stop bleeding of cuts, etc."

Bolivia / W. J. Eyerdam 24661 / 1939 / "Unpleasant foetid odor"

4916   *Hulsea algida* U.S.: Mont. / A. Cronquist 8093 / 1955 / "...strongly aromatic and glandular-viscid"

4917   *H. heterochroma* U.S.: Calif. / S. F. Blake 9939 / 1927 / "Strong and rather pleasant odor. Juice from glands very irritant"

4918   *Hymenoxys cooperi* var. *cooperi* U.S.: Ariz. / R. M. Scott s.n. / 1918 / "'Poisonous to sheep'"

4919   *Helenium amarum* U.S.: Ind. / C. C. Deam 51497 / 1931 / "Hogs do not molest it"

U.S.: Ala. / L. V. Porter s.n. / 1938 / "'Sneezeweed'"

U.S.: Ala. / L. C. Bush s.n. / 1928 / "'Bitterweed'"

4920   *H. hoopesii* U.S.: Utah / Pammel & Blackwood 4258 / 1902 / "Sheep will not eat this plant"

4921   *H. microcephalum* U.S.: Tex. / L. H. Shinners 7935 / 1945 / "Leaves somewhat viscid and slightly aromatic"

4922   *H. autumnale* U.S.: Penn. / S. P. Sharples 145 / 1858–64 / "'Sneeze Weed'"

4923   *H. scorzoneraefolium* Mexico / Dr. Ghiesbreght 527 / 1864–70 / "...employée comme sternutatoire"

4924   *H. quadridentatum* Cuba / E. L. Ekman 16804 / 1923 / "'Romerillo americano'"

4925   *Cephalophora aromatica* Peru / Macbride & Featherstone 2572 / 1922 / "Grown in monastic garden. Used dry as snuff substitute. Odor Pungent" / "'Pelona'"

4926   *Amblyolepis setigera* U.S.: Tex. / G. C. Woolson 1 / 1877 / "Said to be a powerful sudorific"

U.S.: Tex. / F. Lindheimer 302 / 1850 / "(The dry plant has a strong aromatic odor like melilotus)"

4927 *Nicolletia edwardsii* Mexico / I. M. Johnston 7686 / 1938 / "... with powerful odor"

4928 *N. trifida* Mexico / I. L. Wiggins 4402 / 1930 / "Plant with disagreeable odor"

4929 *Nicolletia* indet. Mexico / R. M. Stewart 2639 / 1942 / "'Yerba del Venado'"

4930 *Tagetes erecta* Mexico / G. B. Hinton 1942 / 32 / "'Flor de Muerto'"

Mexico / H. S. Gentry 5321 / 1939 / "Annual with astringent pleasant odor. Vernacul: Sin pual"

4931 *T. filifolia* Mexico / H. S. Gentry 1759 / 1935 / "Vernacular: Anissilla, Mex. Medic. Decoction made from herbage for diverse ailments. Collected in the Sierras and carried below for sale to druggists"

Mexico / Y. Mexia 901 / 1926 / "'Anisillo.' Strongly scented"

Costa Rica / P. C. Standley 414597 / 1925 / "'Manzanilla'... strong-scented"

Argentina / A. T. Hunziker 1716 / 1942 / "'Anis del campo'"

4932 *T. foetidissima* Costa Rica / A. Alfaro 32377 / 1924 / "'Flor de muerto'"

4933 *T. lucida* Mexico / H. S. Gentry 2691 / 1936 / "Vernacular: Yerbanis, Mex. Brewed as a tea (quite savory) for refreshment and for slight ailments"

Mexico / Y. Mexia 838 / 1926 / "Dried and used for smoking out mosquitos" / "'Santa Maria'"

Mexico / G. B. Hinton 1272 / 32 / "'Pericon'"

Mexico / L. C. Ervendberg 94 / 1858 / "... called by the people anis and smells like it"

Mexico / Mex. Biol. Exped. of Students of Univ. of Ill. 985 / 1938 / "Leaves licorice flavored"

Mexico / C. H. & M. T. Mueller 695 / 1934 / "Tea from leaves much used as hot beverage"

4934 *T. micrantha* Mexico / A. Dugès s.n. / 1891 / "'Anisillo'"

Mexico / I. M. Johnston 7476 / 1938 / "plant with strong licorice odor"

4935 *T. microglossa* E. Salvador / P. C. Standley 22780 / 1922 / "Strong-scented..." / "'Flor de muerto'"

4936 *T. patula* Guatemala / P. C. Standley 24309 / 1922 / "Cultivated" / "'Flor de muerto'"

4937   *T. remotiflora* Mexico / G. B. Hinton 2090 / 32 / "'Rosa de muerto'"

4938   *T. schiedeana* Mexico / Dressler & Jones 245 / 1953 / "...strongly licorice scented"

4939   *T. sororia* Guatemala / A. F. Skutch 1045 / 1934 / "Foliage strong-scented"

4940   *T. subulata* El Salvador / P. C. Standley 19286 / 1921–22 / "Plant with strong odor" / "'Flor de muerto'"
Venezuela / H. Pittier 7611 / 1917 / "'Mata-pulgas'"

4941   *T. tenuifolia* Mexico / E. Palmer 748 / 1898 / "purchased in market. City of Zacatecas"
Mexico / Y. Mexia 8788 / 1937 / "'Rosita de Muerto'"

4942   *T. caracasana* Colombia / Pennell & Killip 6481 / 1922 / "Herb, with strong odor"
Colombia / W. A. Archer 695 / 1930 / "Plants used as insect repellant in houses" / "'ruda gallinazo'"

4943   *T. minuta* Brazil / Y. Mexia 4769 / 1930 / "...pungent odor... Leaves crushed and applied to swellings..." / "'Rabo de foquete'"

4944   *T. multiflora* Peru / Mr. and Mrs. F. E. Hinkley 15 / 1920 / "Local name 'Chigchipa.' Use, salads"

4945   *T. terniflora* Colombia / Killip & Smith 19667 / 1927 / "'Ruda cimarena.' Used as an insect poison"

4946   *T. zypaquirensis* Ecuador / A. Rimbach 250 / 1934 / "...strongly aromatic"

4947   *Dyssodia porophylloides* U.S.: Calif. / E. Palmer 131 / 1867 / "strong disagreeable odor"

4948   *D. papposa* U.S.: Ky. / C. W. Short s.n. / 1837 / "This plant is so abundant and exhales an odour so unpleasant, as to sicken the traveller over the Western Prairies of Illinois, in Autumn"
U.S.: Tex. / L. C. Hinckley s.n. / 1937 / "strong odor when fresh"
U.S.: Tex. / L. H. Shinners 8190 / 1945 / "Plants resinous dotted and aromatic, odor suggesting Anthemis Cotula with an infusion of Nepeta Cataria"
U.S.: Tex. / V. L. Cory 50335 / 1945 / "'false dogfennel'"
Mexico / G. Thurber 799 / 1800 or 1802 / "Strong disagreeable odor" / "'Yerba de Santa Gertrudes'"
Mexico / Stanford, Retherford & Northcraft 224 / 1941 / "...heavily grazed by goats"

4949   *D. polychaeta* U.S.: Tex. / B. H. Warnock W588 / 1941 / "Frequent and 'smelly' in Green Valley—Glass Mts."

4950 **D. berlandieri** U.S.: Tex. / R. McVaugh 7704 / 1947 / "... odor faintly resinous"

4951 **D. micropoides** U.S.: Tex. / R. McVaugh 7745 / 1947 / "... plants with lemon odor when crushed"

Mexico or Central America: Monterey / Dr. Gregg s.n. / 1847 / "Used as a poultice for wounds"

4952 **D. aurea** U.S.: Tex. / L. C. Hinckley 1633 / 1941 / "Highly scented"

4953 **D. acerosa** Mexico / I. M. Johnston 9366 / 1941 / "'Matrimonio Viejo'"

Mexico / Dr. Gregg 13 / 1847 / "'Yerba de San Nicholas'"

4954 **D. anomala** Mexico / H. S. Gentry 1300 / 1935 / "Medic. for Catarrh" / "'Mira Sol Chiquita,' Mex. 'Turasali,' W."

Mexico / H. S. Gentry 5621 / 1940 / "... strong pleasant pungent odor. Infused & taken as potion for stomach ailments" / "'Manzinillo'"

Mexico / J. González Ortega 6570 / 1926 / "Cataplasmas para reducir el baso"

4955 **D. anthemidifolia** Mexico / L. Constance 3158 / 1947 / "Scented annual"

4956 **D. montana** Guatemala / P. C. Standley 59671 / 1938 / "'Valeriana'"

4957 **D. pentachaeta** Mexico / R. M. Stewart 1544 / 1941 / "'Limoncillo'"

Mexico / Stanford, Retherford & Northcraft 236 / 1941 / "... heavily grazed by goats"

4958 **D. puberula** Mexico / Dr. Gregg 53 / 1847 / "Paraleña used for indigestion pain in stomach cough Etc. (in decoction)"

4959 **D. setifolia** Mexico / E. Palmer 716 / 1898 / "Market of San Luis Potosi"

4960 **D. speciosa** Mexico / L. Constance 2173 / 1947 / "... creosote-scented..."

4961 **Syncephalantha sanguinea** Guatemala / P. C. Standley 58551 / 1938–39 / "'Flor de muerto'"

4962 **Porophyllum macrocephalum** Mexico / Dressler & Jones 125 / 1953 / "... has strong odor and is used for seasoning" / "'papaloquelite'"

4963 **P. punctatum** Honduras / A. Molina R. 1077 / 1948 / "'Mata piojo'"

4964 **P. viridiflorum** Mexico / G. B. Hinton 5315 / 33 / "Sold in markets for food eaten raw"

4965 **P. nutans** Mexico / G. B. Hinton 2241 / 32 / "Young shoots eaten raw" / "'Pápalo quelite de venado'"

4966 **P. crassifolium** Mexico / I. L. Wiggins 11404 / 1946 / "herbage disagreeable odor"

4967   *P. tridentatum* Mexico / H. L. Mason 1928 / 1925 / "Aromatic"

4968   *Chrysactinia mexicana* Mexico or U.S.: Tex. / F. Lindheimer 77 / 1849 / ". . . strong odor of Pinus"

Mexico / Stanford, Retherford & Northcraft 218 / 1941 / ". . . heavily grazed by goats. Flowers yellow, strong odor"

Mexico / Dr. J. Gregg 103 / 1848–49 / "'Yerba de San Nicolas'"

4969   *Pectis ciliaris* U.S.: Fla. / G. R. Cooley 2651 / 1954 / "Plant has a lemon odor"

Venezuela / H. Pittier 8173 / 1918 / "N.v. 'Cominillo aromatic'"

4970   *P. angustifolia* U.S.: Tex. / M. S. Young s.n. / 15 / "Strongly lemon-scented"

Mexico / A. López 10 / 1947 / "serve para tomar como el cafe"

Mexico / S. S. White 2441 / 1939 / "With lemon-like odor"

Mexico / C. V. Hartman 817 / 1891 / "Plant very aromatic (like Anise)"

4971   *P. papposa* U.S.: Calif. / A. K. & F. E. Harrison s.n. / 1931 / "Chinch weed"

U.S.: Calif. / C. B. Wolf 7596 / 1935 / "Aromatic"

Mexico / I. L. Wiggins 11498 / 1946 / "Odor disagreeable"

4972   *P. prostrata* Mexico / J. Gregg 455 / 1848–49 / ". . . pleasant smell" / "'*Contrayerba*'"

Mexico / A. López 36 / 1948 / "'Herb de la Gallina' good for the stomach"

4973   *P. capillaris* Panama / P. H. Allen 1018 / 1938 / ". . . odor of lemon oil"

4974   *P. elongata* Panama / P. C. Standley 27722 / 1923 / ". . . strong-scented"

Panama / P. C. Standley 28187 / 1923–24 / "'Hierba de alacrán'"

Dominican Republic / H. A. Allard 13524 / 1945 / ". . . orange or lemon scented plant . . ."

Dominican Republic / R. A. & E. S. Howard 9926 / 1946 / ". . . plant very aromatic"

Venezuela / H. Pittier 7023 / 1917 / ". . . the whole plant aromatic"

4975   *P. palmeri* Mexico / H. S. Gentry 4745 / 1939 / ". . . with aromatic odor"

4976   *P. stenophylla* Mexico / H. S. Gentry 1025 / 1934 / "Medic. for catarrh (cold), brewed and the fumes inhaled" / "'Cominillo,' Mex."

Mexico / C. E. Lloyd 402 / 1890 / "Sweet, aromatic"

4977   *P. oerstediana* Guatemala / J. A. Steyermark 32015 / 1939 / "Odor resembling that of 'stink bug'"

4978 *P. schottii* Brit. Honduras / W. A. Schipp 673 / 1930 / "...leaves have unpleasant odor when crushed..."

4979 *P. febrifuga* W.I.: Curaçao / Curran & Haman 196 / 17 / "Medicinal—Fever"

4980 *P. densa* Colombia / H. H. Smith 528 / 1898–1901 / "An infusion of whole plant used as remedy for fevers"

4981 *Pectis* indet. Peru / H. E. Stork 11368 / 1939 / "...lvs. resinous with pellucid dots"

4982 *Athanasia trifurcata* S. Africa / M. A. Pocock S.34 / 1926 / "Medicinal—'found' in every farmhouse about here" / "Infusion of leaves used for stomach complaints" / "'Guarri-sung'"

4983 *Santolina scarriosa* Morocco / E. K. Balls B2954 / 1936 / "Strongly aromatic"

4984 *Anthemis cotula* U.S.: Md. / S. F. Blake 9477 / 1926 / "Very malodorous when crushed"
U.S.: Tex. / Innes & Moon 935 / 1941 / "Aromatic-leaved herb..."
U.S.: Iowa / L. H. Pammel 44 / 1904 / "Dog's Fennel or May-weed"
Peru / Mrs. R. S. Shepard 134 / 1919 / "Used for flavoring. Local name 'Mansanilla'"

4985 *A. arvensis* U.S.: Penn. / S. P. Sharples 146 / 1858–64 / "'Wild Chamomile'"
U.S.: Md. / S. F. Blake 9488 / 1926 / "Slightly ill-scented"

4986 *Anthemis* indet. Mexico / J. N. Rose 14086 / 1910 / "Cultivated for medicinal use"

4987 *Achillea atrata* Denmark / Hall 111 / no year / "Plante tonique; vulnéraire"

4988 *A. nobilis* Denmark / Hall 109 / no year / "Aromatique, stomachique"

4989 *Matricaria courrantiana* Guatemala / P. C. Standley 91288 / 1941 / "'Manzanilla.' Sold in market. Much used in domestic medicine all over Guatemala"

4990 *M. inodora* Peru / R. Kanehira 254 / 1927 / "'Mansanilla,' medicine (bought)"

4991 *Chrysanthemum parthenium* Colombia / Killip & Smith 19671 / 1927 / "'Mañanilla.' Remedy for tropical anemia"
Peru / Macbride & Featherstone 1251 / 1922 / "Plant with a pungent minty odor"
Ecuador / Dr. Rose 6 / no year / "'Santa Maria'"

4992 *C. coronarium* Crete / J. B. Patten C- / 1900 / "Used to keep away fleas and bedbugs? Constituent of dalmation powder?"

4993 *C. indicum* Japan / E. Elliott 135 / 1945 / "Medicinal Use: Dried flower decoction cures boil and mump" / "'Aburagiku'"

China / H. T. Feng 37 / 1924 / "For medicine and decoration" / "'Wild Chrysanthemum'"

China / W. T. Tsang 20663 / 1932 / "...fruit edible. Leaves used for tea making"

4994 *C. sinense* China / H. T. Feng 9 / 1924 / "flowers used for medicine and leaves are also useful"

4995 ***Tanacetum nubigenum*** Nepal / Stainton, Sykes & Williams 9021 / 1954 / "...plant aromatic"

4996 ***Artemisia tilesii*** Alaska / Y. Mexia 2252 / 1928 / "'Wormwood'"

4997 ***A. glauca (dracunculoides)*** Canada: Saskatchewan / A. J. Breitung 1386 / 1941 / "Probably planted as some medical herb. Near a deserted dwelling. One clump found"

U.S.: Calif. / Wolf & Stark 5463A / 1933 / "Herbage dark green and very aromatic"

4998 *A. canadensis* U.S.: Mich. / R. J. Webb 387 / 1899 / "'Canada Wormwood'"

4999 *A. tridentata* U.S.: Ore. / Rev. Spalding s.n. / no year / "Sage or Wormwood of Oregon—Plains"

U.S.: Wash. / F. G. Meyer 1712 / 1939 / "Whole plant very aromatic"

5000 ***Artemisia*** indet. W.I.: Martinique / H. & M. Stehlé 7137 / 1946 / "'Absinthe'"

5001 *A. magellanica* Argentina / Eyerdam, Beetle & Grondona 24398 / 1939 / "...pungent, aromatic odor"

5002 *A. afra* Basutoland / D. G. Collett 486 / 1938 / "'Wormwood'; 'Wilde Als'"

5003 *A. asiatica* Japan / E. Elliott 110 / 1946 / "Besides mixing with mochi (dumpling) young leaves are eaten with barley" / "Young leaves are picked for mixing with the dumpling" / "'Yomogi'"

5004 *A. biennis* Nepal / Stainton, Sykes & Williams 3788 / 1954 / "Plant pungent"

5005 *A.* aff. *eriocephala* Nepal / Stainton, Sykes & Williams 3924 / 1954 / "Plant pungent"

5006 *A. japonica* Japan / E. Elliott 22 / no year / "Young leaves roasted and eaten with soy sauce" / "'Otokoyomogi'"

5007 *A. maritima* Kashmir / R. R. Stewart 17940 / 1939 / "'moru' used for santonen"?

5008 *A. mutellina* Denmark / N. C. Seringe 2264 / no year / "Cette Artemisia, la Spicata & Glacialis, sont employées par les habitants des Alpes comme sudorifiques & fébrifuges"

5009   *A. persica* Afghanistan / E. Bacon 78 / 1939 / "The plant is boiled and the liquid drunk for stomach trouble. Called *tirx* in Puštu"

5010   *A. strongylocephala* var. *cachmirica* f. *angustiloba* Nepal / Stainton, Sykes & Williams 3734 / 1954 / "Plant greyish and pungent"

5011   *A. vallesiaca* Denmark / N. C. Seringe 2274 / no year / "C'est avec cette plante que le grand Haller faisait préparer l'extrait d'absinthe"

5012   *Artemisia* indet. India: Punjab / W. Koelz 8452 / 1936 / "Not eaten fresh by animals but collected for winter"

5013   *A. vulgaris* China / W. T. Tsang 718 / 1928 / "...medicine"

E. ASIA / R. D. Anstaed 132 / 1926 / "...aromatic shrub"

5014   *A. lactiflora* China / W. T. Tsang 20655 / 1932 / "Use as food for pigs"

5015   *Crossostephium chinense* China / F. A. McClure Y-98 / 1929 / "a drug plant"

5016   *Liabum glabrum* var. *hypoleucum* El Salvador / P. C. Standley 22265 / 1922 / "'Espinillo,' 'Papelillo,' 'Palo de San Nicolás'"

5017   *L. pichinchense* Ecuador / Y. Mexia 7697 / 1935 / "Used to kindle fire. 'Santa Maria'"

5018   *L. polymnioides* Argentina / Eyerdam & Beetle 22608 / 1938 / "Very aromatic"

5019   *Neurolaena lobata* Panama / P. H. Allen 239 / 1937 / "...used by natives for a fever remedy; bitter tea made from leaves"

5020   *Erechtites hieracifolia* Mexico / G. B. Hinton 3654 / 33 / "'Borraja'"

Brit. Honduras / H. H. Bartlett 11917 / 1931 / "...aromatic"

El Salvador / P. C. Standley 22213 / 1922 / "'Te del suelo'... Remedy for coughs"

China / W. T. Tsang 27935 / 1937 / "Used for feeding pigs"

5021   *Culcitium canescens* Peru / Macbride & Featherstone 2485 / 1922 / "Highly valued cough remedy" / "'Wida wida'"

5022   *C. reflexum* Ecuador / Y. Mexia 7655 / 1935 / "Used medicinally in infusion; said to be a powerful diuretic" / "'Arquitecto'"

5023   *Arnica acaulis* U.S.: Md. / C. C. Plitt 675 / 1901 / "'Leopard's-bane'"

5024   *A. longifolia* U.S.: Wash. / Suksdorf 568 / 1885 / "Pubescence short and very glandular, emitting a strong peculiar odor..."

5025   *Gynura pinnatifida* China / Fan & Li 546 / 1935 / "...root with medicinal value"

5026   *G. angulosa* India / N. L. Bor 16165 / 42 / "...fed to pigs"

5027   *Gynura* indet. Burma / F. G. Dickason 7709 / 1937 / "Use leaves as greens"

5028   *Crassocephalum crepidioides* Philippine Is. / C. B. Tadeo 6009 / 1948 / "Green leaves is used for food prepared as salad, similar to other vegetable" / " 'Doyan-doyan' " Bukidnon

5029   *C. mannii* Nyasaland / L. J. Brass 17597 / 1946 / ". . . plant aromatic"

5030   *Psathyrotes pilifera* U.S.: Ariz. / H. C. Cutler 3154 / 1939 / ". . . strongly pungent odor"

5031   *Cacalia reniformis* U.S.: Penn. / J. A. Shafer 1478 / 1901 / " 'Great Indian-Plaintain' "

5032   *C. atriplicifolia* U.S.: Ohio / R. J. Webb 380 / 1899 / " 'Pale Indian Plaintain' "

5033   *C. decomposita* Mexico / W. P. Hewitt 68 / 1945 / "This is among the 3 or 4 most important medicinal plants in the Sierra . . . . Tea used for diabetes and kidneys; also powdered or as a wash for skin irritations" / " 'Matarique' "

     Mexico / H. S. Gentry 2815 / 1936 / "The roots are valued for their medicinal properties" / " 'Matariki,' Mex."

     Mexico / H. S. Gentry 1959 / 1935 / "Roots valued for their medicinal properties, are collected and transported to the lowland towns, and sold to druggists" / " 'Matariki,' Mex."

5034   *Senecio amplectens* U.S.: Nev. / Maguire & Holmgren 22647 / 1943 / "(Plants with disagreeable odor)"

5035   *S. obovatus* U.S.: Ohio / R. J. Webb 433 / 1901 / " 'Round-leaf Squaw-weed' "

5036   *S. smallii* U.S.: Va. / S. B. Kovacs 3057 / 1933 / " 'Small's Squaw-weed' "

5037   *S. tomentosus* U.S.: Md. to Del. / no collector s.n. / no year / "Known as 'Yaw leaf'; 'Good for the yaws' whatever that is . . ."

5038   *S. angulifolius* Mexico / E. K. Balls 5633 / 1938 / ". . . whole plant strongly aromatic"

5039   *S. bellidifolius* Mexico / W. P. Hewitt 116 / 1946 / "Taken as a tea for bladder and kidney trouble. Fresh plant ground with olive oil applied as poultice for boils, tumors, infections, changing several times daily" / " 'Chucaca' "

5040   *S. douglasii* Mexico / I. L. Wiggins 5403 / 1931 / "Herbage with strong, disagreeable odor"

5041   *S. hoffmannii* Costa Rica / Dodge & Thomas 6469 / 1930 / "Capitana, used as substitute for quinine by natives"

5042   *S. kermesinus* Honduras / J. B. Edwards P-741 / 1934 / " 'Flor de San Juan' "

     Mexico / C. & E. Seler 1814 / 1896 / " 'flor de niño' "

El Salvador / P. C. Standley 19281 / 1921–22 / "'Flor de colmena'" / "Used as decoration on altars"

Honduras / Yuncker, Koepper & Wagner 8218 / 1938 / "...aromatic"

5043 *S. praecox* Mexico / A. Dugès 452 / 1891 / "'Candelero'"

Mexico / E. K. Balls 4119 / 1938 / "Strongly chocolate scented"

5044 *S. salignus* Mexico / Y. Mexia 2504 / 1929 / "'Yerba de las Animas'"

Guatemala / P. C. Standley 63643 / 1938–39 / "'Chilca'"

Mexico / R. Q. Abbott 68 / 1936 / "'Jarilla'"

Mexico / E. Langlassé 97 / 1898 / "'Yara'"

5045 *S. sartorii* Mexico / H. S. Gentry 5659 / 1940 / "...highly odorous shrubs"

5046 *S. toluccanus* Mexico / G. B. Hinton 3279 / 1933 / "Stalk eaten raw" / "'Conejo'"

5047 *Senecio* indet. Mexico / H. S. Gentry 1411 / 1935 / "Has an odor like an exotic perfume which I have noticed in feminine company"

5048 *Senecio* indet. Mexico / R. M. Stewart 2227 / 1942 / "...strong odor" / "'Jarilla'"

5049 *Senecio* indet. Mexico / H. S. Gentry 7178 / 1945 / "'Peyote'"

5050 *Senecio* indet. Mexico / W. P. Hewitt 16 / 1945 / "ground fresh juice applied directly to fly bites on cattle to kill blow worms hatching in flesh" / "'Sopepare'"

5051 *S. lucidus* W.I.: Martinique / H. & M. Stehlé 3657 / 1939 / "'herbe à pique'"

5052 *S. adenotrichius* Chile / G. Montero 58 / 1922 / "Una yerba fétida pegajosa"

5053 *S. jorquerae* Chile / I. M. Johnston 4889 / 1925 / "'Yerba raton'"

5054 *S. leucus* Chile / I. M. Johnston 4842 / 1925 / "'Yerba de jote'"

5055 *S. pinnatus* S. America / no collector 107 / 1826 / "A strong disagreeable smell"

5056 *S. rhizomatus* Bolivia / W. J. Eyerdam 25057 / 1939 / "Peculiar odor"

5057 *S. saxicolus* Chile / Worth & Morrison 16255 / 1938 / "...strong odor of sheep"

5058 *Senecio* indet. Peru / Y. Mexia 8105 / 1936 / "Leaves stewed and eaten" / "'Col del Monte'"

5059 *Senecio* indet. Peru / C. Vargas C. 1554 / 1939 / "Planta medicinal, para pulmones" / "'Huamanipa'"

5060 *Senecio* indet. Peru / H. E. Stork 11389 / 1939 / "Agreeably fragrant, more so at night" / "'San Juan'"

5061 *Senecio* indet. Peru / Mr. & Mrs. F. E. Hinkley 52 / 1920 / "'Chechera cimarrona del cerro'"

5062 *S. procumbens* Philippine Is. / C. O. Frake 521 / 1958 / "Remedy for rice aphids" / "'Kalambuay'" Sub.

5063 *S. sonchifolia* Philippine Is. / C. O. Frake 781 / 1958 / "Leaves applied to skin eruptions" / "'taket'" Sub.

Philippine Is. / M. Adduru 217 / 1917 / "to cure wounds"

5064 *S. lautus* Australia: Queensland / C. T. White 12692 / 1944 / ". . . completely ignored by stock"

5065 *S. arnicaeflorus* S. Africa / N. Bolus 121 / 1882 / "'False Bay'"

5066 *S. bupleuroides* Africa / Pretoria Herb. 11490 / 1934 / "Sent in by Director of Vet. Services from O/C Alberton Laboratory. Suspected of poisoning in stock"

5067 *S. halimifolius* S. Africa / MacOwan 931 / 1888 / "False Bay"

5068 *S. multicorymbosus* Brit. E. Africa / R. H. Goodwin 11 / 1937 / ". . . rank disagreeable odor"

5069 *Emilia sonchifolia* China / G. W. Groff 79 / 1919 / "Medical use" / "'Yeung T'ai 'Ts'o'"

5070 *Emilia* indet. Philippine Is. / W. Beyer 11 / 1948 / "For treating boils. Masked leaves applied" / "'Hubbu Hubbu'" Ifugao

5071 *Werneria nubigens* Peru / Stork & Horton 9986 / 1938 / "'Cebollin'"

5072 *W. poposa* Argentina / T. Meyer 34205 / 1940 / "Medicinal y comestible" / "'Pupusa'"

5073 *W. pygmaea* Chile / M. Gay? s.n. / no year / "'Romerilla'"

5074 *Gazania pinnata* S. Africa / MacOwan 126 / 1882 / "'False Bay'"

5075 *Carlina?* Afghanistan / E. Bacon 96 / 1939 / "Used for fodder"

5076 *Atractylis chinensis* var. *simplicifolia* China / K. Ling 2539 / 1926 / "medicine"

5077 *A. coreana* China / C. Y. Chiao 2661 / 1930 / "Drug plant"

5078 *A. ovata* China / C. Y. Chiao 2917 / 1930 / "drug plant"

5079 *Cousinia* indet. Afghanistan / E. Bacon 100 / 1939 / "One of few types of plants untouched by grazing sheep. Used as fodder (apparently after it has been dried and broken up)"

5080 *Carduus nutans* Canada: Ontario / J. A. Longman s.n. / 1925 / "Musk or Nodding Thistle"

5081 *Carduus* indet. Mexico / R. M. Stewart 1267 / 1941 / "'Cardo Santo'"

5082 *Cirsium conspicuum* Mexico / G. B. Hinton 13874 / 1939 / "'Cardo santo'"

5083 *C. mexicanum* El Salvador / P. C. Standley 21474 / 1922 / "'Cardo santo'"

5084   *C. subcoriaceum* Mexico / Y. Mexia 9044 / 1937 / "'Cardo-Santo'"

5085   *Cirsium* indet. Mexico / E. Langlassé 970 or 910 / 1899 / "'Can o santo'"

5086   *C. arvense* Afghanistan / E. Bacon 42 / 1939 / "Used for fodder, fuel"

5087   *Serratula chinensis* China / W. T. Tsang 20662 / 1932 / "Roots used as medicine"

5088   *Centaurea americana* Mexico / G. Thurber 748 / 1852 / "'Cardus benedictus'"

5089   *C. chilensis* Chile / G. Montero 65 / 1922 / "'Flor del minero'"

5090   *C. ibirica* Afghanistan / E. Bacon 29 / 1939 / "Used for cooking"

5091   *Cnicus arvensis* U.S. or Canada / A. W. Graham s.n. / 1889? / "Canda or cursed Thistle"

5092   *Chuquiraga floribunda* Bolivia / J. Steinbach 6319 / 1924 / "'Tigrillo' o 'Cujuchi del monte'"

5093   *C. insignis* Ecuador / F. C. Lehmann 469 / 1881 / "...febrifuges..."

5094   *C. microphylla* Ecuador / L. Mille S.J. 753 / 1906 / "Planta amara febrifuga"

       Ecuador / A. Rimbach 41 / no year / "Infusion of twigs is said to have a similar effect as Quinine. Vulg: 'Chuquiragua'"

5095   *C. ulicina* Chile / I. M. Johnston 4799 / 1925 / "...leaves pungent"

       Chile / I. M. Johnston 5669 / 1925 / "shrub (browsed)..."

       Chile / Reed s.n. / no year / "'Yerba buena'"

5096   *Gochnatia hypoleuca* Mexico / Stanford, Retherford & Northcraft 184 / 1941 / "...heavily grazed by goats"

5097   *Cyclolepis genistoides* Argentina / T. Meyer 4054 / 1942 / "famoso como medicinal" / "'Palo azul'"

5098   *Lycoseris crocata* Guatemala / J. A. Steyermark 42188 / 1942 / "'Santo Domingo'"

5099   *Plazia pinifolium* Chile / J. West 3870 / 1935 / "...aromatic succulent foliage"

5100   *Pleiotaxis antunesii* Africa / Mrs. Leona S. Tucker 50 / 1924 / "Ochisipesipe—so named because there is a sweet liquid in the flower. The roots are used as medicine for fever and stomach"

5101   *Barnadesia parviflora* Ecuador / F. C. Lehmann 5239 / no year / "'Palo-Santo'"

5102   *Ainsliaea fragrans* China / S. K. Lau 4554 / 1934 / "...used as medicine"

5103   *Pachylaena atriplicifolia* Chile / Morrison & Wagenknecht 17177 / 1939 / "Used as a medicine" / "'Yerba Santa'"

       Chile / J. L. Morrison 16894 / 1938 / "Used as stomach medicine" / "'Yerba santa,' 'Escarapela'"

5104 *Chaetanthera involucratum* Chile / I. M. Johnston 4855 / 1925 / "'Escarapela'"

5105 *C. spathulifolia* Chile / J. L. Morrison 16943 / 1938 / "'Escarapella'"

5106 *Trichocline auriculata* Argentina / Meyer & Bianchi 33032 / 1940 / "'Contrayerba'"

5107 *T. incana* Argentina / Eyerdam, Beetle & Grondona 23486 / 1938 / "... these herbaceous plants ... were completely grazed out of the surrounding area"
Argentina / T. Meyer 3982 / 1941 / "lo fuman mezclado con tabaco" / "'Coro'"

5108 *Gerbera piloselloides* China / W. T. Tsang 21883 / 1933 / "Used for medicine"

5109 *Gerbera* indet. China / W. T. Tsang 22149 / 1933 / "Used as medicine"

5110 *Chaptalia nutans* Honduras / S. F. Blake 7390 / 1919 / "Root medicinal. N.v. 'Valeriana'"
El Salvador / P. C. Standley 19389 / 1921–22 / "'Valeriana'"

5111 *Polyachyrus glandulosus* Peru / Mr. & Mrs. F. E. Hinkley 49 / 1920 / "Use, medicinal" / "'arubarina cimarrona'"

5112 *Perezia thurberi* Mexico / H. S. Gentry 2012 / 1935 / "Medic. Warihio report use roots for decoction for killing worms in live stock, and for curing human diseases" / "'Mata gusano,' Mex."

5113 *Perezia* indet. Mexico / R. M. Stewart 692 / 1941 / "'Tarantula'"

5114 *Perezia* indet. Mexico / R. M. Stewart 2971 / 1942 / "'Arnica'"

5115 *P. multiflora* Bolivia / J. West 6374 / 1936 / "Used medicinally"

5116 *Trixis radialis* El Salvador / P. C. Standley 21955 / 1922 / "'San Pedro'"
Chile / W. J. Eyerdam 24647 / 1939 / "... lvs. pleasantly aromatic"

5117 *T. wrightii* Mexico / H. S. Gentry 1371 / 1935 / "Medic. mash flowers, and apply to temples and forehead when a person is 'loco'" / "'Yerba del Aigre,' Mex."

5118 *T. antimenorrhoea* var. *discolor* Argentina / W. Lossen 109 / "Jan. 25" / "'Contrayerba d.l. Sierra'"

5119 *Jungia revoluta* Chile / G. Looser 2218 / 1932 / "'Yerba del pasmo'" / "... planta aromática"

5120 *Jungia* indet. Argentina / R. Alaiy? 27 / 41 / "'Zarzaparrilla'"

5121 *Lapsana apogonoides* China / W. T. Tsang 24877 / 1935 / "... used for feeding pigs"

5122 *Hypochaeris brasiliensis* Argentina / P. Jorgensen 1007 / 1915 / "Comestible"

5123 *H. grandidentata* Chile / E. E. Gigoux s.n. / 1894 / "'Palo de Bandera'"

Chile / E. E. Gigoux s.n. / 1885 / "'Renca'"

5124 *H. meyeniana* Peru / Mrs. R. S. Shepard 3 / 1919 / "Local name 'Sicke' (Indian name)"?

5125 *Hypochaeris* indet. Ecuador / J. N. & G. Rose 22480 / 1918 / "'Chicora'"

5126 *H. stenocephala* Peru / Y. Mexia 04191 / 1935 / "'Pilly' or 'Chicoria'"

5127 *Taraxacum mongolicum* China / H. T. Feng 38 / 1924 / "Milky juice in the stem. Used for eating and medicine" / "Pukung Ying"

5128 *T. officinale* China / W. P. Fang 3622 / 1928 / "...medicine"

5129 *Launaea arabica* Kuwait / H. Dickson 125 / 62 / "...eaten by all animals" / "Vern. Name 'Murrar'"

5130 *L. nudicaulis* Arabia / H. Dickson 19 / 61 / "...eaten by all animals" / "Vern. Name: 'Howal Ghagal'"

Kuwait / H. Dickson 120 / 62 / "...grazed by all animals" / "'Howal Ghagal'"

5131 *Sonchus oleraceus* U.S.: Tex. / V. L. Cory 53468 / 1947 / "'sow thistle'"

Mexico / S. S. White 3412 / 1940 / "...juice milky"

El Salvador / S. Calderon 103 / 1922 / "'Lechuga montés'"

Ecuador / L. Mille S.J. 5 / 1922 / "in hortis"

Peru / Macbride & Featherstone 1367 / 22 / "Used fresh for stomach disorders"

Peru / Mrs. R. S. Shepard 30 / 1919 / "Local name: 'Khanachu'"

Japan / T. Tyozaburô 100112 / 1924 / "...stem with milky juice"

Philippine Is. / M. D. Sulit 7673 / 1948 / "...leaves are edible, like cabbage" / "Common name: 'Gatgatang'"

Gambier Is. / F. R. Fosberg 11042 / 1934 / "'Taru'"

5132 *S. asper* U.S.: Mich. / C. A. Davis 150 / 1892 / "'Spiny-leaved Sow Thistle'"

U.S.: Ohio / R. J. Webb 253 / 1897 / "'Spiny Sow-Thistle'"

El Salvador / P. C. Standley 19218 / 1921–22 / "'Valeriana'"

Argentina / J. E. Montes 2226 / 46 / "Nombre vulgar: 'Cerraja'"

5133 *S. arvensis* U.S.: Mass. / E. F. Fletcher s.n. / 1911 / "'Common Sow Thistle'"

U.S.: N.Y. / S. A. Beach 149 / 1892 / "'Sow Thistle'"

China / C. W. Wang 62886 / 1935 / "'Cardus'"

China / W. T. Tsang 22597 / 1933 / "...edible" / "'Ye Fu Mak'"
China / W. T. Tsang 25110 / 1935 / "...fr. edible" / "'Ye Fu Mak'"
Burma / F. G. Dickason 7665 / 1938 / "'Dong Dor'"
India / H. F. Mooney 3464 / 49 / "'Kahalandi'"

5134 *Sonchus* indet. Mexico / R. M. Stewart 2158 / 1941 / "'Borraja'"

5135 *Sonchus* indet. Mexico / R. M. Stewart 2197 / 1941 / "'Borraja'"

5136 *Sonchus* indet. China / W. T. Tsang 22320 / 1933 / "...fr. edible" / "'Fu Mak Tsoi'"

5137 *S. bipontini* Tanganyika Terr. / F. G. Carnochan 345 / 1928 / "'Lusunga'"

5138 *Lactuca scariola* Canada: Alberta / E. H. Moss 238 / 39 / "'Pincher'"

5139 *Lactuca* indet. U.S.: Wis. / I. A. Lapham s.n. / no year / "Called 'Hoosong' or Chinese Asparagus"

5140 *L. intybacea* W.I.: Curaçao / Curran & Haman 191 / 1917 / "Medicinal-throat" / "'Salada andijvie'"
Virgin Is. / J. B. Thompson 1074 / 1925 / "'Anna's Hope'"

5141 *L. diversifolia* China / W. T. Tsang 28814 / 1938 / "'Ye Fu Mak Ts'oi'"

5142 *L. indica* Hainan / W. T. Tsang 337 / 1927 / "'Shan Foo Mak T'soi'"
China / W. T. Tsang 23217 / 1933 / "'Sai Fu Mak'"
Hainan / Tsang & Fung 648 / 1929 / "'Chim Ip Fu Mak'"
Hainan / W. T. Tsang 450 / 1928 / "'Ye Fu Mak'"
China / W. T. Tsang 20941 / 1932 / "Flower...edible" / "'Sai Ye Fu Mak'"
China / W. T. Tsang 24139 / 1934 / "...frt. edible" / "'Chim Yip Sai Fu Mak'"
China / W. T. Tsang 27678 / 1937 / "'Ye Fu Mak'"
China / W. T. Tsang 28199 / 1937 / "...fr. edible" / "'Ye Fu Mak'"
Sumatra / R. S. Boeea 8598 / 1935 / "'Samoer tata'"

5143 *L. laciniata* Japan / K. Uno 379 / 1923 / "Nipp. 'aki-no-nogeshi'"
Korea / Mrs. R. K. Smith s.n. / 1934 / "'Aki n geshi'" Japanese / "'Katulpyungi'" Korean

5144 *L. laciniata* f. *indivisa* Japan / K. Uno 384 / 1923 / "Nipp. 'hosoba-no-aki-nogeshi'"

5145 *L. lessertiana* China / C. W. Wang 66002 / 1935 / "Milky juice"

5146 *L. raddeana* Korea / Mrs. R. K. Smith s.n. / 1934 / "'Miyama aki nogeshi'"
China / W. T. Tsang 25551 / 1935 / "'Ye Fu Mak'"

China / W. T. Tsang 28918 / 1938 / "'Ye Fu Mak Ts'oi'"

Japan / K. Uno 369 / 1914 / "'Ya-nigana'"

5147 *L. sibirica* Japan / K. Uno 19734 / 1937 / "'Yezo-murasaki-nigana' (Nipp.)"

Japan / K. Uno 21636 / 1937 / "'Yezo-murasaki-nagana' (Nipp.)"

5148 *L. sororia* Japan / K. Watanabe s.n. / 1890 / "'Murasaki-nigana'"

5149 *L. squarrosa* Japan / E. Elliott 55 / '46 / "...leaves are eaten....
Food uses: leaves boiled and eaten with soy sauce" / "'Akinonogeshi'"

5150 *L. stolonifera* Japan / K. Uno 3636 / 1931 / "'Iwa nigana' (Nipp.)"

5151 *Lactuca* indet. China / G. W. Groff Herb. 4040 / 19 / "...a cultivated
vegetable...Leaves edible" / "'Fu mak, ts'oi' (Chinese)"

5152 *L. brunoniana* INDIA: Kolung. Lahul / native collector for W. Koelz
9872 / 1934 / "'Katakmur'"

5153 *Ixeria chinensis* Japan / K. Uno 11295 / 1934 / "'Takasago-nigana' (Nipp.)"

Korea / Mrs. R. K. Smith s.n. / 1933 / "'Nigana,' 'sing pai'"

Japan / no collector s.n. / 1894 / "'Kawara-nigana'"

5154 *I. chinensis* ssp. *strigosa* China / K. P. To Herb. 2383 / 1918 / "'Ye
Fu Mok'" Cantonese

5155 *I. dentata* Japan / K. Uno 3633 / 1931 / "'Nigana'" Nipp.

Japan / no collector s.n. / 1888 / "'Niga-na'"

China / W. T. Tsang 28789 / 1938 / "'Sai Yeung Fu Mak'"

5156 *I. denticulata* China / W. T. Tsang 20649 / 1932 / "'Wong Fa Lo'"

Japan / S. Ohkara 360 / 1925 / "'Yakushi-so'" Nipp.

Japan / K. Uno 363 / 1929 / "'Hana-yakushiso'" Nipp.

5157 *I. denticulata* ssp. *typica* China / Cheo & Yen 360 / 1936 / "...with
milky juice"

5158 *I. gracilis* China / W. T. Tsang 20425 / 1932 / "'Sai Fu Mak Tsz'"

China / W. T. Tsang 25116 / 1935 / "'Ye Fu Mak'"

5159 *I. japonica* Japan / K. Watanabe s.n. / 1888 / "'Iwa-nigana'"

Japan / K. Watanabe s.n. / 1891 / "'Ji-shibari'" Japanese

5160 *I. polycephala* Japan / no collector s.n. / 1894 / "'No-nigana'"

5161 *Crepis japonica* China / W. T. Tsang 28668 / 1938 / "'Ye Loh Pak
Ts'oi'"

Japan / K. Watanabe s.n. / 1888 / "'Oni-tabirako'"

5162 *C. rigescens* ssp. *lignescens* Burma / F. G. Dickason 7432 / 1938 /
"'Khi Ha' ('Haka Chin')"

5163   *Youngia bungeana* Japan / K. Uno 23202 / 1938 / "Nipp. 'Chosen-yakushiso'"

5164   *Y. denticulata* Japan / E. Elliott 111 / '46 / "Food use: leaves are eaten after being well-cooked to remove bitterness" / "'Yakushiso'"

5165   *Y. japonica* Japan / S. Ohkara 254 / 1925 / "Nipp. 'Oni-tabirako'"

5166   *Y. japonica* ssp. *elstonii* China / W. T. Tsang 26208 / 1936 / "'Yeung Wo Tung'"

Sumatra / R. S. Boeea 9177 / 1936 / "'Sori man(da) pot'"

5167   *Prenanthes alba* U.S.: Mass. / E. F. Fletcher s.n. / before 1924 / "'White lettuce'" / "'Rattlesnake-root'"

5168   *P. trifoliata* U.S.: Mass. / E. F. Fletcher s.n. / before 1924 / "'Rattle-snake-root'"

5169   *P. altissima* U.S.: Mass. / Herb. G. Gilbert s.n. / '94 / "'Rattlesnake-root'"

U.S.: Va. / P. A. Warren 115 / 31 / "'White Lettuce'; 'Rattlesnake-root'"

5170   *Hieracium venosum* U.S.: N.J. / O. H. Brown 10704 / 1917 / "'Rat-tlesnake weed'"

U.S.: Va. / A. Rayland 3936 / 35 / "'Rattlesnake-weed'"

5171   *H. venosum* var. *nudicaule* U.S.: Mass. / E. F. Fletcher s.n. / before 1924 / "'Hairy Hawkweed'"

U.S.: Mass. / E. F. Fletcher s.n. / before 1924 / "'Rattlesnake-weed'; 'Poor Robin's Plaintain'"

U.S.: Penn. / S. P. Sharples s.n. / 1858–64 / "'Rattlesnake-weed'"

5172   *H. gronovii* U.S.: Ohio / Rood & Barber 913 / 1906 / "'Gronovius Hawkweed'"

5173   *H. jalapense* Guatemala / J. A. Steyermark 51092 / 1942 / "Reputed to cure pains in cattle; dry leaves and put the powdered leaves over wounds of the cattle in treating the pains. 'Yerba de culebra'"

5174   *H. loxense* Ecuador / Prieto & Camp E-2485 / 1945 / ". . . 'Taruga tani' (Taruga-deer; tani-herb, or food)"

5175   *H. aurantiacum* Japan / K. Uno 20008 / 1937 / "'Hohrin-tampopo,' 'Yefude-tampopo'"

5176   *H. japonicum* Japan / M. Furuse s.n. / 1958 / "'Miyama-koozorina'"
Japan / H. Ichikawa 336 / 1919 / "'Miyama-kozorina'"

5177   *H. krameri* Japan / K. Uno 338 / 1931 / "'Sui-ran'"

5178   *H. umbellatum* Japan / K. Uno 21728 / 1937 / "'Yanagi-tampopo'"
Korea / Mrs. R. K. Smith s.n. / 1934 / "'Yanagi tanpopo'" / "'Cho-pap pa mul'"?

Japan / no collector s.n. / 1894 / "'Kozari-na'"

Bibliography

Some of the information in this volume has been reported in occasional papers, which include the following:

Altschul, S. von Reis. "Psychopharmacological notes in the Harvard University herbaria." *Lloydia* 30:192–196 (1967).
—— "Vilca and its use." *Ethnopharmacologic Search for Psychoactive Drugs,* ed. Daniel H. Efron. U.S. Public Health Service Publication No. 1645:307–314 (1967).
—— "Unusual food plants in herbarium records." *Econ. Bot.* 22:293–296 (1968).
—— "Ethnogynecological notes in the Harvard University herbaria." *Bot. Mus. Leaflets Harvard Univ.* 22:333–343 (1970).
—— "Ethnopediatric notes in the Harvard University herbaria." *Lloydia* 33:195–198 (1970).
Hartwell, J. L. "Plants used against cancer. A survey." *Lloydia* 30:379–436 (1967); 31:71–170 (1968); 32:79–107, 153–205, 247–296 (1969); 33:97–194, 288–392 (1970); 34:103–160 (1971).
Kreig, M. B. *Green Medicine. The Search for Plants That Heal.* New York, Rand McNally (1964).
Raffauf, R. F., and S. von Reis Altschul. "The detection of alkaloids in herbarium material." *Econ. Bot.* 22:267–269 (1968).
von Reis, S. "Herbaria: sources of medicinal folklore." *Econ. Bot.* 16:283–287 (1962).
Schultes, R. E. "The role of the ethnobotanist in the search for new medicinal plants." *Lloydia* 25:257–266 (1962).
—— "The widening panorama in medical botany." *Rhodora* 65:97–120 (1963).
—— "De plantis toxicariis e Mundo Novo tropicale commentationes I." *Bot. Mus. Leaflets Harvard Univ.* 21:265–284 (1967).

Index to Families

Index to Genera

Medical Index

NOTE: In a number of instances the medicinal use reported for a particular species was unclear. In these cases, a use has been assigned based upon the best possible evaluation of information in the citations. Many intriguing entries could not be indexed at all.

The *catalogue number* of each species, rather than the page on which it appears, is given under the topic headings below.

Analgesics: back pain, 1338, 2185, 2778; chest pain, 16, 1925, 2816, 3232, 3495, 3525, 3658, 4187, 4322; cramps, 586; headache, 96, 155, 563, 619, 647, 791, 1001, 1083, 1102, 1105, 1242, 1243, 1248, 1365, 1521, 1601, 1681, 1697, 1834, 1894, 1898, 1970, 2143, 2288, 2310, 2327, 2329, 2587, 2672, 2821, 3014, 3098, 3220, 3276, 3348, 3412, 3470B, 3497, 3622, 3643, 3644, 3670, 3678, 3791, 4022, 4128, 4154, 4180, 4186, 4257A, 4258, 4331, 4421; joints, 617, 731, 760, 1109, 1955, 2179, 2310, 2313, 2471, 2819, 3111, 3113, 3158, 3203, 3551, 3658, 3676, 3951, 4024, 4045; labor, 1842; stomach, 304, 555, 652, 683, 694, 723, 762, 791, 813, 920, 1003, 1082, 1086, 1139, 1379, 1442, 1697, 1933, 1995, 2005, 2131, 2138, 2201, 2310, 2396, 2397, 2480, 2486, 2501, 2548, 2585, 2657, 2765, 2786, 3003, 3063, 3206, 3361, 3523, 3616, 3680, 3819, 3822, 3827, 3954, 4200, 4249, 4270, 4271, 4408, 4489, 4573, 4746, 4804, 4891; taken internally, 477, 483, 1628, 1868, 2024, 2494, 2606, 3561, 4314; topical application for internal pain, 409, 423, 616, 666, 727, 988, 1046, 1072, 1078, 1234, 1240, 1781, 1796, 1799, 1957, 2007, 2010, 2024, 2044, 2130, 2181, 2283, 2310, 2473, 2583, 2600, 2819, 2824, 2877, 3032, 3290, 4128, 4200, 4270, 4278, 4299, 4303, 4368, 4902, 5173. *See also* Anesthetics; Emollients

Anesthetics: eye, 757, 1269, 1320, 2328, 2356, 2677, 3227, 3329, 3350, 3353, 4870; local, 536

Antitumor agents, 963, 3303, 3761, 5039

Aromatics, 61, 63, 102, 111, 122, 131, 162, 163, 176, 180, 209, 212, 249, 270, 275, 287, 288, 400, 476, 478–80, 484, 489, 491, 497, 514–16, 521, 539, 549, 551, 552, 556, 559, 561, 564–75, 577, 585, 592, 593, 595, 598, 602, 605, 610, 612–14, 618, 639, 643, 680, 695, 742, 750, 755, 803, 808, 815, 817, 904, 908, 925, 958, 962, 971, 1030, 1032, 1036, 1041, 1047, 1084, 1088, 1094, 1096, 1098–1101, 1103–06, 1108, 1110–16, 1118, 1122, 1125, 1126, 1138–43, 1152, 1154, 1161, 1167, 1172, 1173, 1176, 1178, 1183, 1190, 1192, 1199, 1202, 1204–06, 1208–13, 1215–17, 1219, 1220, 1222–25, 1227, 1229–33, 1237, 1238, 1253, 1255, 1256, 1258, 1259, 1261, 1262, 1265–67, 1270, 1272–74, 1277–80, 1282, 1283, 1285, 1288, 1290, 1292, 1294, 1295, 1298–1308, 1310, 1313, 1331, 1348, 1355, 1358, 1371, 1406, 1456, 1457, 1459, 1470, 1471, 1618, 1707, 1789, 1872, 1877, 1879, 1882, 1891, 1892, 1899, 1902, 1904, 1905, 1907–10, 1934, 1938, 1938A, 1939, 1982, 1984, 1995, 1998, 1999, 2030, 2077–79, 2081, 2204–06, 2209, 2214, 2219, 2220, 2223, 2224, 2229–33, 2236–42, 2250, 2259, 2274, 2275, 2312, 2360, 2419, 2438, 2441, 2472, 2492, 2509, 2514, 2558, 2563, 2643, 2648, 2649, 2651, 2652, 2687, 2689, 2690, 2751, 2840, 2854, 2855, 2857, 2860, 2864, 2879, 2888, 2902, 2904, 2979, 3001, 3002, 3037, 3067, 3068, 3071, 3075, 3092, 3096, 3112, 3115, 3119, 3125, 3128–30, 3132, 3134–36, 3140, 3146, 3147, 3151, 3158, 3160, 3162, 3166, 3174, 3176,

360

891, 899, 937, 940, 947, 948, 951, 977, 1334, 1335, 1337, 1342, 1343, 1345, 1346,
1439–41, 1448, 1453, 1520, 1562, 1566, 1567, 1602, 1605, 1608, 1613, 1614, 1624,
1663, 1667, 1688, 1706, 1715, 1718, 1737, 1747, 1757, 1768, 1769, 1785, 1823, 1850,
1851–53, 1867, 1871, 1873, 1963, 2146, 2147, 2202, 2203, 2234, 2387, 2413, 2415,
2488, 2491, 2549, 2555, 2610, 2665, 2691, 2697, 2742, 2746, 2780, 2784, 2797, 2867,
2874, 2890, 2918, 2976, 3089, 3097, 3104, 3137, 3139, 3150, 3154, 3201, 3228, 3241,
3268, 3294, 3314, 3552, 3562, 3564, 3565, 3575, 3580, 3586, 3591, 3593–97, 3602,
3603, 3610, 3635, 3658, 3660, 4001, 4047, 4059, 4067, 4068, 4077, 4079, 4092, 4101,
4353, 4376, 4387, 4396, 4427, 4486, 4566, 4570, 4573, 4581, 4662, 4710, 4717, 4735,
4741, 4823, 4841, 4849, 4853, 4873, 4896, 4948, 4957, 4968, 5012, 5014, 5020, 5026,
5075, 5079, 5086, 5096, 5107, 5121, 5129, 5130; tenderizers, 4105
Elephantiasis, treatment, 4260
Emollients, 1982, 2351, 2855, 4582, 4692, 4850. *See also* Analgesics; Anesthetics; Cosmetics
Escharotics, 2903, 4034
Eye disorders, treatment, 1319, 2464, 2468, 2678, 2826, 2855, 3074, 3329, 3351, 3643,
3677, 4060, 4279, 4485, 4776, 4829, 4867; conjunctivitis, 399, 606, 2978, 4397; inflammation, 181; "pink eyes," 750, 2573; washes, 180, 2330, 2443, 2447; wounds, 2571,
4360

Febrifuges, 58, 416, 447, 448, 459, 539, 587, 684, 741, 830, 968, 974, 1025, 1051, 1763,
1773, 1897, 1911, 1951, 1965, 1992, 2094, 2112, 2208, 2255, 2277, 2357, 2416, 2584,
2626, 2639, 2748, 2755, 2800, 2969, 3041, 3159, 3220, 3307, 3327, 3366, 3412, 3504,
3589, 3608, 3658, 3662, 3665, 3686, 3723, 4012, 4183A. *See also* Malaria
Fermenting agents, 1026, 1838, 1839, 2020, 4023; alcohol production, 3231
Fertilizers, agricultural, 634
Foot infections, treatment, 3968, 4740
Fungicides, 661, 3444, 3561, 3973, 4106; thrush, treatment, 1786, 1898, 2004, 2134, 2501,
2768, 2958

Gastrointestinal system, agents affecting: antidiarrheics, 184, 264, 1442, 1521, 1950, 2134,
2317, 2486, 2722, 2780, 3313, 3704, 3953, 4371, 4431, 4862; antiemetics, 444, 1442,
2733, 4397; digestants, 539, 3384; emetics, 917, 1341, 2014, 2405, 2433, 2607, 2671,
3427, 3505, 4392; enemas, 1331; purgatives, 167, 304, 705, 792, 1356, 1686, 1697,
1699, 1809, 1932, 1951, 2039, 2178, 2265, 2318, 2355, 2356, 2367, 2374, 2429, 2444,
2447, 2448, 2473, 2476, 2534, 2675, 2741, 2847, 2964, 2971, 2991, 3109, 3187, 3189,
3324, 3375, 3477, 3478, 3486, 3492, 3500, 3608, 3682, 4076, 4121, 4218, 4238, 4251,
4298, 4314, 4320, 4345, 4424, 4759. *See also* Analgesics
Gastrointestinal system, disorders, treatment: abdominal spasms, 1244, 1245, 1820, 1903,
2127, 2582, 3352, 4243, 4258; bowel diseases, 704, 4280, 4640; colic, 8, 347, 871, 3844;
dysentery, 107, 764, 815, 888, 1764, 1780, 2094, 2195, 2215, 2289, 2439, 2550, 2662,
2722, 2996, 3313, 3624, 3752, 4431, 4446; 4881; dyspepsia, 2749, 4632; hemafecia,
587, 2134, 3356; hematemesis, 1070, 1442, 1950, 2730; indigestion and flatulence, 892,
1332, 2123, 2612, 2740, 3513, 3683, 3824, 3951, 4095, 4250, 4672, 4783, 4958; intestinal ailments, 1364, 2973, 3103, 3313; jaundice, 3010; nausea, 1249, 1832; stomach ailments, 10, 310, 393, 435, 443, 495, 498–500, 503, 582, 666, 823, 856, 871, 984, 1002, 1080,
1085, 1208, 1252, 1380, 1427, 1437, 1743, 1753, 1875, 1935, 1986, 2134, 2215, 2412,
2435, 2489, 2648, 2696, 2752, 2962, 2975, 2982, 3087, 3136, 3311, 3313, 3483, 3563,
3607, 3618, 3627, 3811, 3814, 3845, 3942, 4072, 4095, 4200, 4256, 4280, 4307, 4370,
4398, 4418, 4459, 4532, 4632, 4640, 4657, 4682, 4703, 4712, 4743, 4746, 4766, 4802,
4832, 4851, 4866, 4876, 4954, 4972, 4982, 4988, 5009, 5100, 5103, 5131. *See also*
Analgesics
Glue, sources, 145, 2914, 3253, 3546

Hydrophobia, treatment, 927, 1693, 3412

Inflammations, treatment, 513, 630, 1356, 2745, 4392, 4451. *See also* Eye disorders; Skin
diseases
Injuries, treatment: bruises, 721, 911, 960, 2328, 2869, 3622, 3955, 4292; burns, 42, 177,
617, 1458, 2581, 2677, 3234, 3550, 4273, 4684; fractures, 68, 431, 857, 1726, 2025;

Menstrual disorders, treatment: amenorrhea, 179, 264, 2083, 3219, 4816; dysmenorrhea, 4455; excessive menstruation, 4275; irregular menstruation, 397, 426, 1699, 4521

Mumps, treatment, 2036, 4993

Nervous system, agents affecting: anticonvulsants, 1690, 3704, 4325; enhancers of olfactory senses, 2270, 2818, 4397; intoxicants, 589, 2177, 2193, 3183, 3188, 3517, 4019, 4891; narcotic agents and snuffs, 206, 711, 1065, 1199, 1722, 3183, 4925; sedatives and soporifics, 264, 1810, 2282, 3742, 3908, 4297; sternutatories, 3414, 3419, 4923; stimulants, 1097, 2579, 3172, 3473. *See also* Analgesics; Tonics

Nervous system, disorders: hangover, treatment, 3925; insanity, treatment, 1321, 3972, 5117; intoxication, prevention, 887; paralysis, causes and treatment, 180, 750, 2624, 2795; vertigo, causes, 3172

Oils, fixed, source, 92, 1064, 1066, 1068, 1069, 1090, 1107, 1297, 1483, 1492, 1653, 1806, 1807, 1848, 1987–89, 2351, 2552, 2569, 2828, 2830, 2855, 2900, 2904, 3019, 3230, 3369, 3993. *See also* Resins and oleoresins

Oral hygiene: dentifrices, 2187, 4161; gargles, 1852, 4461; gums, care, 840, 1385, 1625, 2087, 2216, 2319, 2526; mouth sores, 4044, 4272; obtundents (dental), 59, 398, 526, 536, 597, 676, 728, 898, 1364, 1385, 1664, 1991, 1993, 2135, 2136, 2216, 2496, 2548, 2856, 3051, 3074, 3128, 3221, 3270, 3437, 3622, 4141, 4282, 4302, 4324, 4857; teeth, care, 108, 921, 1436, 1588, 1710, 1882, 2027, 2310, 2358, 2526, 2653, 3278

Pinworms, treatment, *see* Vermifuges

Poisons: arrow, 506, 507, 529, 535, 925, 1008, 1060, 1972, 2421, 2422, 3198, 3289, 3292, 3327, 3340, 3387, 3392, 3393, 3449B; fish, 415, 496, 929, 988, 1000, 1055, 1318, 1380–83, 1546, 1560, 1714, 1720, 1748, 1749, 1752, 1791, 1793–95, 1798, 1799, 1801, 1802, 1828, 2034, 2088, 2095, 2096, 2098, 2100, 2109, 2188, 2265, 2272, 2399, 2403, 2404, 2412, 2414, 2421, 2431, 2436, 2442, 2473, 2573, 2574, 2576–78, 2728, 2832, 2838, 2839, 2893, 2995, 2998, 3015–17, 3130, 3199, 3207, 3222, 3256, 3257, 3286, 3287, 3291, 3342, 3546, 3561, 4061, 4211, 4217, 4719, 4721, 4723–27, 4813; gaseous, 208; to warm-blooded animals, 35, 39, 82, 115, 144, 146, 148, 150, 156, 210, 215, 225, 419, 421, 424, 429, 505, 538, 712, 843, 923, 931, 963, 965, 966, 970, 978, 1074, 1122, 1214, 1268, 1315, 1318, 1380, 1447, 1463, 1467, 1468, 1493, 1494, 1657, 1699, 1722, 1727–31, 1742, 1751, 1752, 1756, 1777, 1801, 1811, 1846, 1956, 1996, 2012, 2017, 2018, 2052, 2084, 2118, 2161, 2265, 2269, 2273, 2299, 2302, 2306, 2308, 2311, 2336, 2356, 2380, 2392, 2420–22, 2430, 2432, 2434, 2459, 2463, 2470, 2475, 2484, 2510, 2528, 2533, 2577, 2606, 2835, 2916, 2950, 2954, 3018, 3023, 3037, 3145, 3148, 3165, 3180, 3184, 3249, 3303, 3305, 3347, 3360, 3369, 3371, 3391, 3396, 3401, 3407, 3412, 3416, 3433, 3451, 3453, 3456, 3469, 3485, 3744, 3820, 4032, 4040, 4043, 4056, 4089, 4104, 4177, 4184, 4198, 4244, 4264, 4265, 4283–87, 4290, 4313, 4332, 4346, 4362, 4364, 4375, 4378, 4382, 4390, 4393, 4469, 4470, 4567, 4706, 4728, 4752, 4918, 4920, 5066; antidotes, 331, 554, 893, 1271, 1892, 1959, 2036, 2981, 3304, 3359, 4017, 4033, 4044, 4356, 4521, 4801. *See also* Injuries, snakebite

Repellents: insect, 479, 496, 1045, 1049, 1738, 1837, 1951, 1994, 2008, 2213, 2418, 2456, 2559, 2717, 3066, 3657, 3945, 4262, 4933, 4942, 4992; snake, 3428

Reproduction: abortifacients, 2394, 3627; aphrodisiacs, 304, 794, 1831, 2298, 2365, 4252, 4389; contraceptives, 2014, 2075; douches, 3573; fertility promoters, 4796; galactagogues, 1038, 1454, 3388, 3429; hemostatic agents, uterine, 427, 1808, 4033, 4113, 4593; impotence, 1565, 2807; oxytocics, 768, 1038, 1842, 2124, 2925, 3274, 3585, 3978, 4334; pregnancy, plants used in, 28, 58, 973; puerperium, treatment during, 56, 587, 648, 665, 726, 887, 1020, 1077, 1208, 1323, 1372, 1383, 1776, 1844, 1903, 1985, 2083, 2132, 2189, 2573, 2588, 2785, 2794, 3010, 3110, 3220, 3482, 3573, 3972, 4107, 4174, 4176, 4183B, 4258, 4259B, 4269, 4616, 4645; sterility, female, treatment, 1086; vagina, ruptured, 1786; women's diseases, treatment, 2235, 2466, 3572, 4763. *See also* Menstrual disorders; Venereal diseases

Resins and oleoresins, sources, 1659, 1999, 2520, 2863, 2882–84, 2886, 4610, 4617, 4618, 4644. *See also* Oils

Respiratory ailments, treatment, 1942, 1945, 3587; antitussives, use, 69, 297, 435, 450, 616,